The future of much conservation interest, both inside and outside nature reserves, is dependent upon sensible habitat management. This comprehensive volume provides a pragmatic, habitat by habitat guide to conservation management, in which the prescriptions and methods are based upon sound science coupled with practical experience. For each habitat, the book guides the reader through the options and solutions, shows the problems to look out for, and gives good and bad examples of habitat management in the past.

This will be a must for all practising ecologists, land managers, wardens, landscape architects and conservationists, and will provide a valuable reference for students of ecology, conservation and environmental science.

Managing Habitats
for Conservation

Managing Habitats for Conservation

Edited by

WILLIAM J. SUTHERLAND

School of Biological Sciences, University of East Anglia

and

DAVID A. HILL

Ecoscope Applied Ecologists

CAMBRIDGE
UNIVERSITY PRESS

CAMBRIDGE UNIVERSITY PRESS
Cambridge, New York, Melbourne, Madrid, Cape Town, Singapore, São Paulo

Cambridge University Press
The Edinburgh Building, Cambridge CB2 8RU, UK

Published in the United States of America by Cambridge University Press, New York

www.cambridge.org
Information on the title: www.cambridge.org/9780521447768

First published 1995
Eighth printing 2007

Printed in the United Kingdom at the University Press, Cambridge

A catalogue record for this publication is available from the British Library

Library of Congress Cataloguing in Publication data
Managing habitats for conservation / edited by W. J. Sutherland and
D. A. Hill
p.   cm.
Includes index.
ISBN 0 521 44260 5 (hardback) – ISBN 0 521 44776 3 (pbk)
1. Habitat conservation.   I. Sutherland, W. J.   II. Hill, D. A.
QH75.M353   1995
639.9′2–dc20
94-34332   CIP

ISBN-13 978-0-521-44260-5 hardback
ISBN-13 978-0-521-44776-8 paperback

Contents

Colour plates are between pp. 180 and 181.

Contributors

John Andrews, Andrews Ward Associates, 17 West Perry, Huntingdon, Cambridgeshire PE18 0BX

Malcolm Ausden, School of Biological Sciences, University of East Anglia, Norwich NR4 7TJ

Chris Baines, P.O. Box 35, Wolverhampton, West Midlands WV1 4XJ

David Bellamy, David Bellamy Associates, Mountjoy Research Centre, Durham DH1 2UR

Neil Burgess, Royal Society for the Protection of Birds, The Lodge, Sandy, Bedfordshire SG19 2DL

Paul M. Dolman, School of Biological Sciences, University of East Anglia, Norwich NR4 7TJ

Peter Evans, Department of Biological Sciences, University of Durham, Durham DH1 3LE

Christopher L. J. Frid, Dove Marine Laboratory, University of Newcastle, Cullercoats, Tyne and Wear NE3D 4TZ

Robert J. Fuller, British Trust for Ornithology, The Nunnery, Nunnery Place, Thetford, Norfolk IP24 2PU

Barrie Goldsmith, Department of Botany, University College London, Gower Street, London WC1E 6BT

Roger Hanbury, British Waterways, Llanthony Warehouse, Gloucester Docks, Gloucester GL1 2EJ

Juliet Hawkins, The Hall, Milden, Sudbury, Suffolk CO10 9NY

David A. Hill, Ecoscope Applied Ecologists. Crake Holme, Muker, North Yorkshire DL11 6QH

Graham Hirons, Royal Society for the Protection of Birds, The Lodge, Sandy, Bedfordshire SG19 2DL

Richard Hobbs, Norfolk Wildlife Trust, 72 Cathedral Close, Norwich NR1 4DF

Nigel T. H. Holmes, Alconbury Environmental Consultants, The Almonds, Ramsey Road, Warboys, Huntingdon PE12 2RW

Peter J. Hudson, Uplands Research Group, The Game Conservancy Trust, Crubenmore, Newtonmore, Inverness PH3 4EP

Reg Land, Norfolk Naturalists Trust, 72 Cathedral Close, Norwich NR1 4DF

Angus J. Macdonald, Scottish Natural Heritage, 2 Anderson Place, Leith, Edinburgh EH6 8NP

George F. Peterken, Independent Consultant in Forestry and Nature Conservation, Beechwood House, St Briavels Common, Lydney, Gloucester GL1S 6SL

Nick W. Sotherton, The Game Conservancy Trust, Fordingbridge, Hampshire ST1 1EF

William J. Sutherland, School of Biological Sciences, University of East Anglia, Norwich NR4 7TJ

Gareth Thomas, Royal Society for the Protection of Birds, The Lodge, Sandy, Bedfordshire SG19 2DL

Desmond B. A. Thompson, Scottish Natural Heritage, 2 Anderson Place, Leith, Edinburgh EH6 8NP

Jo Treweek, Institute of Terrestrial Ecology, Monks Wood, Abbots Ripton, Huntingdon, Cambridgeshire PE17 2CS

Juliet Vickery, Department of Zoology, University of Edinburgh, The King's Buildings, Edinburgh EH9 3JT

Diana Ward, Andrews Ward Associates, 17 West Perry, Huntingdon, Cambridgeshire PE18 0BX

Consultants

Penny Anderson Consultant Ecologist, Penny Anderson Consultants

Ceri Evans Reserves Ecologist, Royal Society for the Protection of Birds

John Lawton Director, NERC Centre for Population Biology, Imperial College

Alan Stubbs Entomologist, formerly Joint Nature Conservation Committee

Colin Tubbs formerly Senior Officer, English Nature, Lyndhurst

John Wilson Senior Warden, Leighton Moss and Morecambe Bay Reserve, Royal Society for the Protection of Birds

Acknowledgements

We are very grateful to ICI for sponsoring this project and the following: the consultants who read and commented on each chapter, thus greatly improving the book; comments were made on specific chapters by Paul Fisher, Chris Gibson, Su Gough, Liz Muirray, Peter Robertson, David Streeter and Peter Wakely; Greg Poole painted the cover; Tracey Sanderson at Cambridge University Press helped especially with sorting out the illustrations and Carmen Mongillo was very helpful in ensuring that its production went smoothly. A number of people helped in a variety of ways: Ruth Chambers, Robin Cox and Valerie McCrea.

Introduction and principles of ecological management

WILLIAM J. SUTHERLAND

Introduction

This book was born out of the frustration caused by the lack of readily accessible information on habitat management. Our aim in this book is to collate the available information to produce a practical guide to what needs to be done rather than exactly how to do it. Thus, for example, we will describe the conditions in which coppicing or building a boardwalk is a good idea but not actually how to coppice or construct a boardwalk. (A good source for advice on such techniques is the British Trust for Conservation Volunteers; see end of book for address.) The authors of each chapter were asked to imagine someone who has become responsible for managing an area and to provide the sort of information that that person would require to make the necessary management decisions. This book is directed towards conservationists in the United Kingdom but we expect it will be of relevance elsewhere, particularly in Western Europe. After the chapters on the general issues of planning and access, each chapter describes a different habitat. Most sites will include many habitats, for example, a wetland site may include open water, fen, reeds, woods and heaths and could even be an urban site.

The need for such a book is clear from the fact that we have seen the rapid local extinction of numerous species not only due to direct habitat loss, persecution and pollution but also from habitat deterioration through lack of appropriate management. Butterflies are particularly sensitive and many species have shown dramatic declines. For example, High Brown Fritillary *Argynnis adippe* has disappeared from 94% of its previous locations, the Heath Fritillary *Mellicta athalia* from 92% and the Marsh Fritillary *Euphydryas aurinia* from about 63%. Thomas (1991) points out that a number of explanations have been put forward for such extinctions including over-collecting, insecticides, habitat fragmentation, climatic cooling and, nowadays, air pollution. The evidence suggests that these play at most a localised role and that habitat deterioration is actually the overwhelming cause of such extinctions. It has become clear that simply declaring an area a nature reserve is insufficient to ensure that the interest of the site persists. Indeed, for some butterflies the extinction rate inside reserves has been as great as that outside.

Some habitats have lost much of their ecological interest. For example, although much Breckland heath has been lost to forestry and agriculture there has also been a considerable loss of characteristic Breckland species from the remaining habitat (Dolman & Sutherland, 1992). Of the five characteristic birds of East Anglian grass heaths, Ringed Plovers *Charadrius hiaticula* and Red-backed Shrike *Lanius collurio* no longer breed on the heaths while Stone Curlew *Burhinus oedicnemus*, Woodlark *Lullula arborea* and Wheatear *Oenanthe oenanthe* have all declined severely on the heaths. Two species of moth, Vipers Bugloss *Hadena irregularis* and Spotted Sulphur *Emmelia trabealis* have become extinct while the Tawny Wave *Scopula rubiginata* is probably extinct. The loss of populations of many plant species during this century has been dramatic: Small Alison *Alyssum alyssoides*, 75% of populations lost; Field Wormwood *Artemisia campestris*, 81%; Glaucous Fescue *Festuca longifolia*, 39%; Grape Hyacinth *Muscari neglectum*, 29%; Perennial Knawel *Scleranthus perennis* subsp. *prostratus*, 93%; Spiked Speedwell *Veronica spicata*, 78%; and Spring Speedwell *Veronica verna*, 63%. The overwhelming majority of these losses can be attributed to habitat deterioration and relatively few are due to site destruction through agriculture, afforestation and development. The decline in many species of butterfly and the decline in many of the characteristic species of such an important area as Breckland shows the importance of correct habitat management. An objective of this book is to outline ways of maintaining, enhancing or restoring the conservation interest of sites. Balanced against these losses are many habitat management success stories such as the Large Blue *Maculinea arion* and the re-creation and management of wetlands.

In this chapter I will consider the practical implications for habitat management of ecological principles. Although conservation is used as a justification for much ecological research, in reality such research is often of little use. For example, the theory of island biogeography has been widely canvassed as a means of justifying such policies as the acquisition of chains of reserves, or of joining small areas to make larger ones. However, such theory is often of little practical use as decisions are determined largely by factors such as availability, access, cost, control of the water regime and the location of specific species and communities. In practice most managers use very little theory in making management decisions and thus much of the information in this book is based on practical experience.

A knowledge of ecological concepts is nevertheless invaluable. For example, it is well known that chalk grassland should be grazed, but if there is also understanding of the importance of the underlying ecological processes of disturbance, regeneration, nu-

Fig. 1.1. Red Helleborine *Cephalanthera rubra* is dependent upon a complex interaction with carpenter bees, wood-boring insects and mycorrhiza. (J. Geeson.)

trient dynamics and competition, then it will be possible to interpret the response of the site to management and refine the prescription. Thus, this chapter will outline some of the general ecological principles which are common to managing many habitats.

Communities

Much habitat management is a mixture of managing for specific species and managing to maintain communities. In theory it should be possible to manage on the basis of specific prescriptions for each of the important species present. In practice, not only would this be very complicated, but usually we also lack the required information. The ecology of individual species is often complex, for example, studies of Red Helleborine *Cephalanthera rubra* (Fig. 1.1) in Sweden (Nilsson, 1983) have shown that it is pollinated by male solitary carpenter bees *Chelostoma fuliginosum*. The female carpenter bees obtain their nectar from Peach-leaved Bellflower *Campanula persicifolia* and, to a lesser extent, Harebell *Campanula rotundifolia*. The male bees visit flowers in search of females. The helleborine lacks nectar but in the bee visual spectrum the colours are practically identical to that of the bellflowers and this mimicry is sufficient to ensure that it is pollinated by the males. The bees, in turn, are dependent upon nest holes made in dry wood by wood-boring insects. Furthermore, like all orchids, Red Helleborine is dependent upon a mycorrhiza to germinate and survive. As well as managing to maintain the helleborines directly it is necessary to ensure that the carpenter bees, wood-boring insects, bellflowers and mycorrhiza all persist.

Other species may be less bizarre but are also particular in their requirements. The requirements of orchids and butterflies are only just becoming understood and we can only guess at the requirements of less glamorous species. Many invertebrates have very different needs at different life stages. Only by maintaining the integrity of a community is it likely that the complex and unknown demands of its constituent species will be met. Even in the case of the Red Helleborine, in which the requirements are known, the management of the two British sites does not take specific notice of carpenter bees or wood-boring insects.

It is ideal to have an inventory of what is at a site to start with so as to assess both the status of the site and the potential for change. It is important to detail the vegetation structure and relative abundance of species and not just list the species present. Very often management radically alters habitat structure and relative abundance but with only minor effects on the presence or absence of plant species on the site. Such changes in structure, or in factors such as whether plants are able to flower and set seed, have important consequences for animals, especially invertebrates. The nature of the communities present at a site is a good indicator of the underlying geology, water regime history and recent management. The National Vegetation Classification is a means of naming plant communities which is being adopted by most conservation bodies. Identification of the plant communities by using the National Vegetation Classification is an invaluable first step before planning any management. Knowledge of the communities present assists in assessing the potential for change because the success of management in improving the conservation interest of a site will depend upon what is already present or is capable of arriving. If a meadow or woodland is species-poor then introducing haymaking or coppicing will not produce a species-rich community. The possibility of plants persisting in the seed bank needs to be considered: for example, Fen Violet *Viola persicifolia* reappeared at Wicken Fen after being considered extinct for many years.

Finally, the animal and plant community occurring on a site is greatly influenced by physical factors such as the aspect, weather and underlying geology. Although acknowledging the importance of such factors, as they cannot be modified they will not be discussed in detail.

Succession and disturbance

Succession is the progression of an area from being devoid of life to one with a climax community present. In one version, autogenic succession, the first species to arrive may make the habitat more suitable for those that follow due, for example, to the build up of nutrients or organic matter. A change in species composition may also occur as colonist species tend to be good dispersers which grow fast but are sensitive to competition and are thus replaced as time proceeds. Many of the most precious wildlife communities in Britain are of an early successional stage and thus much of the effort of conservationists is in suspending succession. The intensity of management is inversely proportional to the maturity of the habitat being managed.

It is now well accepted that disturbance can alter the species diversity of many communities. Disturbance has the twin role of retarding the dominant competitors and providing gaps suitable for regeneration. The intermediate disturbance hypothesis suggests that with little disturbance the community is dominated by a few species while with considerable disturbance only a few species can persist: an intermediate disturbance level has the greatest diversity. This process is illustrated by studies of intertidal beds of mussels *Mytilus californianus* in the USA (Paine & Levin, 1981). In the absence of disturbance this species may persist as an extensive monoculture. When a gap is created through battering by logs or by storms then a succession of species of algae, barnacles and other species of mussels occupy the site. Within four years the bare space is reduced to just 1% and *Mytilus californianus* is again the dominant species.

In the absence of disturbance few annuals and biennials will survive. Many wildflower mixes consist of colourful weeds of an early successional stage, which will die out without regular disturbance. As a consequence, without appropriate management wildflower mixes are often considered an eventual disappointment.

The immediate and obvious response when protecting an area is to reduce disturbance. Although this can be sensible, most habitats would benefit from some carefully
controlled disturbance. On the Breckland, rotovation of grass heaths has been shown to
be highly beneficial (see Chapter 10). Bare patches of ground are essential for many of
the scarcer invertebrates such as various butterflies, ground beetles, ground bugs and
solitary bees and wasps. Yet despite the ecological importance of bare ground, it is often
considered untidy or a waste of plant space.

It is widely considered that Britain had a climax community of largely continuous
woodland cover, so it is curious how many species of conservation concern are
dependent upon early successional stages. This seems true for a disproportionate
number of plants and butterflies. Some species probably used to occur on naturally
disturbed habitats such as river edges, coastal landslips or saltmarshes. The main
reason is probably that the climax community was less stable than often assumed. In the
Białowieza Forest in Poland (Tomiałojc, 1991), which is pristine forest relatively
unaffected by man, gaps are caused by factors such as falling old trees (which then often
cause others to fall) and fire. Gaps in the forest are kept open by grazing deer and by the
rooting of Wild Boar *Sus scrofa*. These gaps are exploited by a high diversity of annual
plants and butterflies.

Thomas (1991) suggested a historical climatic factor which further accounts for the
disproportionate number of butterflies associated with habitats of an early successional
stage: during the warm period of 10 000 to 4500 BP many of the UK's current rarities
could survive in more shaded vegetation, as they currently do in continental Europe; as
the climate cooled 4500 to 2500 years ago, they became more dependent upon man-
made early successional habitats such as coppiced woodland or closely grazed turf
which have a warmer microclimate. As these man-made habitats have disappeared
(through lack of management in addition to habitat loss) over the last few decades the
numbers of these butterflies have plummeted.

It has been suggested that the high diversity of some permanent plant communities
such as chalk grassland is a consequence of variation in the regeneration niche (Grubb,
1977). Gaps in the vegetation will differ according to numerous factors such as the
season, size of site, presence of herbivores, soil surface, extent of plant litter and the
competitors present. Species will differ in their ability to utilise different gaps due, for
example, to the size and shape of the seed, an important factor in determining the
degree to which the seed is buried and the chance of it taking up sufficient water to
germinate. Gaps in river banks caused by scour in autumn usually become occupied by
Great Willowherb *Epilobium hirsutum* whilst those formed in spring are occupied by
Purple Loosestrife *Lythrum salicaria* (Shamsi & Whitehead, 1977). The willowherb
excludes the loosestrife from autumn gaps by its earlier development but the
willowherb is unable to become established as late as the spring so the loosestrife
succeeds at that time.

Grubb (1977) points out that there may also be variation in regeneration between years which contributes to diversity. Dogs Mercury *Mercurialis perennis* will inhibit the growth of invading seedlings of Ash *Fraxinus excelsior* on boulder clay in eastern England but in an unusually wet spring the Dogs Mercury dies back on marginal sites and the Ash, which is tolerant of waterlogging, grows through. In one example, all the seedlings of autumn-germinating Small-flowered Buttercup *Ranunculus parviflorus* were destroyed during severe and prolonged winter frosts whereas seedlings of Spotted Rockrose *Tuberaria guttata* germinating in the spring avoided the hazard (Salisbury, 1929).

Competition

Any gardener knows that a great range of species can be grown in a garden as long as the competitors are weeded out. Many species seem to be restricted to habitats not because they grow better there but because that is the only habitat in which they escape competition. Tolerance to extreme conditions often carries a sacrifice of growth rate and competitive ability. For example, areas of chalk, serpentine, saltmarsh, high altitude or heavy metal contamination are well known for possessing characteristic plant communities (Fig. 1.2). Many saltmarsh species, however, grow well in the absence of salt, and salt even hinders the germination of many saltmarsh species (but not as much as for non-saltmarsh species!). The exclusion of saltmarsh species from non-saline environments must then be ascribed to an ability to compete that is presumably related to the 'cost' of the paraphernalia of salt tolerance (Davy & Costa, 1992).

Many species typical of chalk grassland grow well on non-calcareous soils, although they tend to suffer from aluminium toxicity on really acid soils. Their existence on the chalk is attributable to their ability to tolerate a dry habitat with levels of available iron, nitrogen and phosphorus that would be deficient for other species. Thus, for most species a major factor determining whether they can persist in a site is not only the direct effect of physical factors such as microclimate or soil type but also the degree of competition from other species.

Competition also is important in determining the species composition of animal communities. The faeces of Common Toad *Bufo bufo* tadpoles are known to contain a parasitic alga which inhibits the growth and reduce the survival of Natterjack Toad *Bufo calamita* tadpoles (Griffiths *et al.*, 1992). Natterjacks can burrow down to moist conditions and thus survive hot areas. Common Toads do not burrow and thus cannot persist in arid sites. The successional changes that are currently occurring on heaths due to lack of grazing favour the Common Toads by providing shade. At many famous sites the Natterjack Toads have now become replaced by large numbers of Common Toads (Beebee, 1992). Unlike Red Squirrels *Sciurus vulgaris*, Grey Squirrels *Sciurus carolinensis* feed on unripe hazel-nuts and acorns and so can harvest the hazel-nuts before they are ready for the Red Squirrels, furthermore their populations can persist

Fig. 1.2. Many species of plants are restricted to specific habitats. For example, Yellow Horned Poppy *Glaucium flavum* is restricted to coastal shingle. Many such species can grow perfectly well elsewhere but are restricted to these habitats by competition (J. Geeson.)

on acorns in years when the hazel-nut crop is depressed (Kenward & Holm, 1989). In such cases, the required habitat management may be to make the area less suitable for the superior competitor and more suitable for the more vulnerable species. In our examples, this may involve removing the more rank vegetation favoured by Common Toads from Natterjack sites and not planting oak trees in Red Squirrel sites.

Grazing, mowing and predation

Herbaceous plants and other dicotyledons grow mainly from aerial meristems and are retarded if the top is cut or grazed. Grasses and other monocotyledons grow from a basal meristem and removing the top will have little effect on subsequent growth rates. Hence grazing or cutting usually increases the relative abundance of grasses. Thus the grazing of saltmarshes by sheep leads to the replacement of broadleaved species such as Sea Purslane *Halimione portulacoides* and Sea Aster *Aster tripolium* by the Common Saltmarsh Grass *Puccinellia maritima*. Some species of herbaceous plants such as rosette herbs are tolerant of grazing and tend to increase when grazing occurs.

There has been a considerable body of research on the consequences of grazing on plant species diversity which has produced seemingly conflicting results. The actual underlying rule is very simple: if the dominant species is palatable then grazing will increase diversity, but if the dominant species is unpalatable then grazing will reduce diversity. Experiments on pastures dominated by two palatable species, Rye Grass *Lolium perenne* and White Clover *Trifolium repens*, showed that sheep grazing resulted in an increase in diversity as the palatable dominants were suppressed. Other grazers such as molluscs and insects can have the same effect on other communities (Rees & Brown, 1992).

A major benefit of grazing is the increase in habitat disturbance. This differs between grazers. Rabbits with their burrows, scrapes and dung piles cause considerable disturbance, cattle with cow pats and trampling cause some disturbance, while sheep cause very little (and humans with mowing machines least of all). As a result of this, combined with the variation in grazing habits, each of these grazing species produces a different plant community.

Although grazing often increases the species diversity of plant communities it may reduce the diversity of invertebrates. Van Wieren (1991) documented the decline in insect diversity with grazing and showed this was attributable to fewer plants flowering, less litter, reduced standing crop and a more extreme microclimate. A few invertebrates clearly benefited from grazing such as dung beetles and the consumers of the algae surviving in the damp cattle footprints. Some breeding wading birds benefit from some grazing as this provides the required vegetation structure but at high stocking levels they suffer considerably from trampling: intermediate levels of grazing are thus preferable.

Fig. 1.3. The Silver-spotted Skipper *Hesperis comma* is restricted to areas which are grazed as the females lay eggs only on Sheep's Fescue *Festuca ovina* adjacent to bare areas. They also avoid areas that have been recently nibbled, so the best management is to graze heavily with cattle over the winter, and then leave ungrazed. (J. Geeson.)

Predation also influences the community present. Attempts to introduce Hedgehogs *Erinaceus europaeus* into an area of their preferred habitat in Oxfordshire, from which they were absent, resulted in the majority of those who did not disperse immediately being eaten by Badgers *Meles meles* (Doncaster, 1992). It is clear that Hedgehogs are only common in areas where Badgers are scarce such as urban settlements. It is also likely that pathogens and parasites have a major role in determining community structure as shown by Dutch elm disease and myxomatosis.

This general rule that removing the dominant species increases diversity has been shown to apply also to other processes such as the role of predators on intertidal communities and the role of hemiparasites and pathogens in plant communities. Although the hemiparasite Yellow Rattle *Rhinanthus minor* is characteristic of species-rich sites, the patches it occupies are less diverse than adjacent patches without Yellow Rattle (Gibson & Watkinson, 1992). Experimental removal of the species usually results in an increase in diversity. The explanation seems to be that Yellow Rattle selectively attacks the less competitive species.

Fertility

The communities present will be greatly influenced by how fertile the site is. Before the arrival of man, fertile areas would be scarce, and restricted to areas such as flood-plains. The few plant species adapted to such sites show features such as rapid clonal growth which occupies bare ground quickly, and fast vertical growth which shades out shorter species. There is thus usually an inverse relationship between the productivity of a site and its diversity. Most species of the British flora are adapted to surviving in relatively infertile conditions. With the invention and widespread use of artificial fertilisers the abundance and geographical range of species adapted to fertile environments has expanded dramatically; an increase in fertility allows these species to flourish and may result in a reduction in species diversity. Many grasslands, waterbodies and heathlands have a higher fertility than formerly due to previous or current management, agricultural runoff or nitrates in rainfall and many management policies are directed at reducing these nutrient levels.

For a few habitats it is possible that a modest increase in nutrient may increase diversity and may increase the perceived value of the habitat. For example, prior to 1800, the Norfolk Broads were much less fertile than today as the high levels of calcium carbonate combined with the phosphorus, iron and manganese to form insoluble compounds unavailable to the plants (George, 1992). The sparse community was dominated by stoneworts *Chara* with a low diversity and abundance of slow-growing waterweeds. Increased nutrients from agricultural runoff resulted in a diverse community of dense water plants and invertebrates and a fish catch of 75 kilos from a rod and line in a day. Further eutrophication has resulted in a phytoplankton-dominated community with few water plants, invertebrates or mature fish. Although nutrient-poor habitats such as oligotrophic lakes may be unspectacular in terms of species diversity, their intrinsic interest and the species and communities that are restricted to such situations should be appreciated.

Life histories

The life history of a species has considerable consequences for the way it will respond to the timing and type of management. It is necessary, when conserving animals, to have some knowledge of the timing of different stages, and for plants to know whether they are annuals or perennials.

Annual plants are generally species of disturbed ground of which a given area may be suitable for germination, growth and seed production in some years but not others; seed dormancy is the device annual plants use to overcome poor years. As perennial plants survive between years dormancy is rarer. There are a few annual species without a seed bank and these live in predictable but harsh environments where suitable conditions for germination and growth occur in every year, e.g. Glasswort *Salicornia europaea* or winter annuals such as Corncockle *Agrostemma githago* (Fig. 1.4) and Dune Fescue *Vulpia fasciculata*. A further example is the hemiparasite Yellow Rattle which is not dependent upon bare ground but on a predictable environment provided by the root system of perennial plants. Species without a seed bank can usually be eradicated relatively easily either inadvertently or intentionally as a result of the failure to set seed in a single year.

Most invertebrates do not have a persistent dormant stage and are similarly susceptible to eradication. There are two excellent books on invertebrate conservation (Fry & Lonsdale, 1991; Kirby, 1992) that should be read by any practising conservationist.

The timing of management has to consider the life cycle of the species under consideration. Removal of vegetation at certain times of the year, for example by haymaking, will destroy the eggs and larva of many invertebrates. Cutting or grazing in the summer reduces or prevents the seed-set of plants. This is clearly a particular problem for populations of annuals, biennials and non-clonal perennials. Consequently meadows cut at different times of the year develop different species composition of plants and invertebrates. Cutting hedges and mowing grass during the bird nesting season can destroy eggs and young birds.

Dispersal

Amongst species there can be considerable differences in dispersal distances. Small Tortoiseshells *Aglais urticae* move northwards in the spring at about 1 km per hour of

Fig. 1.4. Corncockle *Agrostemma githago* is an annual plant without a long-term seed bank. It is thus dependent upon a suitable habitat being present every year. (J. Geeson.)

sunshine but successful colonies of Black Hairstreaks *Strymonidia pruni* only spread by about 100 m a year. Similarly amongst plants many weed species with windborne seeds, or burrs or those which are eaten or hoarded by birds can travel large distances whilst many others rarely disperse as much as a metre.

Most permanent herbaceous communities are dominated by clonal plants which spread by rhizomes, stolons or other means. Detailed studies of the individuals of a range of species such as Bracken *Pteridium aquilinum*, Yellow Flag *Iris pseudacorus*, White Clover *Trifolium repens*, Wood Anemone *Anemone nemorosa* and Reed *Phragmites australis*, and of many other perennial grasses have shown that the general rule is that establishment from seed is exceedingly rare except in areas cleared of vegetation. Such species which are largely reliant upon bulbs, rhizomes or stolons can often only disperse a few centimetres a year and rarely cross barriers such as roads or fields and are thus dependent upon the habitat persisting in a suitable state. Once such species become extinct in a site, they are unlikely to recolonise unless there is another population close by. This is the reason why the indicator plant species of ancient wood communities tend to be poor dispersers for example Wood Anemone *Anemone nemorosa*, Ramsons *Allium ursinum*, Lily of the Valley *Convallaria majalis*, and Sanicle *Sanicula europaea* and the same is true of the indicators of ancient chalk grassland which include Squinancywort *Asperula cynanchica*, Dropwort *Filipendula vulgaris*, Rockrose *Helianthemum nummularium*, Horse-shoe Vetch *Hippocrepis comosa* and Chalk Milkwort *Polygala calcarea* (Chapters 8 and 12).

Most plants are thus either long-lived or have a seed bank and most vertebrates are long-lived and so can tolerate a single poor breeding season. By contrast the populations of many invertebrates are dependent upon individuals surviving to reproduce each year. Those species of invertebrates that are poor dispersers are then particularly prone to extinction. For example, studies of the Heath Fritillary show that it is dependent upon recently cut coppice and is a poor disperser. Once the chain of coppice management has been stopped, even for a few years, the species becomes extinct in the wood. The failure to maintain continuity of management within semi-natural habitats is at the heart of most conservation problems.

Ancient sites continually managed in one way are thus often of high conservation importance especially for plants and invertebrates. The traditional management should be maintained wherever possible. If a habitat has been traditionally managed in a specific manner, for example cut for hay in June and then grazed, then the species present will be those that can survive such management. A switch in method or timing of management is likely to be deleterious to such species while the species that could benefit may be unable to disperse there to take advantage of the new regime. If the traditional management has been long abandoned then species depending upon it may well have disappeared. In which case, automatically reintroducing traditional management is not necessarily the best option.

Patches and mosaics

Recent research has shown that localised extinctions and colonisations can be frequent and it is best to think in terms of a population of populations – a 'metapopulation'. Studies of shrews on islands in Finnish lakes show that extinction occurs regularly (Peltonen & Hanski, 1991) but recolonisation also takes place by shrews swimming to islands in summer or walking across the ice in winter, although the probability of this occurring declines with distance from the shore. Extinction is also much more likely on islands with small populations of shrews. The species vary in their susceptibility to extinction: Pygmy Shrew *Sorex minutus* is much more likely to become extinct than the larger Common Shrew *Sorex araneus*. This difference makes biological sense as the smaller species have much smaller body reserves and are particularly sensitive to variation in food availability. The distribution of the shrews can be explained by these processes. At any one time they are most likely to be present on larger islands and those nearest the shore. Pygmy Shrews tend to be restricted to the larger islands but Common Shrews regularly occur on quite small ones.

Numerous studies of small mammals, birds, spiders, aphids and freshwater snails have also shown both that the probability of extinction is lower in larger patches and that more isolated patches are less likely to be recolonised. If the patches are large or close then the metapopulation can persist even though extinctions occasionally occur. This process of extinction and recolonisation is probably fundamental to many populations of annual plants and invertebrates. A clear example of this is the study of the Silver-studded Blue butterfly *Plebejus argus* in North Wales (Thomas & Harrison, 1992) where it has an irregular distribution occupying ephemeral patches. In heathland it occupies early successional patches which are likely to disappear every 5–50 years. This species was able to occupy virtually all suitable patches less than 1 km away from existing populated patches. The turnover in terms of extinctions and recolonisations was relatively high in small patches. The existence of the metapopulation depends upon suitable habitat being continually present within dispersal distance of previous colonies.

If a block of habitat is small then species which have poor dispersive ability are likely to go extinct at that site and small populations are unlikely to persist. A small and isolated block of wood, scrub, marsh or reedbed will tend to contain little of interest.

It is however, important to create a mosaic within a habitat and not automatically manage the site in a single manner. As an example, in coppice woodland, Pearl-bordered Fritillary *Boloria euphrosyne* and High Brown Fritillary *Argynnis adippe*, require areas that have been coppiced within the last three years, Small Pearl-bordered Fritillary *Boloria selene* requires lush violets *Viola* spp. in areas coppiced 2–5 years previously, while Dormouse *Muscardinus avellanarius* requires older coppice. A rotation in which the entire site was cut at the same time may be suitable for maintaining many of the plants but would eradicate all three species of butterfly and the Dormouse: a mosaic of different age coppice in close proximity is essential to maintain them. The mosaic principle applies to practically all management and applies over a range of scales: it can apply to regional planning policies as well as to management of a small area to ensure a mosaic of microhabitats.

Structure

Many studies have shown that the diversity of groups such as birds is related to the structural diversity of the vegetation and that the general structure is often as important as the fine details of plant species composition. It is thus important to maintain the existing range of physical features, vegetation types and ages including the commonplace. Obsessive tidiness is a great problem. Ensure that you leave room for the fallen tree, poached mud, nettle patch and overgrown hedge.

The physical structure of the ground is also important. Hummocks, banks and depressions regularly contain species not found elsewhere on the site. Irregularities will increase the diversity of microclimates due to local differences in drainage and aspect. The south-facing slope may be used by basking reptiles and butterflies. Being small, invertebrates have high surface area to volume ratios and thus microclimate is important in determining their body temperature. As a result many invertebrates are restricted to particular conditions of sun or shade.

Most species during each day or during their life cycle need to use a range of habitats. For example, the first four instars of Wart-biters *Decticus verrucivorus* live in short turf where it is thought the warm microclimate is important; older stages occur in tussocks perhaps to avoid predation; and the females visit shorter swards to lay eggs (Cherrill & Brown, 1990). Wading birds of wet pastures tend to require tussocks in which to nest but open areas in which the chicks can feed. It is therefore essential to maintain a diversity of microhabitats in close proximity. Even in habitats that are predominantly monocultures such as reedbeds it is important to have a diversity of habitats such as ditches, pools, patches of scrub and blocks of different ages of reed.

Research

There has been far less progress in understanding habitat management than there should be and the responsibility for this is shared between practising conservationists often failing to monitor and document their work and ecologists often claiming to be studying conservation whilst not actually doing so. Management should always be followed by monitoring (see Chapter 2).

The requirements of a given species are often little understood; this applies even for some birds, the most studied group. There is a clear need to carry out further observations to determine species' natural history and requirements. Thomas's (1991) work on butterflies is a good example of the type of research necessary to enable informed effective management. As described in Chapter 2 the common approach is to determine the habitats required and the means of establishing these. There are great opportunities for research into many groups: for example, the larval stages of most flies have yet to be discovered and the habitat requirements of most species of plants and animals are little known.

A major hindrance in the accumulation of knowledge is that management experience is rarely quantified, documented and disseminated. The need for controlled experiments cannot be stressed too strongly. A common problem is that managers alter the management of the entire site. The changes that then occur in the populations or communities present may be either due to the alterations in management or to changes in external factors such as the weather. Thus if you wish to determine whether mowing improves a site, as a very minimum mow one area and leave a comparable area as a control then monitor and compare the two blocks. As described in Chapter 2, such experiments are greatly improved if these control and treated areas can be repeated a number of times within the site. Also, too often site managers make a total commitment and then do not get the result they hoped for. A step-by-step approach allows for gaining experience in how a site will respond to treatment, giving the opportunity to modify or abandon ideas that are not proving beneficial.

Practically all management should be followed by some form of monitoring. This may range from a detailed analysis to a subjective view as to whether the management has been a success. Such monitoring needs to be well documented.

Ethics and species introductions

Although this book is about practical habitat management, it is often necessary to consider why an area is being conserved. Decisions often need to be based upon subjective value judgements. As Rackham (1986) stresses, it is important to consider the meaning of the habitat. An ancient woodland is much more than a diverse community. Its shape, structure and species composition give it a history and meaning which cannot be recreated elsewhere any more than Stonehenge can be recreated merely by balancing stones on top of each other.

Selecting the best form of management is not straightforward as any management action will favour some species and hinder others: it is necessary to think about the objectives. The management recommendations thus depend upon which species or groups of species you wish to conserve.

The process of moving individuals of a species between sites is highly contentious. Amongst many conservationists there is deep concern over introductions. The Joint Committee for the Conservation of British Insects (1986) gives the following definitions:

Re-establishment: the release and encouragement of a species in an area where it formerly occurred but is now extinct

Introductions: an attempt to establish a species where it did not previously occur

Reintroduction: an attempt to establish a species in an area where it had been introduced but the introduction has been unsuccessful

Reinforcement: attempting to increase population size by releasing additional individuals

Translocation: the transfer of individuals from one site to another (for example to boost the population or save individuals which would otherwise be destroyed).

The Conservation Committee of the Botanical Society of the British Isles suggest that if the species was previously known within a 1 km radius of the site, then it is referred to as a reintroduction, re-establishment or reinforcement while species introduced at a greater distance are deemed introductions.

Introductions are well known for causing chaos on isolated islands such as New Zealand but they also cause problems within the UK. Species such as Sycamore *Acer pseudoplatanus*, Mink *Mustela vison* and Grey Squirrel are the bane of British conservationists. Even the apparently innocuous Ruddy Duck *Oxyura jamaicensis* from North America has now spread from Britain, where it was introduced, to Spain where it is can hybridise with the White-headed Duck *O. leucocephala* which is already a globally threatened species. Even extending a species outside its natural range within the British Isles can cause problems. Loch Lomond is one of only six sites in Britain for the Powan *Coregonus lavaretus*, but the introduced Ruffe *Gymnocephalus cernua* has become abundant and is eating the eggs of the Powan (Maitland, 1991). Numerous introductions of fish have resulted in the introduction of parasites which can affect other species.

In deciding whether an introduction is acceptable, much may depend, for example, upon whether the aim is to maintain a semi-natural habitat or to landscape a site such as a new gravel pit or park or motorway verge. Introducing new species by tree planting, adding seeds or plants, introducing fish or waterweeds may be appropriate for landscaping but usually not for semi-natural habitat. Tree planting is considered synonymous with conservation yet it is often unnecessary in semi-natural habitats, natural regeneration usually being preferable. Many attractive pieces of meadow and road verge have been destroyed by well-meaning grant-aided tree planting schemes.

Re-establishments seem much less contentious. If the species used to be present in the site then there are less likely to be ethical objections to re-establishing it. In practice many re-establishments are unsuccessful as they often do not consider why the species went extinct initially or whether the required habitat has been restored. Such re-establishments are often glamorous but time-consuming and readily attain a higher priority than less glamorous but often more useful habitat management.

Any re-establishment project implies a management commitment to suit the re-established species. In the enthusiasm for the project it is possible to overlook the consequences for the existing flora and fauna. Only carry out re-establishments if they are compatible with accepted best management of existing communities. It should also be remembered that re-establishment may obscure trends in population range size and thereby frustrate the use of such species as biological indicators.

The Joint Committee for the Conservation of British Insects (1986) and the Conservation Committee of the Botanical Society of the British Isles have made a series of recommendations concerning re-establishment:

1. Consult widely before deciding to attempt any re-establishment. Consult the Joint Committee for the Conservation of British Insects or the BSBI panel before translocating any rare insects or plants.
2. Every re-establishment should have a clear objective.
3. The ecology of the species to be re-established should be known.
4. Permission should be obtained to use both the receiving site and the source of material for re-establishment. The effect on the donor population needs to be carefully considered. The 1981 Wildlife and Countryside Act (and subsequent amendments) list those species which require prior permission before they can be moved from the statutory conservation agency. If the site is a Site of Special Scientific Interest then the statutory bodies need to be consulted.
5. The receiving site should be appropriately managed.
6. Specific parasites should be included in re-establishment (the logic is that if a species is rare, its parasite is even rarer!).
7. The numbers released should be large enough to secure re-establishment.
8. Details of the release should be meticulously recorded.
9. The success of re-establishment should be continually assessed and adequately recorded.
10. All re-establishments should be reported to the Biological Records Centre and to the Joint Committee for the Conservation of British Invertebrates (for invertebrates) or the Conservation Committee of the Botanical Society of the British Isles (for rare plants), the local office of the statutory conservation agency and the local wildlife trust.

Legal considerations

If the land you are thinking of managing is a Site of Special Scientific Interest, then the owner or occupier of the land is legally obliged to consult the relevant statutory country conservation agency (Countryside Council for Wales; English Nature; Department of the Environment, Northern Ireland; Scottish Natural Heritage; see addresses at the end of the book) before carrying out any 'potentially damaging operation'. The site may also be one identified as important by the local wildlife trust (see addresses) who should be contacted before any significant changes in management are implemented.

References

Beebee, T. (1992). Trying to save the Natterjack Toad – a case study in amphibian conservation. *British Wildlife* 3, 137–45.

Cherrill, A.J. & Brown, V.K. (1990). The habitat requirements of adults of the Wart-biter *Decticus verrucivorus* (L.) (Orthoptera: Tettigoniidae) in Southern Britain. *Biological Conservation* 53, 145–57.

Davy, A.J. & Costa, C.S.B. (1992). Development and organisation of saltmarsh communities. In *Coastal Plant Communities of Latin America*, ed. U. Seeliger, pp. 179–99. San Diego: Academic Press.

Dolman, P.M. & Sutherland, W.J. (1992). The ecological changes of Breckland grass heaths and the consequences of management. *Journal of Applied Ecology* 29, 402–13.

Doncaster, C.P. (1992). Testing the role of intraguild predation in regulating hedgehog populations. *Proceedings of the Royal Society* 249, 113–17.

Fry, R. & Lonsdale, D. (1991) *Habitat conservation for insects: a neglected green issue.* Orpington, Kent: Amateur Entomologist's Society.

George M. (1992). *The Land Use, Ecology and Conservation of Broadland.* Chichester: Packard.

Gibson, C.C. & Watkinson, A.R. (1992). The role of the hemiparasitic annual *Rhinanthus minor* in determining grassland community structure. *Oecologia* 89, 62–8.

Griffiths, R.A., Edgar, P.W. & Wong, A.L.-C. (1992). Interspecific competition in tadpoles: growth inhibition and growth retrieval in natterjack toads, *Bufo calamita. Journal of Animal Ecology* 60, 1065–76.

Grubb, P.J. (1977). The maintenance of species-richness in plant communities: the importance of the regeneration niche. *Biological Review* 52, 107–45.

Joint Committee for the Conservation of British Insects (1986). Insect re-establishment – a code of conservation practice. *Antenna* 10, 13–18.

Kenward, R.E. & Holm, J.L. (1989). What future for British red squirrels? *Biological Journal of the Linnean Society* 38, 83–9.

Kirby, P. 1992. *Habitat Management for Invertebrates: a Practical Handbook.* Sandy, Bedfordshire: Royal Society for the Protection of Birds.

Maitland, P.S. (1991). Conservation of fish species. In *The Scientific Management of Temperate Communities for Conservation*, ed. I.F. Spellerberg, F.B. Goldsmith & M.G. Morris, pp. 129–48. Oxford: Blackwell Scientific Publications.

Nilsson, L.A. (1983). Mimensis of bell flower (Campanula) by the red helleborine orchid *Cephalanthera rubra. Nature* 305, 799–800.

Paine, R.T. & Levin, S.A. (1981). Intertidal landscapes: disturbance and the dynamics of pattern. *Ecological Monographs* 51, 145–78.

Peltonen, A. & Hanski, I. (1991). Patterns of island occupancy explained by colonisation and extinction rates in three species of shrew. *Ecology* 72, 1698–708.

Rackham, O. (1986) *The History of the Countryside.* London: J.M. Dent.

Rees, M. & Brown, V.K. (1992). Interaction between invertebrate herbivores and plant competition. *Journal of Ecology* 80, 353–60.

Salisbury, E.J. (1929). The biological equipment of species in relation to competition. *Journal of Ecology* 17, 197–222.

Shamsi, S.R.A. & Whitehead, F.H. (1977). Comparative eco-physiology of *Epilobium hirsutum* L. and *Lythrum salicaria* L. IV. Effects of temperature and interspecific competition and concluding discussion. *Journal of Ecology* 65, 71–84.

Thomas, C.D. & Harrison, S. (1992). Spatial dynamics of a patchily distributed butterfly species. *Journal of Animal Ecology* 61, 437–46.

Thomas, J. (1991). Rare species conservation: case studies of European butterflies. In *The*

Scientific Management of Temperate Communities for Conservation, ed. I.F. Spellerberg, F.B. Goldsmith & M.G. Morris, pp. 149–97. Oxford: Blackwell Scientific Publications.

Tomiałojc, L. (1991). Characteristics of old growth in the Białowieza Forest, Poland. *The Natural Areas Journal* 11, 7–18.

Van Wieren, S.E. (1991). The management of populations of large mammals. In *The Scientific Management of Temperate Communities for Conservation*, ed. I.F. Spellerberg, F.B. Goldsmith & M.G. Morris, pp. 103–27. Oxford: Blackwell Scientific Publications.

2 Site management planning

GRAHAM HIRONS, BARRIE GOLDSMITH
AND GARETH THOMAS

Introduction

Land-use changes this century have caused the disappearance of many wildlife habitats in Britain and reduced the distribution of others to mere remnants. At such a time of rapid change, nature reserves have an important role: in retaining viable blocks of important ecosystems or habitats holding natural assemblages of plants and animals; as centres for research, education and public enjoyment; and for demonstrating the importance of conservation and the part played by protected areas in an intensively used countryside. Other chapters in this book give the important management practices which can be employed to enhance the conservation interest of them. However, all these management techniques should be applied within the framework of an appropriate management plan.

Conservation management of a site involves knowing what species and communities are present; understanding the ecology of the site; identifying the broad goals for the site (the objectives); identifying the management needed to achieve them (management prescriptions and work programmes); the means to determine progress towards achieving the objectives and showing that resources of cash, labour and skills are being used efficiently (monitoring prescriptions). The management plan is a convenient mechanism for bringing these elements together while the process of producing and implementing it should ensure that the necessary management is widely approved and carried out in an agreed manner to an agreed time-scale.

The purpose of this chapter is to demonstrate the need for management plans, to encourage their production by providing a worked example and to give guidance on monitoring. The latter is important because having undertaken the management, the plan should also outline the monitoring which is required to measure its effectiveness. Otherwise how are we to learn from our successes and failures?

Many conservation organisations have been reluctant to embrace management plans.

This has meant that many nature reserves are managed without them. Without planning how can the objectives for a site be decided, on what basis can management decisions be made and, if there are no objectives, how can the usefulness of reserve establishment be gauged? Most conservation organisations have limited resources and to justify expenditure on reserve acquisition and management sites must be managed cost-effectively with tangible benefits. This alone should be sufficient justification for planning.

Occasionally a site may need no management but it will still need a management plan. This is because it will need an evaluation of its features of interest, extraneous features which might cause degradation will need to be considered, objectives will need to be formulated and agreed, specified limits will need to be identified for highly valued or problem species, legal and other constraints will need identification, monitoring should be planned, and in some situations, for example Areas of Outstanding Natural Beauty, agencies will need to be co-ordinated and promotion may need to be considered.

Functions of the management plan

a) To describe the site by collating all available physical and biological information
b) To identify the objectives or purpose of managing the site
c) To anticipate any conflicts between, and problems achieving, the objectives for the site and suggest the best means of resolving them
d) To identify and describe the management necessary to achieve the objectives
e) To identify the monitoring needed to measure the effectiveness of management
f) To organise manpower and funding
g) To act as a guide for new staff, i.e. to guarantee continuity of effective management
h) To link with national species and habitat action plans
i) To demonstrate the effectiveness of management
j) To ensure that site management objectives and operations reflect the policies of the parent organisation
k) To facilitate communications between sites and organisations

These functions allow the conservation value of the site to be evaluated, providing a baseline enabling recognition of any subsequent changes to the site and showing where information is lacking thereby pointing the need for further surveys, etc. Setting objectives is the most fundamental part of the planning process and yet is frequently neglected. The various management options should be listed and the most appropriate selected in order to meet the objectives. Management plans should also encompass periodic ecological audits of sites in order to assess the degree of conformity to the plan and the extent to which the objectives set for the site are being achieved, e.g. the degree to which sites and species have been safeguarded.

Monitoring is an essential component of management and planning. Even where habitat or species management objectives can be achieved without direct intervention, it is essential that the habitat or species is monitored to confirm that this is so.

Conformity with the objectives for nature conservation at a national and international scale is important if duplication of effort is to be avoided and efficient use of resources is to be achieved. Furthermore, management techniques are constantly being developed and refined and we often have to begin the management planning process from a position of limited knowledge of the best techniques to use. Sharing this management information between sites increases the effectiveness of conservation bodies and is made easier by the use of common planning and recording systems.

The management plan also enables organisations to plan effective conservation in the context of and in relation to other sites. For example, in order to pursue the organisation's objectives it may be important to create a series of reserves for educational purposes which fit with existing reserves of the organisation or of those of others. This requires a Site Management Framework and each reserve is seen as one jewel in the crown. Other jewels may be Sites of Special Scientific Interest (SSSIs), County Wildlife Trust reserves, or properties of the National Trust, English Heritage or the local authority. A procedure equivalent to management planning needs to be conducted for each series of sites.

Management plan formats

The management plan is the framework within which all future management of the site is planned and carried out. The plan should enable any person involved to understand how and why decisions are taken. It should indicate what, when, where, why and how management is to be conducted, at what cost, and who pays. The plan, therefore, has to be a comprehensive and unambiguous document but at the same time it has to be very easy to use. A plan, however elegant and sophisticated, is only successful if it is translated into appropriate action.

Differences in management plan format arise because the aims and objectives of the organisations and the rank of staff preparing them differ and the time considered to be appropriate for their preparation and production varies. Although there is no universal format, most are based on, and compatible with, what has become known as the 'NCC format'. Provided the guidelines given in Box 2.1 are adhered to, it does not matter what framework is used. There are also pitfalls. Organisations with limited resources may find it difficult to produce full plans. Plans can easily be too large and may appear too academic – they need to be interpreted on the ground by practical individuals. Some plans are written in the absence of adequate baseline data and may be damaging to the conservation of the site if they are implemented on the ground.

Producing the management plan

Unfortunately management planning has tended to become shrouded in mystique, too bureaucratic and too bogged down by worries over detail or format. This has meant that many organisations and individuals have lacked the confidence to produce plans. A range of approaches is possible, from very wide consultation involving statutory bodies and other national and local conservation bodies for large sites of major conservation importance at high spatial scales, i.e. national and international, to more limited consultation involving local authorities and local wildlife groups, and less detail for small local sites. The most important considerations at the outset are agreeing the objectives, the management prescriptions to achieve them, the legal and/or other constraints on management (e.g. boundary problems, etc.) and a good, self-explanatory summary. It is sometimes argued that the priorities of management, which determine the objectives, are to safeguard the least numerous species at the cost of reducing the abundance of commoner ones. However, over-zealous adherence to this kind of policy can lead to excessive small area management or 'gardening'.

Management plans are usually prepared one to two years after acquisition when the warden or person responsible for the site has had time to assemble all the necessary background information. For some sites outline plans may be produced first as a point for discussion with interested parties. The components of the plan are summarised in Box 2.1.

Box 2.1 Management plan format: an example using the format of the Royal Society for the Protection of Birds.

Management plan formats essentially have four main features which can be followed in the same way as someone writing a business plan:

Preamble: general policy statement on behalf of organisation responsible for the management of the site with reasons for acquisition and an outline of the objectives for the site.

Description: location, tenure, map coverage, aerial photographs, physical description (geology, geomorphology, climate, hydrology, soils, vegetation, fauna), cultural information, land use history, archaeology, public interest, past management, ecological interrelationships, bibliography and register of scientific research.

Policy and prediction: reasons for establishment, evaluation of features, objectives, factors influencing management (legal, constraints of tenure and access), obligations, man-induced trends, externalities, specified limits.

Prescription: the part of the management plan which sets down what work needs to be carried out and precisely how and when to do it; project register, project groups (the assembly of management tasks into groups that occur in the same compartments at more or less the same time and with similar objectives, e.g. management of species-rich grassland, sea-bird recording in their breeding sites), work schedule, budget, manpower, and maps showing the desired state and the management required to achieve it.

THE RSPB FORMAT

0. Summary

A concise précis of the content and important features in the plan and hence objectives, what needs special attention, what time-scales are required for different communities and what monitoring is required. The summary should provide all the necessary information for someone new to the site to gain an overall impression of objectives and importance. This is valuable not only for managers, organisation directors, local press, media and the public, but also for fundraisers.

1. General information

Much of this information relates to maps which can be given as appendices. This information is descriptive and, apart from boundaries which may change with future acquisitions, should remain the same for many years.

1. Location
2. Site status
3. Tenure
4. Site definition and boundaries
5. Legal and other official constraints and permission
6. Main fixed assets

2. Environmental information

2.1 Physical
1. Climate
2. Hydrology
3. Geology
4. Soils

Again this is descriptive and should remain relatively unchanged for ten years or so and will require less updating than sections later in the plan. Hydrology may be a factor which could change depending on management of the site and its surrounds, the latter of which may be outside the responsibility of the site-owner.

2.2 Biological
1. Habitats
2. Flora
3. Fauna

These environmental data may need more regular updating. Biological communities tend to change more readily than physical ones. Habitats may change as a result of management, etc. Use standard descriptions for flora and fauna. Seek help from local archives, local sources of data. Has any local person been collecting information from the site in the past? Have there been any published documents relating to the ecology of the site? National Vegetation Classification categories can become degraded through time but are still attributed the same NVC code.

2.3 Cultural
1. Commercial use
2. Recreational use
3. Research, survey, monitoring
4. Conservation management already achieved

Visitor access, information and control must be built into management plans, considering the integration of habitat/landscape, business of the organisation, attitudes and values of land managers, views of other people with an influence on the land managers. Consult the book 'Practical Conservation' by Tait, Lane & Carr (1988). Rights of way and established use obviously have to be respected and occasionally will have to be controlled to avoid disturbance in order not to adversely affect the conservation value of the site. But it is essential to engender support for nature conservation amongst the general public and so usually it is better to think in terms of how to accommodate the needs of visitors rather than how to justify turning them away. Education is a vital part of the promotion of nature conservation and a positive approach should be adopted such as directing school groups to areas where their impact will be minimal (away from steep slopes, wet and muddy areas, woodland herb layer, etc.). More detail on visitor management is given in Chapter 3 and this will help in completing the management plan.

3. Evaluation

Here it is important to describe the attributes of the site in relation to a set of common criteria used as a standard in British conservation ('A Nature Conservation Review', Ratcliffe, 1977). It is necessary to indicate the status of each feature in terms of its international, national, regional and local status. The authority for each statement should be indicated, e.g. Nature Conservation Review, White Paper Command No. 7122, EC Directive on the conservation of wild birds (79/409/ EEC), EC Directive on the conservation of natural habitats and of wild fauna and flora (92/43/EEC). The criteria are as follows:

1. Size
2. Diversity
3. Naturalness
4. Rarity
5. Fragility
6. Typicalness
7. Recorded history
8. Position in ecological/geographical units
9. Potential value for future development:

   Habitat and species action plans
   Visitor services
   Education
   Marketing, retail
   Marketing, recruitment
   Public affairs
   Demonstration and advisory work

10. Intrinsic appeal
11. Other criteria
12. Identification/confirmation of important features
13. Operations likely to damage the special interests
14. Main factors influencing the management of the site
15. Land of conservation or strategic importance in the vicinity of the reserve

4. Management policy

This is the intention part of the plan. Have a path-finder meeting to discuss the policy for the site. By reference to the site's main features what should be the underlying policy for management? Have you considered non-intervention? Include as wide a range of interested parties including site staff, managers, fundraisers, statutory conservation agency personnel, local experts, etc., depending on the size and conservation importance of the site. Write down notes under the following headings and use to compile the policy statements.

1. Major management goal and supporting rationale
2. Main management objectives for:

   Habitat and species management
   Visitor services, interpretation and education

Estate services and major machinery
Public relations and administration
Research, survey, monitoring

5. Management prescriptions and operations

This is the action part of the plan. What needs to be done over the next five years in order to meet the objectives of the plan? When you have decided on the most appropriate management in relation to the priorities for it, go back to verifying that this work relates directly to the objectives of the plan (rationale), then list the work projects which will be undertaken in order to carry out the management. It is a good idea to prioritise: do the most important things first, and relate them to the amount of time required and the resources available and required. What estate services and major machinery will be needed? Is it likely that the resources will not stretch to these and if so how damaging will it be to the management objectives to put off this aspect of management until later? If so when will it be done? Have you listed all the appropriate points necessary to achieve good public relations? Have you identified the major items of research, survey and monitoring needed to quantify what is on the site and how some species/communities will change as a result of targetted management? Does this include methods for analysing the data? Have you considered fully appropriate survey or experimental design which will account for normal variation and bias? In the research and survey have you built in randomisation or stratified randomisation? If you are setting up an experiment have you considered replication?

5.1 Habitat and species management
As an example on RSPB reserves this would include prescriptions for grazing, mowing, strimming, burning, ditch clearance, hydrological considerations, patrolling to maintain refuges, coppicing, tree planting and felling, ride creation and maintenance, island creation, nest box erection, predator and competitor management, non intervention.

5.2 Visitor services, interpretation and education
As an example, on RSPB reserves this would include prescriptions for laying out nature trails, walks, open days for local people, provision of displays and leaflets, facilities for school educational programmes, membership recruitment and retail sales.

5.3 Estate services and major machinery
As an example, on RSPB reserves this would include prescriptions for siting and erecting hides, work stations and other buildings, maintaining or erecting fences, gates, paths, roads, buildings, sluices and cattle-handling units, acquisition or maintenance of tractors and other agricultural and forestry equipment and boats.

5.4 Public relations and administration

As an example, on RSPB reserves this would include prescriptions for renewal of leases, formal meetings with neighbouring landowners and local authorities, action to combat nearby developments, information to be sent to other organisations, volunteers and their tasks, minimum staffing requirements, identification of possible reserve extensions, liaison meetings with other conservation organisations.

5.5 Research, monitoring, survey

As an example, on RSPB reserves this would include prescriptions for surveys and monitoring of biotic groups, contributions to national monitoring programmes, assistance to on-site biologists, monitoring effects of management programmes, need for autecological species studies, attitude surveys amongst visitors.

Further, the major classes, reasons, range and scope of research, survey and monitoring projects are given below using examples of work undertaken on RSPB reserves, which can be a useful guide to those setting up such projects:

5.5.1 Research
A. Management experiments on reserves
Reasons: to determine the optimum (and cost-effective) management for important taxa (particularly key bird species) and communities on reserves. To enable management techniques to be refined and exported to other reserves or conservation managers.

Range and scope: Ecological management on reserves should be viewed as experiments with monitoring of experimental and control areas.

5.5.2 Monitoring
A. Red Data and other key bird species on reserves
Monthly means and maxima or fixed date counts, size of breeding populations and productivity

Reasons: to evaluate the effectiveness of reserve management for Red Data Book (RDB) and key species (targets identified in Management Plans). To compare the effectiveness of the reserve mechanism against other potential measures, e.g. SSSIs, Environmentally Sensitive Areas (ESAs) to conserve important species. To advertise the role of the reserve's mechanism in conserving important species. To contribute to national species monitoring needs and systems. To enable within- and between-sites analysis of factors affecting important species.

Range and scope: all reserves, RDB, candidate and other key species identified in individual Management Plans. Data collected by compartment where appropriate.

B. Monitoring of key non-bird species/taxa on reserves
Reasons: to detect effects on key non-bird taxa of habitat management designed to

benefit important bird species. To determine the most effective management for key non-avian taxa. To demonstrate to outside bodies, e.g. Broads Authority, Ministry of Agriculture Fisheries and Food, Internal Drainage Boards (IDBs), etc. the wider conservation effectiveness of RSPB reserve management.

Range and scope: all non-avian RDB species, key nationally scarce plants and key indicator species (e.g. ditch flora) identified in the reserve Management Plan, monitored at appropriate intervals (1–5 years depending on taxa).

C. Vegetation monitoring on reserves
Reasons: to ensure that management designed to benefit key bird species does not damage important plant communities. (N.B. Many reserves are designated as SSSIs for their botanical interest.) To help understand the response of key species and communities to management in order that management techniques can be refined.

Range and scope: key indicator communities or where Management Plan prescriptions have been defined in terms of vegetation manipulation or which are likely to damage important plant communities. Mostly annual.

D. Monitoring the effects of reserve management on the physical environment
Reasons: to quantify and evaluate the effects of management on reserves, i.e. what was done and what happened? To relate performance of key species to management in order that management techniques can be refined. To provide evidence of damage to reserve interests by external influences. To enable management techniques to be exported to other reserves or conservation managers.

Range and scope: parameters, to which management can respond, identified in the Management Plan as likely to affect key bird species on the reserve (e.g. water levels, water quality, soil nutrient levels, etc.). All managed reserves. The frequency of measurement will depend on the parameter.

E. Reserve visiting figures by site, month and type
Reasons: to optimise allocation of visitor resources among reserves. To prioritise, develop and justify capital investment programmes on reserves. To identify and rationalise marketing opportunities on reserves.

Range and scope: recorded on a monthly basis for wardened reserves, sampled occasionally for others. Information should be reviewed annually.

5.5.3 Survey
A. Species lists for reserves

Reasons: to determine whether important species are present on the site and which might be sensitive to management intended to benefit key bird species or merit management in their own right. (N.B. Site safeguard imposes wider conservation responsibilities on RSPB than management for birds alone.) To detect long-term ecological changes which might need attention.

Range and scope: as a minimum, higher plants, vertebrates and other groups, particularly priority insect groups, for which the reserve is likely to be especially important (as identified in the Management Plan). Information should be updated every five years.

B. Baseline habitat and vegetation surveys on reserves
(a) Habitat area (NCC/RSNC codes) by individual reserve and collectively

Reasons: to record the long-term effects of management and other influences on the extent of different habitats on reserves. To assess whether habitat-based objectives set in reserve Management Plans have been achieved. To determine areas of particular habitats managed by RSPB (for use by various departments, e.g. Habitat Action Plans and to allocate resources towards priority habitats). To place value of RSPB reserves for habitat safeguard in a national context.

Range and scope: all reserves, mapped and updated every five years but with significant changes reported annually.

(b) Mapping of classified plant communities (NVC classification)
Reasons: to record and evaluate in a national context important plant communities on reserves in order to assist management planning decisions and to demonstrate major changes. (NVC provides a standardised, adaptable and cost-effective method for classifying and mapping vegetation; survey quadrats can be converted into monitoring quadrats to record future changes in relation to management.)

Range and scope: reserves with nationally important plant communities and reserves where management for key bird species has/would result in gross changes to important plant communities.

6. Five-year work programme

Detail all work to be undertaken over the five-year period which fits with the prescriptions and hence to fulfil the objectives of this five-year term of the plan.

6.1 All projects and years of operation listed

7. Appendixes

1. References
2. SSSI notification sheet
3. List of Potentially Damaging Operations (PDOs)
4. Maps and plans

In many cases for larger sites a 'path-finder meeting' is arranged and can be a key stage in the plan production process at which site managers, ecologists, land agents and a representative of the statutory nature conservation organisation participate. In the Royal Society for the Protection of Birds for example, other staff and even outside specialists also take part as appropriate in order to consider such factors as research requirements, education and marketing initiatives. The main purpose of a path-finder meeting is to decide the main policies or objectives for the reserve and to anticipate and resolve any problems or conflicts involved in achieving them.

It can be useful to produce a draft management plan, usually within six months of initial consultations, in which the detailed prescriptions or methods of operation are presented. To aid the preparation of the plan, the person responsible for carrying out management of the site should be issued with guidelines with worked examples indicating the way in which information should be presented and the level of detail required, although the latter will vary depending on the size and complexity of the site. The draft plan should then be reviewed by the original team involved in the path-finder meeting.

The collection of background data and the processes of external and internal consultation are considered to be especially important in the plan production sequence. In particular the involvement of a multi-disciplinary team ensures that there is a common understanding of the plan's purpose.

Biological research, survey and monitoring in relation to site management

Many organisations place high emphasis on managing sites for their intrinsic nature conservation value. For example, RSPB strives to integrate management for birds with other important nature conservation goals. To ensure that the conservation value of sites is maintained or enhanced considerable resources (in relative terms) should be put into research, survey and monitoring of sites.

Research

Much of the habitat and species management undertaken on reserves should be regarded as experimental and should be designed as such with appropriate monitoring of experimental and control areas. In any research it is necessary to account for two sources of error: normal variation and bias. Bias is associated with poor study design, individual observer bias, inadequacy in the experimental design, lack of randomisation, lack of replication, surveying in bad weather or inappropriate times of day or year, comparing habitats where detectability of, say, birds, is different because of different habitat structures. Two important considerations are precision and accuracy. Results with minimal amounts of normal variation and bias are known as precise and accurate (or unbiased), respectively. We might measure the girth of coppice stools with considerable precision down to the nearest centimetre. But it would be inaccurate for us to say that the mean girth of all coppice stools was x cm if we had not sampled the 'population' of stools randomly and without bias. We might be very precise in our measurements but inaccurate in that our predicted mean may be far from the 'true' mean for reasons of our biased sampling programme.

An experimental approach enables us to learn about habitat management in a structured way. Often 'natural' experiments are conducted to take advantage of habitat change. For example, bird communities might be studied in two forests, one with many trees felled by wind, the other unaffected by wind-blow. However, this is not a true experiment. Likewise counting birds in one plot before and after management and attributing any resultant change to the effect of management is erroneous. If the counts had been made in successive years how would we tell whether the change in, let's say, a particular species, was due to habitat management (the experimental factor) or to an extraneous factor (e.g. low rainfall)? We need to know what would have changed on the plot if no habitat alteration or management had occurred, i.e. to be a formal experiment the observed site must receive a purposeful 'treatment' with another site acting as a control. A well-designed study, therefore, needs its own control with the same level of research effort in the 'before' and 'after' stages of the trial as is given to the experimental plot itself. In the wind-blown forest example above, the study would constitute a natural experiment if data had been collected from both sites before the wind-blow occurred. Box 2.2 shows the use of an experimental control to test the effects of thinning conifer plantations on populations of Wren *Troglodytes troglodytes* (Bibby *et al.* 1992). If Plot A had been studied alone the decline in numbers between Year A and Year B could equally have been attributed to extraneous factors such as heavy winter snowfall. By including a control plot in the study, on which numbers of wrens did not decline, we can conclude that the decline in numbers on the experimental plot was probably due to thinning.

To increase the confidence we can place in an experiment it is important to replicate the blocks in a trial. For example, if we studied the effect of coppicing on tree-dwelling invertebrates, the coppiced plot might differ in some other, unmeasured way, from the control plot. The invertebrates in the coppiced plot might change more than in the control but for extraneous reasons. If the coppice 'treatment' was replicated by having two blocks each receiving a treatment of coppicing and a control, and the same differences occurred in both blocks, we could be more confident that the effect was due to coppicing. Blocks can be replicated a number of times. Each additional replicate increases the confidence placed on the observed effect being the result of the treatment.

In nature, few things are distributed randomly. However, in order to have an effective sample, it is important to sample randomly from the population under study. If we place

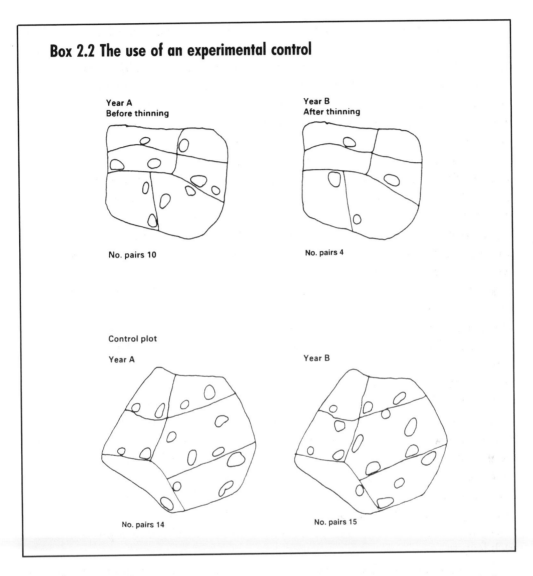

Box 2.2 The use of an experimental control

Year A
Before thinning

No. pairs 10

Year B
After thinning

No. pairs 4

Control plot

Year A

No. pairs 14

Year B

No. pairs 15

quadrats in a woodland in order to measure something about the ground flora communities, it is better to place them randomly within a grid placed over the woodland rather than simply placing them in certain areas. It is a good idea to generate a series of random numbers on a calculator which are then used to find the intersection of the grid placed over a map of the woodland. Quadrats will then be placed at random. If the objective of the research is to study, say, the population density of small mammals on a reserve prior to management, a number of plots can be selected at which to conduct the sampling. The areas should be randomly located within the total area over which we want the answer to apply, i.e. the whole reserve. The critical feature of randomness is that, as each plot is chosen, it should have equal probability of falling anywhere within the study area and its location should not be influenced by any prior knowledge. This requirement is essential to ensure that the resulting answers are representative of the study area without bias. A further development of random sampling is stratified random sampling. In some cases it may be advantageous to recognise the effect of habitat variation on the populations of study organisms. If we wished to sample the abundance of hoverflies on a site but also in different habitats, some of which cover larger areas than others, plots would be chosen with knowledge of the distribution and abundance of the various habitats. In some cases it may be desirable to put more effort into the less frequent habitats than would be the case if samples were taken at random. All possible plots on a grid placed over the site are attributed to a stratum (e.g. habitat) and the sample squares are picked randomly from within each stratum rather than for the site as a whole. Where the habitat variation is not sharply defined but is more continuous, plots could be spread along the range of variation, vegetation cover measured and the analysis performed by regression methods.

More detailed methodologies are given in various books on quantitative plant ecology and in Krebs (1989) and Bibby *et al.* (1992).

Monitoring

For the purposes of this chapter we will call the repeated recording of site, habitat or species '*monitoring*' and the regular updating of the management plan '*review*'.

Monitoring is important for all species and habitats but especially rare and endangered species, all of which should have 'specified limits' indicated and justified within the management plan. Specified limits are defined as the acceptable ranges of values for the main species and habitats. These limits should include high value species and those which might become too numerous, i.e. the maximum and minimum values for population size or extent. They can act as criteria for commencing or terminating management or monitoring. They should be shown in a table which indicates the present status, lower limit, upper limit, trend, controlling factors (where known), for each feature.

Monitoring is also important for fast-growing or aggressive plant or animal species, e.g. *Spartina* on estuaries, Silver Birch *Betula pendula* and Bracken *Pteridium aquilinum* encroachment on heathland, gull colonies which compete for space with terns and Avocets *Recurvirostra avosetta*. It is important to know change in species abundance in relation to a change in habitat and to be able to separate natural and man-made changes. Often, annual changes in a long-term trend may be quite small but there may be substantial year-to-year variation which could conceal a long-term trend which is important for the manager to understand.

Monitoring is not the same as recording or survey. It should have clear objectives which need to be agreed and set down in the plan (Goldsmith, 1991). It should have unambiguous methods including data analysis and interpretation, e.g. to count the number of avocet nests each year and perhaps their distribution. This should include an indication of the thresholds which should trigger action, such as if the population falls below a certain number contact a particular person, or carry out a specific study or control measure. In turn, these too will need monitoring.

In monitoring vegetation it may be desirable to identify all species of grasses and perhaps mosses and lichens as important indicators but it is important to use appropriate methodology. For example, use percentage cover rather than Domin scale and fixed and aerial photography are often used to monitor change cost-effectively. For some species the whole population can be counted, e.g. breeding terns nesting on a shingle spit but other, commoner species which make up a particular community, may need to be counted from a series of samples. The monitoring methods employed need to be repeatable from year to year. Repeatable counts do not need to be accurate in the sense that the population numbers recorded are the actual population figures. If the numbers are constantly wrong for any reason, the *changes* from year to year can still be measured accurately. One of the most often quoted monitoring schemes of British birds, the Common Birds Census (CBC), relies on observers visiting the same area year upon year. All individual birds registered are recorded onto maps from a series of 10–12 visits spread through the breeding season. The method is often employed by reserve wardens to monitor birds on a site. The CBC, when analysed for all plots counted by volunteers, avoids the problem of some of this bias by calculating a population index based only on plots counted in the same way by the same observer in successive years. Provided an observer with poor hearing detects twice as many Goldcrests *Regulus regulus* if numbers double, it does not matter if only six and twelve pairs are recorded when in fact there were ten and 20. This method is very costly in terms of the volume of data produced, however. If the reserve is especially large (>20 ha) the CBC method is not the most appropriate and other methods such as transects, point counts, etc., supported by look-see searches for rarer species, should be used (see Bibby *et al.* 1992, for birds). The most important point about population monitoring is not that the methods be accurate but that they be similar from year to year so that systematic inaccuracy does not have an adverse effect.

Finally, 'fulfilment' or 'stopping rules' need to be considered. As monitoring is time-consuming and therefore expensive it needs rules for termination. The problem here is that once monitoring has been set up it is difficult to stop it. Often the longest runs of data are the most useful although some examples could be quoted which use poor methods with few or no products but which continue to be undertaken simply because they have been done for many years. In the end such schemes can only be said to have wasted money. The review of the management plan should be used to review monitoring schemes and their products on the basis of their cost-effectiveness.

Recording management undertaken

It is particularly important to record the habitat and species management undertaken on reserves (including type, location, cost, manpower) and to monitor the results of this. Only by so doing is it possible to document and evaluate the ecological and cost-effectiveness of the management carried out. Recording the management undertaken also enables the cost-effectiveness of reserve acquisition and management to be compared against other potential conservation measures with similar aims. Many site managers fail to record adequately the management undertaken on reserves and some may not even record it at all! Alternatively management may be undertaken that is not prescribed in the management plan and bears no relationship to achieving the objectives set out within it.

A computerised management planning and recording system, the Countryside Management System (CMS), has been developed for the British government's conservation agencies in order to record management undertaken on reserves and to ensure relevance to the management plan. CMS provides the framework to convert prescriptions (to meet objectives of the plan) into 'projects'. Projects are clearly defined units of work which form the basis for the production of work plans. Advantages of the system include standardisation across reserves and organisations, provision of reports with greater ease, encourages systematic recording of management and species responses to it through various co-ordinated research and survey projects, provides a reporting structure to managers through to the parent organisation, and links resource requirements such as manpower and costs to the management plan, enabling priorities of work to be set. A disadvantage is that the system, when first implemented, appeared bureaucratic, too theoretical and gave a sense of 'managing conservation on paper' to the site wardens who are charged with carrying out the practical conservation work on the ground. For small sites and for organisations with limited resources, it is unlikely that they should need or desire to implement such a sophisticated system, though it is as important for them to monitor the results of their management as it is for the larger organisations.

Monitoring the implementation of management plans

A workable, basic, standardised recording and reporting (preferably annual) system is essential to allow the progress towards attaining the ecological objectives of the management plans, and the costs of achieving them, to be assessed but few organisations operate in this manner. Some do require annual reports from their site managers, but often these are not standardised across the organisation and the information they contain is often difficult to retrieve. The main vehicle for monitoring progress towards implementing the management plan should be a reserve Annual Report.

As an example, the R S P B Annual Report format for reserves mirrors that of Section 5 of the Management Plan format (see Box 2.1) with reports organised under the same headings. It consists mainly of a series of standard proformas which have to be completed by the warden. These are used to summarise bird numbers and productivity and to record habitat and species management projects undertaken, staffing levels, visitor numbers and marketing information.

The project recording form is almost identical to the planning and recording form employed in the C M S system described above and incorporates the standard N C C project codes. An example of a completed project recording form is shown in Box 2.3.

Periodic review of management plans

The management plan with its successes and failings should be reviewed regularly, say every five years. The descriptive part and formulation of objectives may remain unaltered for a decade or more whilst the prescriptive part should be seen as being flexible and evolving. Plans need to be flexible and require regular updating to take into account new information, successional and other changes that occur, new extrinsic threats and new ecological information. The type of information that needs to be included are:

Extent of priority habitats and changes over five years

Population levels of rare and conservationally important species for the period of the existing plan

Population trends of any other key/target species

The work achieved against the previous plan (major species and habitat prescriptions undertaken, and whether they achieved the plan's stated objectives)

Education, visiting and marketing information (more relevant to larger bodies)

New information provided by other conservation and statutory bodies

Box 2.1 should be used as a guide in constructing a management plan. At appropriate places questions are asked. Answering these questions and by following the guidance in this chapter should enable anyone with a need to produce a management plan to do so in a standardised way.

Box 2.3 Example of a project recording form. (The one on the right has been left blank so that it can be copied.)

SITE CODE	MINS		
PROJECT CODE	MH76	PROJECT NO 01	FINANCIAL YEAR 92-93
PROJECT TITLE	MANAGE HABITAT, COASTAL, BY SCRUB CONTROL		
QUALIFYING PHRASE	MANAGE DUNE VEGETATION		
SHORT DESCRIPTION			
MANAGEMENT OF SCRUB TO IMPROVE DUNE VEGETATION COMMUNITY			

SECONDARY PROJECT CODE		SECONDARY PROJECT NO	
COMPARTMENTS	49, 58, 59		
MONTHS ACTIVE	JAN/ FEB/ MAR/ APR/ JUN/ JUL/ AUG/ SEP		
PROJECT PRIORITY	1	PHOTOGRAPH	N
FILE NUMBER	MH72	RECORDER	GEOFF WELCH

MAN DAYS			
WARDEN	2	ESTATE WORKER	VOL WARDEN 5.5
OTHER STAFF		OTHER STAFF	
SUBHEAD	7100MI	EXPENDITURE 10	INCOME 0

REPORT

SEP:
Gorse and grass bordering the dune footpaths was cut back and burnt on site to widen selected existing paths in an attempt to reduce erosion of the vegetation by trampling

MAR:
Further work carried out behind East & Public Hides to complete this task.

This is the start of a more intensive scrub control and mowing programme to maintain an existing dune vegetation community.

Work recommended by English Nature.

SITE CODE			
PROJECT CODE		PROJECT NO	FINANCIAL YEAR
PROJECT TITLE			
QUALIFYING PHRASE			
SHORT DESCRIPTION			

SECONDARY PROJECT CODE		SECONDARY PROJECT NO	
COMPARTMENTS			
MONTHS ACTIVE			
PROJECT PRIORITY		PHOTOGRAPH	
FILE NUMBER		RECORDER	

MAN DAYS			
WARDEN		ESTATE WORKER	VOL WARDEN
OTHER STAFF		OTHER STAFF	
SUBHEAD		EXPENDITURE	INCOME

REPORT

References

Bibby, C.J., Burgess, N.D. & Hill, D.A. (1992). *Bird Census Techniques*. London: Academic Press.

Countryside Commission (1986). *Management Plans: a Guide to their Management and Use*. CCP 206, Cheltenham: Countryside Commission.

Countryside Commission (1992). *AONB Management Plans: Advice on their Format and Content*. Technical report. Cheltenham: Countryside Commission.

Goldsmith, F.B. (1983). Ecological Effects of Visitors and the Restoration of Damaged Areas. In *Conservation in Perspective*, ed. A. Warren & F.B. Goldsmith, pp. 201–14. Chichester: Wiley.

Goldsmith, F.B., ed. (1991). *Monitoring for Conservation and Ecology*. London: Chapman and Hall.

Krebs, C.J. (1989). *Ecological Methodology*. New York: Harper Collins.

Nature Conservancy Council (1983). *Handbook for the Preparation of Management Plans*. Peterborough: Nature Conservancy Council.

Nature Conservancy Council (1988). *Site Management Plans for Nature Conservation: a Working Guide*. (Shorter format). Peterborough: Nature Conservancy Council.

Ratcliffe, D.A. (1977) *A Nature Conservation Review*, vols. 1 & 2. Cambridge: Cambridge University Press.

Roberts, K.R. (1991). Field monitoring: confessions of an addict. In *Monitoring for Conservation and Ecology*, ed. F.B. Goldsmith, pp. 179–212. London: Chapman and Hall.

Tait, J., Lane, A. & Carr, S. (1988). *Practical Conservation: Site Assessment and Management Planning*. Milton Keynes: Open University Press.

Warren, A. & Goldsmith, F.B. (eds) (1983). *Conservation in Perspective*. Chichester: Wiley.

3 Access

JULIET VICKERY

Introduction

In recent years there has been a growing interest in outdoor recreation, 'green' tourism and wildlife conservation. This has resulted, in turn, in an increase in the number of people visiting the countryside in general and reserves in particular, and a growing awareness among conservation organisations of the need to accommodate these interests by providing easy access to reserves. There is no better way to excite and interest people in wildlife and the countryside than through first-hand experience. Reserves provide invaluable opportunities to educate and inform the public concerning nature and nature conservation. However, visitors to reserves are also a potential cause of considerable habitat damage and disturbance and the provision and management of access to sites requires careful planning.

The first stage in considering 'access management' on a reserve is to decide whether access should be encouraged at all and if so to what extent? This will be determined by the degree to which the nature conservation value of the site may be put at risk by increasing levels of human disturbance. The major concern is physical damage to plant communities and disturbance to breeding or feeding birds or mammals. For birds, human-induced disturbance during the breeding season may lead to nest desertion and increased predation or, outside the breeding season, to a reduction in the level of foraging activity and ultimately the use of a site. The need to prevent access completely to a site is unusual except perhaps in the case of fragile plant communities such as raised peatland bogs where a footprint on the *Sphagnum* surface may take 20–30 months to disappear. In practice it is almost always possible, through careful site layout and design, to provide for access without damaging the conservation value of a site.

Assuming access is to be allowed the next question is to what extent? Once again this will depend primarily on the nature of the site, as discussed below, and the resources available. However, there are a number of general considerations worth bearing in mind at the outset. The development of facilities, such as a visitor centre, hides or toilets on the site will require not only capital outlay but considerable running costs in terms of maintenance and insurance. It is also worth remembering that a good insurance policy, with public liability cover, will be necessary for any reserve and the cost of this will increase with the number of facilities such as hides that are provided on the site. A visit by a health and safety inspector and/or a safety audit will highlight any hazards and may reduce the insurance premium.

Vandalism is frequently a problem, particularly on sites relatively close to urban areas. In general, on such reserves it is often best to keep the use of man-made artifacts to an absolute minimum, even down to avoiding the use of stakes and whips for trees. Vandalism can often be held in check by rectifying the damage immediately it occurs, e.g. cleaning off graffiti as soon as it appears. The culprits are usually local and experience suggests that it is best to 'make no enemies'!

The maintenance of any facilities, from repairing general wear and tear or damage from vandalism to the emptying of litter bins, will of course be much more difficult on unwardened reserves. Whilst most of the principles of managing access apply equally to wardened and unwardened sites it is best to keep facilities to a minimum at the latter.

Remember also, when considering the extent to which to develop a site, that many people visit reserves to escape their fellow countrymen and seek the solitude of the countryside. Some areas should be retained where this experience is possible and throughout the site the visual impact of any facilities from hides to litter bins should be minimised. Maintaining official opening hours does have the advantage of ensuring that some periods of peace and quiet are retained both for neighbours and wildlife, but the collection of visitor fees on entry leaves little time to do much else on a reserve.

A final general point concerns access for visitors with special needs. Features and facilities designed for people with special needs cannot be seen in isolation but rather as part of the entire management plan – there is little point in providing the odd handrail or ramp if, to reach a site, a person with special needs must negotiate steps or stiles beforehand. Few of the features and facilities involve major expenditure; it is usually a case of adapting to cater for a wide cross-section of the community. Detailed advice on dimensions, designs and standards of facilities are given in 'Informal Countryside Recreation for Disabled People' (Countryside Commission, 1982) and Countryside Commission for Scotland Information Sheets (CCS, 1988a).

Zoning

There are some broad principles of visitor management that must be considered before the details of site layout and design can be developed. Nature conservation objectives may be better satisfied by allowing the bulk of recreational use to remain concentrated in certain areas (see Fig. 3.1). This can reduce the impact on species that are sensitive to damage or disturbance and it may be more economical to provide facilities in localised areas. The main concern, as for access in general, is the trampling of vegetation and disturbance of birds. Mapping the areas where vulnerable vegetation communities and sensitive bird species occur is often a useful first step towards zoning. In the latter case this should indicate the location of possible nesting, roosting and feeding sites for sensitive species and, where possible, the minimum areas required by a species. The ideal zoning system grades use in terms of noise and activity levels, for example: car parks, picnic and recreation areas, a bankside path and hides and an area of no public access. The subsequent positioning of hides and paths will depend on the zoning, for example hides should not be placed too far into the centre of the reserve if this is to serve as an undisturbed area.

Zoning may be adopted on either a permanent or a seasonal basis. It is often used on waterbodies where visitor access is restricted to one area of a large lake or river, or to one or two smaller lakes in a series such as those that comprise disused gravel pits. The main threat is disturbance of waterbirds. Tolerance distances for most waterfowl lie

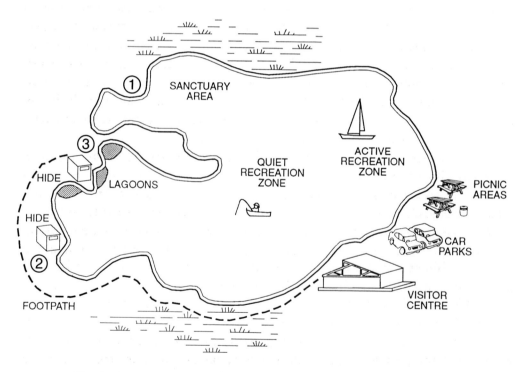

Fig. 3.1. An example of a zoning scheme for a large inland waterbody. (1) sanctuary areas should be designed according to the habitat requirements of the species present: wildfowl will use open shores for loafing, dabbling ducks require shallow water (< 1 m) for roosting and feeding, whilst diving ducks require deeper water (1–2.5 m) for feeding. Shingle islands provide ideal nesting areas for waders and terns. (2) Hides should be set well back (< 50 m) from the edge or built up to see over screening vegetation. (3) Where footpaths run adjacent to the sanctuary areas, they should be well-screened with dense and/or evergreen cover. (After Sidaway, 1991 and Ward & Andrews, 1993.)

between 100 m and 200 m and sites should be designed such that birds can withdraw at least 200 m from sources of disturbance or into cover (Andrews, 1991). The creation of roost sites on islands and spits of gravel in the pits at Snettisham, Norfolk has proved successful in attracting up to 70 000 Knot *Calidris canutus*.

In offshore areas access and visitor pressure, mainly visiting pleasure boats and scuba divers, can be regulated by providing moorings at suitable locations, e.g. near to seal haul-outs (so they can be observed with minimal disturbance) and at the most popular diving sites. Backed by high profile patrolling this can do much to reduce damage to seabed communities and breeding birds and mammals. Such a system of regular patrolling and user zonation has been successfully employed at Skomer Marine Nature Reserve (Fig. 3.2).

Zoning can also be effective in fragile terrestrial habitats such as sand dune systems. Exposed dunes near the sea are particularly sensitive to pressure from public use; trampling destroys the characteristic sparse dune vegetation and accelerates erosion particularly on the foredunes. Foreshores and beaches can, however, withstand high densities of people and slacks, stable backdunes and dune grasslands may tolerate more intensive use than the exposed dunes. By allowing relatively free access to the seashore and to inland buffer zones, pressure can be reduced on the more sensitive areas between them. Cars can be kept off dune grassland by digging roadside ditches or building low timber barriers. Providing facilities in inland zones, such as picnic tables, will also help to encourage people to remain in these areas.

The Royal Society for the Protection of Birds (RSPB) adopt a zoning system in Abernethy Forest, Speyside, where there is concern over the effects of human disturbance on Capercaillie *Tetrao urogallus* for which Abernethy is one of the few remaining suitable pinewoods. Here a 'honeypot' has been created at Loch Garten where successive pairs of Ospreys *Pandion haliaetus* have been shown to the public from a hide equipped with binoculars, telescopes and closed-circuit television. The hide also serves as a visitor centre and there are waymarked walks in the pinewoods nearby. This part of the reserve is well publicised with large car parks and large signs to welcome the visitor. However, the rest of the reserve, i.e. the forest itself, is not promoted by the RSPB. Media features, location interviews and glossy photographs of the site are kept to an absolute minimum and in all such publicity the fragility of the site is stressed. Within the forest people are quite welcome to explore the area using the forest tracks but no new ones are being created and a number of existing ones are being removed to reduce the extent of forest penetrated by tracks. Vehicular access is not prevented but the track to the Forest Lodge is not tarmaced.

Waymarked trails are often a good way of concentrating visitor use away from sensitive areas. Ideally trails of different lengths should be provided in order to allow some freedom of choice in how far people wish to walk, but keep all distances relatively short. A figure of eight provides an easy way of creating both a short walk (one loop) or a

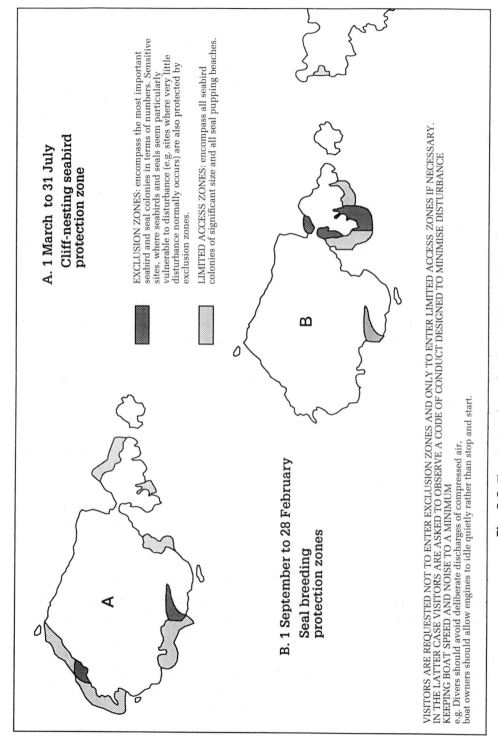

A. 1 March to 31 July
Cliff-nesting seabird
protection zone

EXCLUSION ZONES: encompass the most important seabird and seal colonies in terms of numbers. Sensitive sites, where seabirds and seals seem particularly vulnerable to disturbance (e.g. sites where very little disturbance normally occurs) are also protected by exclusion zones.

LIMITED ACCESS ZONES: encompass all seabird colonies of significant size and all seal pupping beaches.

B. 1 September to 28 February
Seal breeding
protection zones

VISITORS ARE REQUESTED NOT TO ENTER EXCLUSION ZONES AND ONLY TO ENTER LIMITED ACCESS ZONES IF NECESSARY. IN THE LATTER CASE VISITORS ARE ASKED TO OBSERVE A CODE OF CONDUCT DESIGNED TO MINIMISE DISTURBANCE KEEPING BOAT SPEED AND NOISE TO A MINIMUM
e.g. Divers should avoid deliberate discharges of compressed air, boat owners should allow engines to idle quietly rather than stop and start.

Fig. 3.2. The zoning scheme for Skomer Marine Nature Reserve.

longer one (both loops). Recommended distances for such walks are less than one and less than two miles respectively.

Seasonal zoning may be required in addition to or instead of geographical zoning particularly if there are practical limitations to spatial zoning. In some peatland sites it may be necessary to restrict access to midsummer when the peat is at its driest and/or midwinter when the peat is also relatively dry and the vegetation dormant and so less vulnerable to damage. This seasonal zoning has been suggested for the dome of the raised bog, Cors Fochno in Wales.

Seasonal zoning may also be usefully adopted in coastal habitats since colonies of breeding seabirds and seals and roosts of wintering waders are often particularly sensitive to disturbance. On the Farne Islands, off the Northumbrian Coast, access to the colonies of breeding seabirds is regulated by issuing a fixed number of licences to operate boats transporting visitors to Staple and Inner Farne. In April, August and September boats are permitted to land an unlimited number of times, between the hours of 10 a.m. and 6 p.m. at recognised jetties. However in May, June and July each boat may land only once between 10.30 a.m. and 1.30 p.m. on Staple and once between 1.30 p.m. and 5 p.m. on Inner Farne. In this way the seabird colonies remain undisturbed for 21-hour periods during the breeding season. Similarly access is not permitted to the huge colonies of Gannets *Sula bassana* on Grassholm, off the Pembrokeshire coast, during the breeding season.

A special case of zoning is that required for rare species. In this case providing access to one site may serve to reduce the pressure and therefore potential damage elsewhere. Classic examples of this are orchids such as the Monkey Orchid *Orchis simia* which occurs at a number of sites in Kent. Access has been developed at one of the sites where it was already relatively easy. Individual monkey orchids on the site are caged to prevent accidental damage but these cages may be removed, for photographs, by the warden. Creating conditions that allow some plants to be photographed may reduce the likelihood of unwelcome attention elsewhere. This principle also applies to less rare, but photogenic species, within any reserve – if some specimens are easily photographed there is less chance of adjacent plants being damaged through uncontrolled 'gardening'.

The provision and promotion of access to a limited number of nest sites of rare bird species may also serve to reduce disturbance of this species elsewhere. This is a policy that has been adopted by the RSPB both for Ospreys at Loch Garten and Slavonian Grebes *Podiceps auritus* at Loch Ruthven. Here publicity is given to particular nests that can be guarded and to which access can be controlled.

The sudden appearance of a rare bird on a reserve may be a mixed blessing. The dramatic increase in the popularity of 'twitching' means that a rare bird is likely to attract huge numbers of people to the site very quickly. The decision as to whether to report a rarity depends on the personal choice of the warden and the nature of the site and the bird species concerned. On a reserve where the vegetation is extremely sensitive to trampling or if the bird is a potential breeding species it may be better kept quiet. A rarity can, however, give many birdwatchers a great deal of pleasure and generate considerable income through entry or car park charges. At Holme in Norfolk, in 1992 alone, each of four rarities generated around £1000 through entry charges for non-members and voluntary contributions from members. On one of the four occasions at Holme a temporary car park was established from which access to areas where the vagrant Ruppell's Warbler *Sylvia rueppelli* could usually be seen without causing disturbance elsewhere on the reserve. The huge numbers of visitors, attracted to the reserve, effectively 'queued' to be taken in small groups for a limited period of time. If the rarity was not seen during this time then it was back to the queue to try again later! Although this requires a good deal of organisation it is particularly appropriate if the bird is a secretive or skulking species and there is usually no shortage of volunteers to assist where rare birds are concerned.

Insects are probably less sensitive to disturbance than birds and mammals. The major concern for these species is that of collection. For example, the Large Blue butterfly *Maculinea arion*, which is the subject of an ongoing reintroduction programme at four sites in Britain, may suffer from this form of exploitation. Although currently the location of these sites is a well-kept secret since the reintroduction is so recent, access will be promoted to some sites in the future. The favoured habitat of this species is unfertilised, well-grazed, thyme-covered grassland and is not vulnerable to trampling. The presence of visitors on a site may effectively guard the species against the activities of potential collectors.

Car parks

Control of access begins at the car park. Its location and capacity set the approximate maximum number of visitors to the reserve during peak periods. It should ideally be on free-draining soils, to reduce the cost of surfacing, and in an area that minimises the visual impact of the site. Where possible the car park should be screened by existing areas of natural vegetation. Tree planting, with species typical of the area, can improve the appearance of car parks and the shade provided will make it more pleasant for users. An orderly layout allows capacity use whilst minimising congestion and confusion. Large formal car parks with rows of cars often look intrusive in rural areas and there may be considerable cost in preparing the site; clearing it of vegetation, improving the surface and creating internal barriers. Small enclosed bays are visually less intrusive but the area is less easily policed and can encourage theft or vandalism.

Five per cent of spaces should be for people with special needs, located near facilities such as toilets and visitor centres and in more scenic parts of the car park. The parking spaces must be wide enough to allow easy movement in and out of the car (minimum width of 3.0 m, preferably 3.6 m – a standard bay is 2.4 m) and with level access to footpaths.

Footpaths

Footpaths should be visually attractive but unobtrusive, be easily maintained and have minimum liability to erosion. Minimising the visual impact of a path is relatively easy in woodlands but more difficult in grasslands, heaths and montaine habitats. In general it is best to follow any natural visual boundaries of the terrain and vegetation and make maximum use of topographic shelter. Direct ascents of hills and straight-line paths should be avoided.

People will, however, frequently cut corners and there are various ways of generally 'encouraging' walkers to keep to paths. Obligatory methods such as fences and barriers are not recommended since they are frequently expensive, unattractive and often ineffective unless totally unclimbable! Screens of evergreen shrubs or hedges of thick or thorny vegetation such as Hawthorn *Crataegus monogyna* and Blackthorn *Prunus spinosa* are a good deterrent to wandering off the route. Screening species should, however, be in keeping with the habitat and are usually unsuitable in open habitats. They can prevent or reduce corner cutting by serving as physical barriers or by obscuring the view so that the way ahead is not immediately obvious. Boulders have proved effective at keeping walkers on the paths on long-distance routes such as the nature trail up Ben Lawers. However, they should be carefully placed and look natural, e.g. set into the ground with the weathered side up and surrounding vegetation replaced. In woodlands a few small, well-positioned logs at the edge of the path can be effective in reducing wandering from the path. Paths into dark woodlands, such as conifer plantations, may pose very different problems since these can be intimidating to some people. One solution to this problem is to have wide and light grassy rides with a mown central strip. Paths that do have to go under a canopy should be clearly defined by surfacing such as woodchips to encourage use and ensure people are not concerned about losing their way.

The sensitivity of species of plants to trampling and birds to disturbance varies between different community types and species. Heaths and moorlands are particularly sensitive to damage from trampling. Few of the plant species are resistant to this, the underlying peat and soil tends to erode rapidly once the vegetation is removed and, due to the naturally low fertility of such habitats, pioneer vegetation is slow to establish on bare areas. By comparison grass species are relatively resistant to trampling but there is a need to judge the sensitivity and importance of plant species before encouraging free-ranging access. Ground-nesting birds such as Golden Plover *Pluvialis apricaria* (Yalden

& Yalden, 1989), Lapwing *Vanellus vanellus* and Redshank *Tringa totanus* (Andrews, 1991) on moorland and grassland sites may be vulnerable to disturbance especially when with eggs and young. It has been suggested that in areas suitable for breeding Golden Plover, adjacent paths should be a minimum of 400 m apart to ensure birds are not disturbed in the breeding season (Yalden & Yalden, 1989).

The location of the path in terms of the gradient of the slope and the nature of the soil and vegetation is critical in reducing the risk of erosion and therefore the cost of maintenance. For this reason a physical survey is advised before planning paths and routes on a reserve. This should focus particularly on drainage and topography (slope), and soil and vegetation type, all of which will influence the susceptibility of the route to erosion. Erosion becomes more damaging as a slope steepens. The optimal maximum for the slope of a path is considered to be 7°; paths of this gradient are comfortable to climb and can be constructed with narrow ditches at the side to prevent water running down and eroding the surface. Where the slope is steeper a choice has to be made between a steeper ascent or the construction of zig-zags. Where erosion is likely a gradual ascent with zig-zags should be chosen if at all possible. Downhill use is more damaging than uphill and routes should be designed so that users ascend steep slopes and descend gentle ones. Steps are difficult and costly to build and should be avoided unless absolutely necessary, for example where this is the only way to prevent further extreme erosion or the path is dangerous as a result of being wet and slippery.

Soils vary greatly in their resistance to wear; surfaces with a high proportion of coarse particles (rocks or stones) are generally least affected by walkers, followed by sandy soils, loams and silts, with clay and peat soils being most easily affected (Bayfield & Aitken, 1992). Most soils have lower resistance to wear under wet conditions and drainage is therefore critical in determining the suitability of sites for paths. Excess water is indicated not only by the presence of surface water but also by the colour of the soil. Permanent waterlogging results in grey or blackish soil and seasonal waterlogging leads to mottled soils with paths of a rusty brown colour (Agate, 1983). Where possible paths on permanent or seasonally waterlogged soils should be avoided.

Boardwalks are often deployed where terrain is impassable, such as on deep marshes, or to protect fragile habitat such as bogs, marshes or sand dunes. Boardwalks are visually intrusive and should be kept to a minimum. They need to be carefully and tidily constructed and well maintained. There is often no physical reason for walkers to stay on boardwalks, except in very wet marshes and bogs, and the width and line must be such that walkers are not tempted off them. Suggested widths are: 0.75 m to 0.9 m for one-way use and 0.9 m to 1.2 m for two-way use with minimum widths of 1.2 m and 1.7 m respectively for wheelchairs. Boardwalks designed as nature trails must have large passing places perhaps with information boards. Viewing platforms and passing places should be provided with handrails and ideally with a seat or perch to allow leaning or resting for the handicapped visitor. Where handrails are considered unacceptable, boardwalks with an edging kerb at least 10 cm high may still be safe for wheelchair users. Such kerbs will also be valuable tapping rails for the blind or partially sighted visitors. Detailed specifications concerning the construction of boardwalks are available from the Countryside Commission for Scotland.

On peatland special attention must be paid to the type of timber used since preservatives from treated timber board walkways can cause damage to vegetation on peatland bogs. Suitable woods include hardwoods, such as Beech *Fagus sylvatica* and Sweet Chestnut *Castanea sativa*, which have high endurance and will last a long time whereas locally available rough sawn timber may be adequate for less demanding uses (Rowell, 1988). Piles in deep peat must be long enough to ensure the structure does not sink. At Oxwich NNR access to floating fen has been provided by laying standard boardwalk sections over old car tyres (Rowell, 1988).

Boardwalks in sand dune systems should not be sited where sand levels are likely to change and result in their being undercut or buried and so are often unsuitable for traversing dunes. Access routes should follow the more resistant lower ground and should enter the shore at an oblique angle to the prevailing wind. On fixed dunes boardwalks can be successful, good examples of such boardwalks are those at Holkham, Norfolk, and Ynyslas, Dyfed (Agate, 1986).

Paths should, wherever possible, be constructed of natural materials that are in keeping with the surroundings and do not contrast in colour with the surrounding ground. In some instances, where use is low level, vegetation that is relatively resistant to trampling may provide an adequate surface, e.g. ruderals such as Annual Meadow Grass *Poa annua* (Bayfield & Aitken, 1992). This will rarely be sufficient and where paths need to be surfaced stone or wood are preferable to materials such as concrete or steel. Unfortunately the latter are ideal for a person with special needs, on foot or in a wheelchair, since they provide smooth surfaces. However, the areas used most by these visitors will be around car parks and facilities where formal footpaths are more acceptable. Well-compacted crushed rock, gravel or hoggin may also be suitable provided chippings are not too large (and therefore uneven) or too small and uncompacted (resulting in wheels, sticks and callipers becoming embedded).

Full details of practical methods of path clearance, drainage and surfacing, bridge and boardwalk construction are given in manuals and handbooks produced by the British Trust for Conservation Volunteers (Agate 1983, 1986) and by the Institute of Terrestrial Ecology (Bayfield & Aitken, 1992).

Hides

Hides are not always necessary in order to provide the visitor with better views or a better understanding of the wildlife on the reserve. It is worth considering at the

Fig. 3.3. A poorly sited and poorly maintained boardwalk across a peat bog. (J. Vickery.)

Fig. 3.4. A badly designed visitor centre, constructed in materials which are not sympathetic with the surroundings, painted in a visually intrusive way and having a well-hidden entrance that will not serve to attract visitors. (J. Vickery.)

planning stage whether these interests could be equally well served by adapting natural viewpoints or screens, particularly since there is a general presumption against erecting new buildings on reserves. In deciding on the location of any construction consider not only what will be seen from it, in the case of hides for example, but also from where it will be seen. Ideal locations are bird oases such as pools and lagoons or a major feeding area or high tide wader roost.

Feeding stations can sometimes be created by using simple bird feeders. These are often more appropriate in relatively artificial situations such as country parks. But they can also be of considerable value in areas where many visitors will find it difficult to see birds such as in woodlands. Simple peanut feeders in carpark clearings attract many of the commoner woodland bird species and are often of particular interest to children. The feeders do of course have to be regularly filled – use peanut feeders rather than grain which blows away easily. These small-scale feeding stations rarely warrant the construction of hides.

When hides are needed they should be unobtrusive and positioned in cuttings into banks or against background vegetation rather than on skylines. Hides should ideally face north, away from direct sunlight and, at roost sites, east rather than west into the setting sun. Hides should be constructed from natural materials that blend in with the habitat, a good example are the hides overlooking lakes and reedbeds at Pulborough, Sussex, which are built in a style that mimics those of nearby farm buildings (Fig. 3.5). However, vandalism of hides is a common problem on reserves near urban areas. This problem should not be underestimated; it may vary from hides simply being defaced to, at worst, being burnt down. The use of fire-resistant paint, although sufficient to prevent accidental damage is no protection against determined vandals. At the RSPB reserve at Barons Haugh near Motherwell, Scotland successive wooden hides have been burnt to the ground and are now being replaced with hides built of local stone. At such sites viewing is through unprotected windows since glass is smashed and perspex scratched. Displays in hides are often defaced and the solution adopted at this reserve is that of 'bolt-on' portable displays set up on days when, for example, large parties are visiting the reserve and taken down at the end of the day.

Approaches to hides should be screened either between earth banks or vegetation. In the latter case, the width of the vegetation must be sufficient to ensure its value in winter as well as summer and to allow the regular coppicing required to keep basal vegetation thick. Screens of reeds or fencing such as larch-lap or chestnut paling have little sound-absorbing capacity and their use is best restricted to regularly used sites where wildlife may habituate to noise levels (Andrews, 1991). Disturbance of wildlife may be reduced by setting the hide well back from the area to be viewed and raising it on stilts – ensuring the hide is high enough to command a view over tall summer vegetation.

Visitors with special needs require level or ramped approaches of no more than 1:12 gradient and ample space outside the door to manoeuvre a wheelchair. Inside the hide, space for one or two wheelchairs can be provided by omitting some of the fixed benches or installing removable ones (Countryside Commission, 1982). Excellent examples of specially designed hides exist at Minsmere Nature Reserve, Suffolk, Newborough Warren, Anglesey and Rutland Water Nature Reserve. Guidelines concerning the construction of a standard birdwatching hide are published by the Royal Society for the Protection of Birds (RSPB, 1986).

Other facilities

The other facilities most commonly provided on reserves include litter bins, toilets, benches and picnic tables and the principle of 'minimising the visual impact' applies to all these features. Details of the range and construction of all these facilities and others are given in Countryside Commission for Scotland Information Sheets (CCS, 1988b). Litter bins may actually attract litter and emptying them regularly requires a good deal of work. They should, however, be provided near outlets such as shops and visitor centres. Elsewhere litter problems may be reduced by simply putting up a sign asking

(a)

(b)

Fig. 3.5. Two new hides at Pulborough Brooks RSPB reserve, both carefully designed to match typical Sussex hovel barns. The red tiled-roof effect is achieved through the use of tile sheeting (rolled pressed steel) textured to catch lichens, and the black walls are of pressure-treated wood coated with a water-based black preservative. The small hide (a) is well hidden; sited under a major oak and sunk into the bank. The large hide (b), although in a more open location, is sited at the corner of an old hedgerow and willow belt, and will be less obvious when the recently planted screening vegetation becomes established. (T. Calloway.)

visitors to take litter home and by picking up any litter quickly since this encourages people to keep a site clean.

Many wardens consider the provision of toilets to be a 'necessary evil'. They are often demanded by the public but require a good deal of time and money to maintain. The cost of maintenance may be reduced by investing initially in high quality, robust products. The type of hand dryers and soap dispensers used in motorway service stations are often good examples.

Signs

Visitors should be welcomed to a reserve with large and obvious signs, whereas signs elsewhere on the reserve should be unobtrusive and kept to a minimum. A standard 'welcome' sign usually includes a simple map of the area, a contact name and address, a short list of do's and don'ts and words to the effect of 'we hope you enjoy your visit'. Clear signposting from the car park to the reserve is essential.

A visitor centre provides an opportunity to give much more detailed information, drawing attention to ways to enjoy a visit, for example, giving information about hides, pools, nature trails, facilities such as shops and toilets and local ecology or any local

Fig. 3.6. An information board incorporating both a bench and a shelter. (J. Vickery.)

traditions (e.g. reed cutting). Interpretative displays are a good way of interesting and exciting the visitor. For example, at Loch Gruinart an important overwintering site for geese in Scotland, displays give a picture of the geese all year round. In the Wood of Cree in Wigtownshire, Scotland, displays of cross-sections of wood from different trees accompany information about the biology of the species and the use made of it by man in the past. Additional information concerning other local attractions, places to visit, places to stay and Bed and Breakfast all serve to enhance the sense of welcome for visitors. It is of course also an ideal place to encourage people to join an organisation!

Elsewhere on the reserve signposting must be strategic: warnings of any kind, from 'please do not drop litter' to the warnings about the occurrence of species such as adders, must be positioned at all entry points to ensure visitor safety and guard against any legal problems should accidents occur. 'Portrait' signs of A4 size are commonly used. A good general rule is not to include more than 150 words of text. Signs should convey the uniqueness and importance of the site since this is why many people visit a particular reserve. Portrait signs can be positioned at points of interest around the reserve. For example, on entry into a new habitat type or near management features such as coppicing or reed cutting. Signs at viewpoints of a lagoon might include information about bird or insect species that can be seen. Information about the typical plants of a habitat are best sited near distinctive species. If the feature is a seasonal one, such as flowers or migrant birds, then signs should be so constructed that they can be removed easily and reinstated, for example, as boards that bolt onto permanent posts.

Signs should not be visible from any considerable distance. Their visual impact can be reduced by positioning them close to natural features such as trees and bushes. Signs can be constructed such that they do not break the skyline by having them at waist height and tilted at an angle. On reserves where there is a great deal of bird interest it is a good idea to have a whiteboard on which the species that have been seen on the site within the last few days can be listed and altered regularly. Footpaths, bridleways and nature trails can be unobtrusively marked with either 'fingerpost' signs, consisting of a post branded with the relevant symbol or a colour-coded arrow, or waymarking discs that should, where possible, be secured to existing posts, gates or stiles.

Natural materials such as wood should be used wherever possible. However fibreglass will be required for any detailed maps and has the advantage of being more durable than wood. If the risk of vandalism is high signs may need to be constructed of steel to avoid them being burned down and may even need a bullet proof coating as protection against target practice with air rifles! The text can be an integral part of the sign or be printed on sticky-back vinyl which can then be mounted onto enamel signs and easily replaced if damaged or defaced. Another option is to use signs that can be bolted into the ground during the day and removed at night.

Further information and interpretation may be provided through guided walks. Although these require a large time commitment they frequently attract visitors to a site and result in a captive audience as far as education and conservation are concerned. Guided walks are often best done on an *ad hoc* basis, advertised by a simple sign put out on the day, rather than regularly and at fixed times throughout the year. This allows flexibility in terms of the other management needs on the reserve and the weather. Guided walks on a specific theme are often very successful: for example, the dawn chorus or a fungus foray, or for specific groups of people such as a walk for children that involves beachcombing or pond dipping.

References

Agate, E. (1983). *Footpaths. A Practical Conservation Handbook*. Reading, Berkshire: British Trust for Conservation Volunteers.

Agate, E. (1986). *Sand Dunes. A Practical Conservation Handbook*. Reading, Berkshire: British Trust for Conservation Volunteers.

Andrews, J. (1991). Planning for visitors on unwardened nature reserves. *British Wildlife* 2, 206–13.

Bayfield, N.G. & Aitken, R. (1992). *Managing the Impacts of Recreation on Vegetation and Soils: a Review of Techniques*. Banchory, Kincardineshire: Institute of Terrestrial Ecology.

Countryside Commission (1982). *Informal Countryside Recreation for Disabled People*. Cheltenham, Gloucester: Countryside Commission.

C.C.S. (1988a). Countryside Commission for Scotland Information Sheet Numbers 1.1.1–6.19, 13.1, 16.1. Battleby: Countryside Commission for Scotland.

C.C.S. (1988b). Countryside Commission for Scotland Information Sheet Numbers 17.1–17.6.1. Battleby: Countryside Commission for Scotland.

Rowell, T.A. (1988). *The Peatland Management Handbook*. Peterborough: Nature Conservancy Council.

RSPB (1986). *The Design and Siting of a Standard Birdwatching Hide*. RSPB Advisory sheet. Sandy, Beds.: RSPB.

Sidaway, R. (1991) *Good Conservation Practice for Sport and Recreation*. London: Sports Council.

Ward, D. & Andrews, J. (1993) Waterfowl and recreational disturbance on inland waters. *British Wildlife*, 4, 221–229.

Yalden, D.W. & Yalden, P.E. (1989). The sensitivity of breeding Golden Plovers *Pluvialis apricaria* to human intruders. *Bird Study*, 36, 49–55.

Coastal habitats

4

CHRISTOPHER L.J. FRID AND PETER R. EVANS

Introduction

This chapter encompasses a range of habitats including the open sea, rocky shores, shingle, estuaries, sea cliffs, dunes and saltmarshes. There are 19 336 km of coastline in the UK. Within mainland Britain, the coastal habitats contain many of the most important areas for conservation and are of international importance. For many of these habitats, the opportunities for management may be restricted. Coastal areas are popular for a wide range of human activities including sailing, fishing, shooting, diving, and the study of natural history, such that minimising the conflict between access and conservation is one of the major issues.

Land ownership may be complex. Legislation, conventions and site declarations vary in the extent to which they cover the intertidal area. Below the low water mark the seabed in UK territorial waters is the responsibility of the Crown Estate Commissioners, while passage through the overlying sea for the purpose of peaceful navigation is guaranteed by international convention. Many organisations and international conventions operate in coastal waters, including estuaries and intertidal areas.

The manager of a coastal habitat which includes any area below the high water mark is immediately faced with a range of legal and traditional rights to take into account in the management plan. Voluntary marine nature reserves can work, in spite of the wide range of organisations with jurisdiction or interest in the coastal zone, but their establishment can be an extremely slow process (Gubbay, 1986).

Coastal communities

The open sea and subtidal habitats

The open sea is important to all coastal communities. The plankton provide food for many filter feeders, such as Mussel *Mytilus edulis*, which are important in the productivity of coastal systems. It also redistributes non-living organic matter, an important food source for deposit feeders such as Ragworm *Nereis* spp. and *Nephtys* spp.

and Lugworm *Arenicola marina*. It also provides the mechanisms of dispersal of young, such that the continued presence of a particular species at a given site may be dependent on the water quality and the biological communities present some distance away. The open sea is therefore of considerable conservation value, but due to its extensive nature, the complex legal regime which operates on the high seas and the dynamic nature of the system it is not readily amenable to management in the way coastal habitats or the seabed are.

Subtidal areas of seabed often contain rich communities, which in turn are important in supporting fish populations. These are in turn exploited by man, marine mammals and seabirds. The benthic communities are under direct threat from activities such as the use of destructive fishing gear (dredges, trawls, etc.), dredging (maintenance and for aggregates) and waste disposal and dumping. In addition, the large biomass removed by the fishing industry is likely to have resulted in more subtle changes in the ecology of those areas exposed to heavy fishing pressure.

The conservation interest of marine subtidal communities results from the fact that the UK lies at a boundary between marine biogeographic provinces. In the north, i.e. the Shetlands, NE Scotland and the northern North Sea coasts, we have many species with essentially arctic or subarctic distributions. The warming influence of the North Atlantic Drift (Gulf Stream) results in the presence of Lusitanian and southern species in the south west and west of the UK, as far north as the Western Isles. The unusual conditions arising in the sea lochs of Scotland and Ireland have lead to the development of specialist communities of high conservation value in these environments.

The subtidal habitat is interrelated to other habitats, for example, as a source of sediment, nutrients, and recruits for species such as Lugworm and, therefore, there is a need to protect from adverse impact (e.g. pollution, gravel extraction) and disturbance (e.g. suction dredging).

Rocky shores

Although rocky shores show less diversity than subtidal habitats they are much more accessible and thus of considerable educational importance. In the UK, approximately 34% of the coastline comprises rocky shores, which range from the wave-exposed headlands of ocean coasts to the algae-covered shores of sea lochs which receive relatively little wave action. The former are dominated by a few hardy species, mussels and barnacles typically, which have evolved to deal with the periodic disturbance caused by the high degree of wave action, and will quickly recolonise following disturbance. In contrast, sea loch shores are species-rich, with a fully developed algae canopy and an understorey of algae and sessile animals. Rocky shores are important for species such as Purple Sandpiper *Calidris maritima*, Turnstone *Arenaria interpres* and Rock Pipit *Anthus spinoletta* and are used as temporary resting places by Grey Seal *Halichoerus grypus*.

Fig. 4.1. Natural current activity in estuaries can create extensive sandbanks that when exposed, are used traditionally by seals such as here at Blakeney Point in Norfolk. People enjoy trips to see the seals in the summer months, but access must be managed to avoid undue disturbance. (D.A. Hill.)

The harsh physical environment precludes many of the more delicate subtidal forms from exploiting this habitat, but the extreme lower shore, especially on south western coasts, will often contain individuals of sublittoral species and a variety of unusual red algae and invertebrates have been recorded from shores in Devon and Cornwall in particular.

The dense fucoid canopy that develops on all but the most wave-exposed shores in the UK is important for sheltering invertebrates and smaller algae beneath it, and in the overall productivity of coastal systems. It is a major contributor to the detritus that is eventually deposited in coastal and nearshore sediments, where it is utilised by infauna such as polychaete worms.

Sandy shores and estuaries

As a generality, coarse-grained sands tend to occur in the north and west of Britain, while fine-grained muds predominate in the south and east, where tidal ranges are less. There are, however, considerable local variations due to differences in wave energy. In areas sheltered from wave action, fine particles settle out where water currents are slow, towards the upper tidal levels, and saltmarshes develop at the transition to land. At lower tidal levels near the main channels, or on areas exposed to wave action, coarser particles are deposited, giving rise to sands and shingles (Fig. 4.1). Intertidal mudflats are very vulnerable to erosion or relocation by chance meteorological events, as well as to human activities such as dredging of river channels.

Sandy shores, which comprise about 12% of the UK coastline, contain a rather restricted fauna of burrowing bivalved shellfish, polychaete worms and strand line crustaceans, such as sandhoppers, which are utilised by a number of bird species, such as Sanderling *Calidris alba*. Invertebrate densities are relatively low so that sandy shores are unable to support the high densities of birds seen on more muddy estuaries.

British estuaries form a high proportion of those in western Europe and comprise 28% of the entire estuary area of the East Atlantic and North Sea coastal states, more than any other European country. Their bird populations are of international importance, especially in the non-breeding seasons. In January approximately 0.6 million wildfowl and 1.2 million waders are present on British estuaries (Davidson *et. al.*, 1991) (Fig. 4.2).

Mudflats provide more attractive habitats than sand beaches to invertebrates living within them, because they contain a greater food supply of organic particles, microscopic plants and bacteria, and their temperature and salinity fluctuations during the tidal cycle are less extreme. They are, however, less well oxygenated, so that only those animals that can cope with this, such as burrow-dwelling worms and bivalves, are abundant. The variety of invertebrates is often highest in sandy muds in the lower half of the intertidal zone. Their abundance and biomass per unit area is often higher there than in the permanently submerged subtidal areas nearby. The high-productivity intertidal zones are attractive to shorebirds and some wildfowl as feeding areas at low tide and to fishes including commercial species such as Sole *Solea solea*, Plaice *Pleuronectes platessa*, Whiting *Trisopterus luscus* and Cod *Gadus morhua* at high tide.

Coastal lagoons

It is important to distinguish between natural or semi-natural lagoons and entirely man-made lagoons. Natural and semi-natural coastal lagoons are permanent bodies of brackish (or salt) water that are separated from the adjacent sea by a barrier of sand or shingle, but which nevertheless exchange water with the sea through an inlet or, more usually in Britain, by tides occasionally overtopping the barrier or by percolation through the barrier via the salt or brackish water table. They are highly variable in character, depending on the extent and regularity of saline and freshwater inflows and on their depth, which influences the degree of stratification or mixing of the waters. Semi-natural coastal lagoons are not strictly naturally formed lagoons as defined by Barnes (1989), but nevertheless have similar environmental variables and species composition (Sheader & Sheader, 1989). The severity of the physical environment in natural and semi-natural coastal lagoons is comparable to that of estuaries and their restricted varieties of flora and fauna show similarities to those of estuarine habitats. Many animal species exploit food webs based on detritus rather than benthic plants or

(a)

(b)

Fig. 4.2. Estuaries in the British Isles hold internationally important populations of wintering waders and wildfowl due to their large intertidal invertebrate populations. (*a*) Here Lugworms are seen living at high density at Lindisfarne, Northumberland. (*b*) Sediment type and hence invertebrate communities determine which populations of wintering waders exploit estuaries. In this case Dunlin and Ringed Plover are feeding on a sandy intertidal area at Studholme, Dorset. Management of estuaries involves minimising impacts from the huge pressures for reclamation and development, through an integrated conservation management policy for the whole estuary, together with reducing disturbance from, for example, bait-digging and dog-walking particularly in the winter period. ((*a*) J. Andrews, (*b*) D.A. Hill.)

phytoplankton. During the summer, when evaporation rates are high, and tidal ranges smaller, lagoons may become isolated as coastal ponds. Even when open to the sea, water currents within a lagoon are weak and the flushing time relatively long. An inventory of natural and semi-natural lagoons in Britain has been compiled (Sheader & Sheader, 1989) and an overview and conservation appraisal is given in Barnes (1989).

Many artificial lagoons, including some of the country's most popular bird-watching sites have been created by conservation organisations by building embankments and flooding the enclosed land (e.g. Titchwell, Norfolk), pumping water onto depressions in coastal grazing marshes (e.g. Elmley, Kent), or excavating shallow depressions with a bulldozer and leaving a network of islands (e.g. Minsmere, Suffolk) (Hill, 1989). These sites attract many migrant waders and some wintering wildfowl as well as being used as breeding sites by a number of wader and wildfowl species. The majority (80%) of the British Avocet *Recurvirostra avosetta* population breeds on artificial coastal lagoons.

Sand-dunes

Sandflats dry out at low water, and the surface particles are then able to be picked up by the wind. Sand-dunes develop where the flow of sand-laden wind is interrupted initially by detritus. Small embryo dunes are then colonised by vegetation, stabilising the sand, and allowing bigger dunes to develop. Larger, more extensive dunes are characteristic of the west coast of the UK where prevailing winds are onshore. In some areas the sea may move eroded deposits of mollusc beds to the shore and the resulting dunes are calcareous and of greater ecological diversity and interest. Sand-dunes are widely recognised as important ecological sites and in addition many are important as sea defences. Many of the UK's dune systems have been given some form of statutory protection and are of European importance. They are of considerable botanical interest and along with heathland are one of the two habitats of the Sand Lizard *Lacerta agilis*.

Dune slacks, which form where troughs between dunes go down to the water table, are usually the areas of greatest botanical diversity. Soluble nutrients leached downwards by rain enrich the water table. It is common to find heath on higher dunes and diverse plant communities typical of base rich soils in the dune slacks. Natterjack Toads *Bufo calamita* breed in wet dune slacks.

Saltmarshes

Saltmarsh habitats develop on alluvial sediments bordering saline water bodies whose water level fluctuates, usually tidally. Their vegetation is comprised of halophytic grassland and dwarf brushwood species which form a buffer between terrestrial and aquatic ecosystems. They defend the land against erosion and trap sediments and pollutants. Saltmarshes are a scarce resource in Britain, covering only 44 370 ha, less than sand-dunes and heathland (Burd, 1989). Over 80% are designated as Sites of Special Scientific Interest.

Marshes which develop in sheltered areas, where tidal flow is the dominant influence, are usually dissected by a system of natural drainage creeks, and the whole marsh slopes gently towards the sea. In situations where wave action plays a more active role in marsh formation, a system of terraces develops parallel to the coastline, each terrace sloping up towards the sea. The relative amounts of sand and silt present in the sediments during accretion control the topographic relief: the high levels of silt in sheltered sites produce a very level marsh.

Two zones may be distinguished, the lower marsh, pioneer or submergence zone, between mean high water of neap tides (MHWN) and mean HW of all tides, and the upper marsh or emergence zone which develops between mean HW and mean HW of spring tides (MHWS). Accretion rates are lower in the emergence than the pioneer zone. Most saltmarsh plants are perennials, except in the pioneer zone and along the strand line.

Several types of saltmarsh can be distinguished around the British coast (Burd, 1989). Those around the Irish Sea are generally sandy and dominated by grasses, and have been used traditionally as grazing marshes; those around the North Sea and the English Channel are less often used in this way because they contain more clay and silt and less grasses amongst the vegetation. However, much of the large expanses of coastal grazing marsh in the south east of England is converted saltmarsh. Coastal grazing marsh is dealt with in Chapter 8.

The character of the south-coast marshes has already been altered by the spread of the Cord-grass *Spartina anglica* since the end of the nineteenth century and this problem has spread north although die back is now apparent at many sites. Thick swards of *Spartina* prevent the usual communities of annual glassworts *Salicornia* spp. from maintaining themselves on the seaward edge of the marsh. This also has important detrimental effects on invertebrate communities and can reduce use by shorebirds and wildfowl by reducing the food supply and increasing vegetation height (Doody, 1984).

Saltmarshes hold a surprisingly diverse arthropod fauna, perhaps because of the wide range of physical environments between the upper and pioneer marshes, but are probably best known in conservation terms for their breeding birds, as feeding areas for wintering wildfowl and as high-water roosting sites for large numbers of gulls, terns, waders and wildfowl. In summer, they hold more than half of the Redshank *Tringa totanus* breeding in Britain and several per cent of the breeding Black-headed Gull *Larus ridibundus*. In winter they provide grazing habitat for internationally important populations of wildfowl, such as Brent Geese *Branta bernicla*, Barnacle Geese *Branta leucopsis* and Wigeon *Anas penelope*.

Shingle

Coastal shingle banks are unstable habitats which occur in areas of fairly high wave energy where a suitable source of small pebbles or shells is available. Continued physical disturbance, and a lack of nutrient and fine particles, are necessary if the shingle is to remain in its open state, with little or no vegetation. This habitat may occur as simple beaches or, if the waves strike the shore at an angle and causes lateral movement of material, as linear features such as spits up to several kilometres long (e.g. Orfordness, Suffolk), islands (e.g. Scolt Head, Norfolk) or parallel ridges (e.g. Dungeness, Kent). These can protect features lying to the landward, such as coastal lagoons.

Vegetated shingle beaches are scarce both in Britain and in Europe as a whole. Plant cover is more dense in areas with smaller stones. They have a characteristic community including Yellow Horned Poppy *Glaucium flavum*, Sea Kale *Crambe maritima*, Sea Beet *Beta vulgaris* and Sea Campion *Silene uniflora*. Rare species such as the Oyster Plant *Mertensia maritima* and Sea Pea *Lathyrus japonicus* are also characteristic of this habitat. Seeds of most of these plants are relatively large so that they wedge between the shingle particles and provide winter food for some passerine birds such as Snow Bunting *Plectrophenax nivalis* and Shore Lark *Eremophila alpestris*.

Open or sparsely vegetated shingle areas can hold important concentrations of breeding birds such as terns *Sterna* spp. and provide a major coastal nesting habitat for Oystercatcher *Haematopus ostralegus* and Ringed Plover *Charadrius hiaticula*. Outside the breeding season, they are often important high-water roosting sites for sea- and shorebirds.

Sea cliffs

Sea cliffs vary considerably in character. Those formed from hard rocks such as sandstones or basalt tend to be very stable, even if near vertical, but others comprised of soft materials such as chalk and boulder clay may erode continuously to produce unstable, often sloping, surfaces. The edges and ledges of unstable cliffs may sustain important butterfly populations, such as the Glanville Fritillary *Melitaea cinxia*. The seepages on clay cliffs and soft rocks often contain a rich invertebrate fauna especially of flies, beetles and a number of rare woodlice.

Sea cliffs are notable chiefly for nesting birds, and their attractive and often unusual plants. Important seabird sites are restricted to the steeper cliffs, but occur also on isolated stacks and islands. Rock Dove *Columba livia*, birds of prey and corvids often nest on cliffs. The species of plants found on the tops, slopes and ledges of cliffs are influenced greatly by the amounts of salt spray that reach them, as well as by the composition of the underlying rocks and soil. Rock Samphire *Crithmum maritimum*, Scots Lovage *Ligusticum scoticum* and Sea Spleenwort *Asplenium marinum* are restricted largely to cliffs. Because prevailing winds blow from the south west in Britain, especially in winter, the influence of salt reaches further inland and higher up the cliffs in western than eastern Britain. Heavy rainfall leaches salt out of the soil and washes it off leaves with the result that in areas such as south west Scotland the maritime vegetation is found only close to the sea, whilst in drier areas maritime vegetation extends further inland (Mitchley & Malloch, 1992). Gorse *Ulex* spp. can be used as an indication of the extent of maritime heath as it is particularly salt sensitive.

A handbook for the practical management of cliff vegetation for nature conservation has been produced by Mitchley & Malloch (1992). Much of this concerns the vegetation of the cliff tops. The management of maritime heath largely involves the techniques discussed in the chapter on lowland heathland (Chapter 10).

Sea walls

Sea walls topped by soil or clay are often the only remaining areas of old coastal grasslands although they may also be reseeded with imported seed mix after construction. They are important for plants such as Sea Clover *Trifolium squamosum*, Slender Hare's-ear *Bupleurum tenuissimum* and Sea Barley *Hordeum maritimum* and hold rare species such as Least Lettuce *Lactuca saligna* and Hog's Fennel *Peucedanum officinale*. They are also important for some invertebrates such as Roesel's Bush-cricket *Metrioptera roeselii*. If ungrazed, they may hold high densities of small mammals and are popular hunting areas for Short-eared Owl *Asio flammeus*, Barn Owl *Tyto alba* and Hen Harrier *Circus cyaneus*.

Sea walls are important habitats chiefly along the lowland coasts of south east England and East Anglia, and around estuaries in western Britain where tidal ranges are large and lowland marshes have been reclaimed from the sea.

Conservation prescriptions

For most of the habitats in this chapter there is little active management that can be carried out beyond maintaining the natural functioning of the system and, where necessary, reducing disturbance, erosion and over-collecting. Although largely outside the remit of this book, there is a need for a national integrated approach to coastal management taking into consideration sediment movement.

With the envisaged rise in sea level there will be considerable opportunities for imaginative conservation. The best management in some areas may be a planned retreat either to new sea walls or to naturally raised ground behind. There is the opportunity here for imaginative collaboration between conservationists and the National Rivers Authority to find ways of creating and managing these habitats.

Erosion

Erosion is a considerable problem on sand-dunes (see Chapter 3 for ways of alleviating human erosion). The movement of exposed sand can affect considerably greater areas than the area originally eroded. It is necessary to restrict the erosion caused by overgrazing, trampling, pony trekking, off-road vehicles, motor bike scrambling and mountain bike riding. Creating pedestrian walkways from the car parks to the beach is essential in heavily used areas. Fencing prevents wandering off such tracks and causing erosion. Access routes should follow the more resistant lower ground and should enter the shore at an oblique angle to the prevailing wind. Clear signposting reduces the numbers wandering through the dunes. Cars can be kept off dune grassland by digging roadside ditches or building low timber barriers.

On shingle, human access, both on foot and especially by off-road vehicles, can cause destruction of the characteristic sparse vegetation, initiate erosion and lower ridge heights, thus making nests and vegetation more vulnerable to flooding by high spring tides. Even on rocky shores algae and sessile animals are easily eroded by trampling. On wave-swept shores recolonisation is rapid and may be complete in 1–2 years; on sheltered shores 7–10 years may be required. Subtidally, in the sheltered conditions of a Scottish sea loch, where the kelp forest is normally exposed to little physical disturbance, it is now possible to observe paths eroded through the algae on the shore extending through the kelp forest, the result of SCUBA divers pulling on the kelp plants or damaging them as they swim. Many of the colonial/semi-colonial invertebrates which attach themselves to hard substrata are slow growing. Ross Coral

Fig. 4.3. Much of sand-dune management involves balancing the conflicting demands of visitor access and maintaining sensitive habitats. (A. Smith.)

Pentapora foliacea 2 m across, or erect sponges 30 cm tall may be the result of decades of growth, and can be destroyed by a single careless fin stroke (Picton, 1991). Education is the most sensible option here.

The major problem faced in managing a cliff-top or rocky islet habitat is the control of erosion. Although natural phenomena, such as nest-burrowing by seabird species such as Puffin *Fratercula arctica*, or haul-outs or breeding aggregations of Grey Seals, can kill vegetative ground cover and allow gales to blow-off the top soil, much more serious are human activities. Climbers can cause considerable erosion to sea cliff vegetation. Cliff-top paths often develop very close to the edge of slopes and lead to trampling and death of vegetation and gradual erosion of soil back from the edge. Since most maritime plant species occur only in a narrow strip along the coast, and because agricultural usage has often claimed the inland borders of this vegetation band, it is often squeezed and fragmented between these two pressures. Restriction of public access to a footpath some distance inland from the cliff-edge, with the provision of occasional viewing points over the cliffs, is a solution.

Sea defences or cliff defences may reduce erosion to the detriment of the habitats downdrift which are dependent on the deposition of material, for example, the Humber and the Wash which obtain some of their fine material from the eroding Yorkshire coast.

Areas of shingle are often dynamic and unstable as a result of erosion and accretion. Attempts to stabilise shingle coastlines with groynes or sea walls may allow colonisation by thick vegetation which may improve the interest in many ways but remove the features which make them attractive to nesting seabirds and shorebirds. If tern nesting sites become overgrown it is best to pull up the plants rather than use herbicides or attempt to lay polythene sheeting under the surface.

Predator control

Colonial nesting birds are particularly susceptible to a wide variety of predators. For example tern colonies in Britain have suffered from predators including Badger *Meles meles*, Hedgehog *Erinaceus europaeus*, Grey Squirrel *Sciurus carolinensis*, Fox *Vulpes vulpes*, humans, Oystercatcher and Kestrel *Falco tinninculus*. Exclusion of mammalian predators by fencing has been successful, including across shingle spits, though to be sure of success an area should be surrounded. The fence should also be set several inches into the substrate to prevent predators burrowing underneath. Electric fencing is particularly efficient but it will short circuit if subject to immersion in salt water.

Brown Rats *Rattus norvegicus* have eliminated many established colonies of burrow-nesting birds and are extremely difficult to eradicate. Control measures need to be implemented chiefly during those times of year when birds are absent and when natural foods are scarce. i.e. in late winter.

Fig. 4.4. Rocky slopes and cliffs provide colonial breeding grounds for many seabirds, for which the British Isles are internationally important. Here Puffins are breeding on the Skelligs in South-west Ireland. (D.A. Hill.)

Predation by birds is even more difficult to control and may require imaginative management. The predation of Roseate Tern *Sterna dougallii* chicks by Peregrines *Falco peregrinus* was reduced by providing nest shelters.

Collecting and hunting

SCUBA diving enthusiasts in the UK have traditionally taken marine life for consumption as well as Sea-urchins *Echinus esculentis* as 'trophies'. Codes of practice issued by the various diving organisations generally discourage this (e.g. BSAC, 1985, p. 246). In voluntary marine nature reserves considerable success has been achieved through education (mainly talks and information sheets) and notices in order to reduce collecting further. Within statutory reserves legal backing is given to these measures.

A good way of controlling wildfowling is to allow a responsible club to use and police the area. This works well to control poaching, but may not always be considered acceptable. Preventing poaching on accessible sites can be very time consuming. If wildfowling takes place it should be incorporated within a zoning policy making sure roosts are protected and at least some parts of a large site must be designated as refuges.

Many areas in recent years have seen massive mortality of both target and non-target species as the result of bait harvesting (Jackson & James, 1979; Cryer *et al.*, 1987). The bait digging itself may cause disturbance. As bait digging is a highly visible activity it is easy to control with sufficient wardening, but management plans need to critically address the need for closed areas and rotation of digging effort around a number of 'sacrificial areas'. At the Lindisfarne National Nature Reserve, Northumberland digging is allowed in an area adjacent to the causeway to Holy Island, but excluded from an area of Budle Bay. Until the early 1980s, digging was allowed in Budle Bay but during the coal-miners' strike the area was dug out in under 8 weeks. It is estimated that 4 million lugworms were removed in that short period. Olive (1993) considers some management recommendations.

The consequences of bait digging are dependent on the site-specific population dynamics of the exploited worms, such that at some sites sustainable exploitation would seem to be impossible, while at others rotation of exploited areas and the provision of permanently closed areas is feasible. The size of the areas and the frequency of rotation again is site-specific and dependent on the biology of the local population: for example, its growth rate and whether or not it is self-recruiting. The organisation of bait diggers into a voluntary self-policing arrangement has proved to be beneficial in Essex.

On cliff-tops control of grazing may also be important. On less-exposed cliff-tops and slopes, scrub woodland may develop in the absence of grazing, thereby reducing the abundance of some of the rarer flowering plants but stabilising the cliff edge and upper face in the process. Conversely, overgrazing by feral goats and rabbits has led to soil erosion.

Access and visitor pressure are often the most important impacts that the reserve manager must regulate. In offshore areas these take the form of visiting pleasure boats and SCUBA divers. In general the provision of moorings at suitable locations, e.g. near, but not too close to, bird roosts or nesting colonies and seal haul-outs (so that they can be observed with minimal disturbance) and the most popular diving sites, backed by high-profile patrolling staff and user zonation has done much to reduce casual damage to sensitive seabed communities and disturbance to breeding birds and mammals as, for example, around the Skomer Marine Nature Reserve, Dyfed.

Coastal lagoons

The conservation importance of natural coastal lagoons is as much for their rare physiography as for their invertebrate faunas (Barnes, 1991). However, a number of typically lagoonal species (Foxtail Stonewort *Lamprothamnium papulosum*, Ivell's Sea Anemone *Edwardsia ivelli*, Starlet Sea Anemone *Nematostella vectensis*, Lagoon Sand Shrimp *Gammarus insensibilis*, Lagoon Sand Worm *Armandia cirrhosa* and Trembling Sea Mat *Victorella pavida*) are protected under schedules 5 and 8 of the Wildlife and Countryside Act 1981. The shallow edges of natural lagoons may be used by migrant waders feeding on the often abundant invertebrates, but these should not be seen as the main management objective of these natural habitats. Lagoons and coastal ponds are extremely sensitive to inputs of nutrients, e.g. sewage and fertilisers which lead to eutrophication problems, and of pollutants or industrial wastes which may accumulate to toxic levels. Management thus requires strict control of the quality of any freshwater inflows. Planting of Reed *Phragmites australis* beds at the points of entry of non-saline water may assist in removal of nutrients and some pollutants. Artificial coastal lagoons are one of the few coastal habitats in which there are considerable opportunities for management and thus they occupy a large part of this chapter.

Management of coastal lagoons for invertebrates should be based on a knowledge of the particular environment of a lagoon and particularly on continuing monitoring of invertebrate populations. Benthic invertebrates can be sampled using standard core samples sieved through a set of sieves (2 mm, 1 mm and 0.5 mm mesh size). Twice yearly samples (e.g. pre- and post-summer) would provide sufficient information on which to base management decisions. In some situations the invertebrate community of artificially created lagoons can become impoverished over time. If monitoring is conducted regularly this can be detected and management actions can be initiated to improve the situation.

Salinity

Salinity is the most important factor determining the invertebrate species present and their densities in a coastal lagoon; it is measured as parts per thousand of salt (‰S). Salinity should be monitored with a hydrometer weekly through the summer and then monthly over winter. Maintaining salinity within fixed limits is an important part of managing for the invertebrate fauna of an artificial coastal lagoon. Evaporation in hot dry summers and rainfall in wet winters can have a very marked effect on salinities in shallow lagoons. A range of salinities from 2‰ to 95‰S has been recorded for one lagoon over the course of a year: little survived. The degree of control of salinities that is possible depends on the amount and salinity of water available. Even if only a single water source is available, salinity change can be buffered by frequent addition of more water at the original salinity.

The community present will change with the salinity. In more freshwater lagoons (0–10‰S) the dominant invertebrates in the benthos are the larvae of chironomid flies, and, in the water body, Lesser Water Boatmen *Corixid* spp. and the Oppossum Shrimp *Neomysis interger*. The optimum salinity for lagoons with this characteristic invertebrate fauna is about 6‰S, with an upper limit of about 15‰. In more brackish lagoons (>10‰S) the dominant benthic invertebrates are usually Ragworm *Nereis diversicolor* and the amphipod *Corophium volutator* with the shrimp *Palaemonetes varians* predominating in the water body. These invertebrates are characteristic of estuaries and can tolerate a wide range of salinities up to seawater (35‰S).

Evaporation in hot summers can lead to hypersaline (>35‰S) conditions in the more brackish lagoons. Although *N. diversicolor* has some tolerance of hypersaline conditions these can lead to a large reduction in invertebrate biomass depending on their severity and duration. During wet winters a lower salinity can also reduce *N. diversicolor* and *C. volutator* biomass. *N. diversicolor* fails to breed at salinities lower than 5‰S, and the overwinter survival of *C. volutator* is greatly reduced at salinities lower than 3‰S.

Predators of invertebrates

Fish populations can have a major impact on invertebrate densities in coastal lagoons. In more freshwater lagoons (0–10‰S), Three-spined Stickleback *Gasterosteus aculeatus* and Ten-spined Stickleback *Pungitius pungitius*, can reach densities of up to 5 per square metre, and can have a major impact on chironomid populations. In more brackish lagoons (over 10‰S) particularly where some of the water supply comes from an estuary, saltmarsh or the sea, densities of Sand Goby (*Pomatoschistus minutus*) can be as high as 20 per square metre which can totally remove *C. volutator* and reduce populations of *N. diversicolor*. Experiments have shown that a density of one stickleback per square metre or two Sand Gobies per square metre is sufficient to significantly reduce invertebrate densities.

Although fish populations may be an important food source for birds such as Red-breasted Merganser *Mergus serrator*, Goldeneye *Bucephala clangula* and Spoonbill *Platalea leucorodia* there may be cases in which it is considered that their predation is deleterious to the conservation interest. Fish populations could be controlled, if monitoring of fish and invertebrate populations suggest that it is necessary, by drying out a lagoon and reflooding with a fine mesh over the water inlet. Drying out for 2 or 3 days in mild autumn or early spring weather would minimise the impact on the benthic invertebrate community. Recolonisation of chironomids in suitable salinity conditions in spring or in early autumn (before October) would be rapid as *C. volutator* and *N. diversicolor* are tolerant of limited drying and would be very little affected providing it is not too hot.

The shrimp *Palaemonetes varians* can act as a predator and at high densities may also affect invertebrate densities. *P. varians* can tolerate a wide range of salinities and can be controlled only by drying out a lagoon and restricting recolonisation with a fine mesh net over the water inlet.

Organic enrichment

Many artificial coastal lagoons have been created by bulldozing and hence removing the topsoil and surface vegetation. Although this is an effective way of creating shallow lagoons with a series of islands, much of the organic matter, which is at the base of the lagoonal food web, is removed. There is an optimum organic content for a lagoon substratum as higher levels of organic matter reduce oxygen levels, thus creating anoxic conditions and reducing invertebrate species diversity. The optimum organic content of the substrate depends on the invertebrate fauna, as different species have different tolerances of anoxic conditions. Generally chironomids are tolerant of anoxic conditions and only very peaty substrates are likely to become anoxic enough to limit chironomid biomass. Although *N. diversicolor* is reasonably tolerant of anoxic conditions *C. volutator* is sensitive to anoxia. Adding organic matter to the substrate of a lagoon with a *N. diversicolor/C. volutator* fauna would preferentially benefit *N. diversicolor* which feeds directly on coarse organic matter but might be detrimental to populations of *C. volutator*, both through more anoxic conditions, and through increased predation by *N. diversicolor*. It is probably worthwhile adding organic matter to such a lagoon only if a high biomass of *N. diversicolor* is the main aim (for example, as a food source for pre-breeding Avocets) or if the original organic content of the substrate is very low.

Rotovating

Rotovation can be carried out along hardened edges following the lowering of water levels over summer or when a lagoon has been dried out for major management work, and during the creation of a new lagoon. Rotovation and flooding of natural depressions would often be preferable to bulldozing away the top soil as a means of lagoon creation because the original organic matter is retained.

Experiments have shown that rotovating a hard substrate on the base of shallow artificial lagoons can increase the biomass of chironomids. No significant effect on *N. diversicolor* has been shown although the trend was for increased densities in rotovated plots. The effect of rotovation on *C. volutator* has not been tested, but it may be beneficial in that it would increase aeration of the substrate and might assist burrowing.

Flooding regimes and management work

It is sometimes necessary to dry out a lagoon to allow access for heavy machinery for major management work such as creating or modifying islands, building or repairing sluices, and repairing sea walls or boundary banks. Chironomid larvae are very sensitive to drying and it would be impractical to try to drain a lagoon without killing them. However, lagoons with a chironomid fauna will be recolonised rapidly by ovipositing chironomids provided that the lagoon is reflooded during the period of peak oviposition from April to September. Some oviposition will take place in March and October, but from November to February there will be negligible recolonisation. Without it, use by passage waders will be reduced.

N. diversicolor and *C. volutator* are much more tolerant of drying out of lagoonal muds. The period of emersion they can withstand depends on the temperature. Laboratory experiments show that at 15 °C, 65% of *C. volutator* survive for 18 days, while *N. diversicolor* can survive for 30 days without significant mortality. At 25 °C there will be 100% mortality of *C. volutator* after 3 days and 100% mortality of *N. diversicolor* after 11 days. Freezing will also rapidly kill exposed *C. volutator* and *N. diversicolor*. Thus management involving draining of a lagoon, particularly if requiring several weeks' work should be done in mild spring or autumn weather if mortality of these two important prey species for birds is to be avoided. Note that rotovation of the dry lagoon mud would kill the invertebrates sealed in their burrows.

Following a major mortality of invertebrates, for example as a result of prolonged drying out, recolonisation of *N. diversicolor* and *C. volutator* will be slow and biomass increase and recruitment depends largely on breeding *in situ*. Complete recovery of *C. volutator* densities will take a year or more, and recovery of *N. diversicolor* densities will be even slower.

Rehfisch (1994), from a study of brackish (less than 10‰ salinity) lagoons at Blacktoft Sands, Humberside, suggests that water levels should be kept at depths of at least a metre to maximise biomass and productivity of Chironomidae, which are important foods of both ducks and waders. The latter, however, rarely feed in water depths greater than 15 cm, so water levels should be lowered during peak passage periods for migrating waders. An ideal low-salinity brackish lagoon would have steep sides to allow deep flooding during winter, to prevent freezing to the bottom; then a shallow slope towards the centre so that gradual emptying would continually expose a new 'shoreline' of suitable depth for feeding waders; and finally, a deeper lagoon centre which would never dry out and so would provide a reservoir from which chironomid larvae would recolonise when the lagoon was reflooded. Because of the oviposition periods of chironomids in England, a lagoon emptied and refilled during late autumn (October) would not build up a large enough biomass to make it attractive to waders if emptied again the following May. But a lagoon emptied and refilled during April and May would build up sufficient biomass during the summer for a second period of use by birds if water levels are lowered in September and October.

Sand-dunes

On the foredune, fencing will act to trap sand and is frequently used to aid dune establishment or to stabilise eroding parts. A single line of fencing with a ratio of fence material to gaps of about 50% is the most cost effective. At higher wind speeds a double line of fencing traps sand at a higher rate. The best materials for fencing are natural, such as brushwood, wood pilings or forestry offcuts, as these will eventually degrade and add organic material to the dune system. The fence should be at least 1 m and preferably 2 m above mean high water to avoid most storm damage. Further details of fence construction and general sand-dune management are given in Brooks (1979) and Ranwell & Boar (1986).

In the event of a blow-out (Fig. 4.5), steps should be taken to stabilise the newly

Fig. 4.5. Where the wind catches parts of dunes, dune blow-out as seen here can result. In such circumstances any remaining vegetation becomes drowned with sand and the dune system remains unstable and liable to further erosion. (Fraid Head, Northern Scotland; J. Andrews.)

exposed sand at the earliest opportunity. This could involve fencing, laying porous mats or the planting or transplanting of dune-building grasses, particularly Lyme Grass *Elymus arenarius* and Marram *Ammophila arenaria*. When planting in eroding locations the Marram rhizomes should be planted horizontally to produce rapid stabilisation, while in accreting sites, such as foredunes, they should be orientated vertically. On more sheltered and flatter parts of a dune system where accretion is minimal it is possible to replant eroding areas with Creeping Fescue *Festuca rubra*, Meadow Grass *Poa pratensis*, Rye Grass *Lolium perenne* and Birdsfoot Trefoil *Lotus corniculatus*. Adding a stabiliser, mulch or seaweed, is essential.

For all restoration programmes work inland from the coast, work from the top of slopes to the bottom and work upwind to downwind. Organic enrichment helps considerably and thus this can usefully be combined with control programmes of Sea Buckthorn *Hippophae rhamnoides* or Gorse *Ulex europaeus*.

Planted areas need to be fenced to prevent trampling or covered with fishing nets supported on stakes 0.3 m above the ground. Fences 1 m high are best for increasing the dune width and those 2 m high are best for building height. Such fences can accumulate 0.9–1.8 m of sand a year and sometimes even more. To prevent terminal scour, the ends of the fences should always be graded into adjoining dunes. It is essential that fencing is followed by planting dune grasses or the sand may be lost in winter storms.

On steep slopes thatching with brushwood is the most effective way of restricting wind erosion; 90–100 tons per hectare are required. It reduces windspeed, encourages deposition and provides a more humid and organically rich environment for seedlings. It must be pegged down, wired down, or partly embedded. The thatching should start at the top of the slope.

Mulching is useful in large scale restorations. Adding organic compost at 1 kg m^{-2} produces marram 4–5 times higher after 4 years (Wilcock & Carter, 1977). Chopped straw, seaweed and manure have been applied at 6 tonnes per hectare (Brooks, 1979). Sewage needs to be checked for heavy metals but has the advantage that the smell ensures that it is not necessary to fence an area in order to restrict people! The absence of vegetation on back shores is often due to human trampling as shown by fencing areas and planting with Sand Couch Grass. Such management seems acceptable to the public as long as there are alternative areas and the policy is explained on notice boards.

Dune turf will support annual grazing rates of 0.5 cattle/ha or 4 sheep/ha. Sheep reduce the sand-trapping capabilities of the turf more than cattle as they crop it more closely. Sheep also range more widely through the dune than cattle, and are therefore potentially more damaging. Fencing can be used to restrict grazing to certain areas. Damage to turf can be extensive in the vicinity of rabbit warrens.

While trampling and overgrazing by domestic livestock is a detrimental factor in dune management, grazing is essential in dune pastures to prevent them becoming dominated by scrub and rank tall growing forms. Grazing also reduces the fire hazard. If natural grazers are absent and domestic stock are thought to be too limiting then mowing can be considered. Cuttings must be left to help build the soil structure and nutrient status.

The main invasive species in UK dunes is the Sea Buckthorn. This species grows well from the foredunes through to the pastures and once established can produce a blanket cover in a few years, with profound changes in the other flora and fauna. The only effective control method is eradication as soon as it is perceived to be a problem. This is best achieved by cutting in autumn or winter. The cut material will dry out in the following weeks and by spring is ready to be burnt on site, or used in the production of barriers/fences. In summer, any new seedlings or regenerative growth should be sprayed (using knapsack sprayers). A dilute solution (1 part per 120 parts of water) of SBK2,4-D/2,4,5T has been very successful in controlling this species at Braunton Burrows, Devon – every leaf must be covered as there is no translocation. Glyphosate (e.g. Roundup) can be used effectively on regrowth, this is translocated and readily degrades in the soil.

With increasing demands on ground water, and potential changes in rainfall patterns as the climate system changes, reserve managers need to ensure that sufficient water remains in the system to keep the dune slacks wet. This may require the use of water drained from nearby areas and linked into the dune system by ditches, etc. or excavation of the slack to lower it to the water table. Natterjack ponds regularly dry out; it is however a mistake to deepen these ponds so that they become permanent because this can encourage a completely different community to develop which is likely to include predators of the tadpoles.

Saltmarshes

On saltmarsh overlying sandy sediments, light grazing by cattle (one to every 2–3 ha) or sheep (2–3 per ha) increases plant diversity although it alters species composition, for example by eliminating Sea Purslane *Halimione portulacoides*. Mowing for haymaking also enhances plant species richness, but not so effectively, as it is more uniform in its effects than grazing, which affects vegetation by selective feeding, trampling and manuring (Long & Mason, 1983). On marshes overlying silt, moderate intensity of grazing on the higher marsh but only light grazing on the low marsh produces the highest vegetation diversity (Dijkema, 1984). Moderate grazing implies 4–6 sheep or 0.6–1 full-grown cattle per hectare, light grazing about half these densities, with livestock present between April and October. By retaining a more diverse and architecturally heterogeneous stand of plants, grazing promotes invertebrate diversity at the same time; but heavy grazing can lead to preferential loss of dicotyledons, upon which many invertebrate species depend, rather than of grasses. It may also allow the marsh to become more vulnerable to invasion by *Spartina anglica* if bare patches form.

Fig. 4.6. Extensive saltmarsh areas on British estuaries support large populations of breeding waders, especially Redshanks. (D.A. Hill.)

Halimione- and *Limonium*-dominated plant communities become scarce if grazing is heavy.

Moderate grazing of the upper marsh can benefit breeding birds, particularly waders whose chicks feed on the invertebrates associated with livestock dung, provided that nests and eggs are not subject to excessive losses from trampling. Restricting grazing to after mid June helps considerably. In this respect, young cattle and sheep do more damage than full-grown cattle. Cattle need to gain experience of how to leave the flooded areas of the saltmarsh during extreme high tides and keeping some experienced animals from one year to the next is considered a good idea to maintain this expertise.

Where the special interest of an area is wildfowl, requiring the provision of winter grazing for wildfowl, heavy grazing by domestic animals in summer is appropriate to keep the sward height suitably short for the autumn. This level of grazing, at about 1–2 cattle per hectare, is far greater than that appropriate for maintaining a diverse marsh and may also be undesirable for breeding birds especially if there are no restrictions on grazing in the breeding season.

The consequences of grazing need not be straightforward. On an eroding saltmarsh, grazing particularly at times of spring tides can lead to local erosion of the marsh surface and thus precipitate more widespread marsh erosion. Furthermore the high species-richness resulting from grazing need not be valuable if, for example, it is largely due to the colonisation of bare patches by low marsh species such as glassworts *Salicornia* spp. and Annual Seablite *Suaeda maritima*.

A major management problem on saltmarshes is the control of invasions by *Spartina anglica* (Fig. 4.8), the fertile hybrid cord-grass first recorded in 1870. This species colonises gaps caused by disturbances such as turf-cutting, promotes rapid accretion (10 cm/yr) in the pioneer zone and produces a level marsh of poor diversity of plants and breeding bird species. It impedes drainage and accumulates organic debris as well as sediment. It may lead to losses of silt from the lower tidal levels of the shore, but at the very least it reduces the intertidal feeding areas available for use by large numbers of waterfowl at low tide and fish at high tide. On some sites such as Langstone Harbour, Hampshire, *Spartina* is undergoing natural dieback. The 'cyclical' aspect of its ecology is poorly understood.

Methods of effective control of *Spartina* spread have been the subject of much debate. Removal of *Spartina* permits recolonisation of sediments by marine invertebrates and leads to renewed use of such feeding areas by shorebirds (Doody, 1984; Evans, 1986). Spraying with herbicide is best achieved from a vehicle with caterpillar

Fig. 4.7. Intertidal flats are also of major importance for wintering wildfowl such as Brent Geese shown here. Management again involves an integrated approach with other uses of the estuary in order to reduce temporally, or spatially, the impact of disturbances from wildfowling, sailing or dog-walking near roosts or feeding concentrations of birds. (D.A. Hill.)

Fig. 4.8. *Spartina anglica* seen here encroaching on a *Salicornia*-dominated saltmarsh is a relatively recent invader and can destroy large areas of marsh, causing eventual silting-up of the middle and upper marsh. Management involves chemical or mechanical removal of *Spartina* before it reaches extensive coverage, although there is evidence that natural die-back can take place after many years. (J. Andrews.)

tracks or 'balloon' tyres that can negotiate soft mud without sinking appreciably. It is essential to spray under calm conditions, particularly if a spray-boom is used, and during the growing season when temperatures are highest; July and August are optimum months. Herbicide should be applied on receding (ebb) tides during the period of neap tides when the maximum numbers of hours of contact between herbicide and plants is achieved before the *Spartina* is covered again by incoming tide. In areas subject to dieback, rapid foreshore lowering has been noted. If this occurred following spraying, it might negate some of the benefits, although it would increase the intertidal areas below MHWS available to shorebirds to forage.

Sea walls

Sea walls require management in order to reduce invasion by trees and shrubs whose roots may damage the integrity of the wall. They are usually mown, and this should be left as late as possible as many characteristic plants are late fruiting. This, however, may often lead to the dominance of Sea Couch *Elymus pycnanthus*. Grazed walls can result in diverse communities, but overgrazing will affect invertebrates. Low intensity grazing should be aimed for to produce tussocky vegetation which is of importance to invertebrates. Such a habitat should also benefit small mammals and hence owls and harriers. A problem is that if stock are likely to damage saltings, long expensive fencing is needed. One solution is to use flexinetting to control stock which graze the wall section by section.

References

Barnes, R.S.K. (1989). The coastal lagoons of Britain. An overview and conservation appraisal. *Biological Conservation* **49**, 259–313.

Barnes, R.S.K. (1991). Dilemmas in the theory and practice of biological conservation as exemplified by British coastal lagoons. *Biological Conservation* **55**, 315–28.

Brooks, A. (1979). *Coastlands*. London: British Trust for Conservation Volunteers.

BSAC. (1985). *Sport Diving*. London: Stanley Paul.

Burd, F. (1989). *The Saltmarsh Survey of Great Britain*. Peterborough: Nature Conservancy Council (obtainable from Joint Nature Conservation Committee, Monkstone House, City Road, Peterborough).

Clark, J.R. (1977). *Coastal Ecosystem Management*. New York: Wiley.

Cryer, M., Whittle, G.N. & Williams, R. (1987). The impact of bait collection by anglers on marine intertidal invertebrates. *Biological Conservation* **42**, 83–93.

Davidson, N.C., d'Alaffoley, D., Doddy, J.P., Way, L.S., Gordon, J., Key, R., Drake, C.M., Pienkowski, M.W., Mitchell, R. & Duff, K.L. (1991). *Nature Conservation and Estuaries in Great Britain*. Peterborough: Nature Conservancy Council.

Dijkema, K.S. (Ed.) (1984). *Saltmarshes in Europe*. Strasbourg: Nature & Environment Sciences, No. 30. Council of Europe.

Doody, P. (Ed.) (1984). Spartina anglica *in Great Britain*. Peterborough: Nature Conservancy Council.

Evans, P.R. (1986). Use of the herbicide 'Dalapon' for control of *Spartina* encroaching on intertidal mudflats: beneficial effects on shorebirds. *Colonial Waterbirds* **9**, 171–5.

Gubbay, S. (1986). *Conservation of Marine Sites*: a Voluntary Approach. Ross-on-Wye: Marine Conservation Society.

Hill, D.A. (1989). Manipulating water habitats to optimise wader and wildfowl populations. In *Biological Habitat Reconstruction*, pp. 328–43. Lymington: Belhaven Press.

Jackson, M.J. & James, R. (1979). The influence of bait digging on the cockle, *Cerastoderma edule*, populations in north Norfolk. *Journal of Applied Ecology* **16**, 671–9.

Long, S.P. & Mason, C.F. (1983) *Saltmarsh Ecology*. Glasgow: Blackie.

Mitchley, J. & Malloch, A. J. (1992). *Sea Cliff Management Handbook*. Lancaster: University of Lancaster.

Olive, P.J.W. (1993) Management of the Lugworm, *Arenicola marina*, and the Ragworm *Nereis virens* (Polychaeta) in conservation areas. The implications of population structure and recruitment processes. *Aquatic Conservation.*, **3**, 1–24.

Picton, B.E. (1991). The sessile fauna of sublittoral cliffs. In *The Ecology of Lough Hyne*, ed. A.A. Myers, C. Little, M.J. Costello & J.C. Partridge. Dublin: Royal Irish Academy, pp. 139–42.

Ranwell, D.S. & Boar, R. (1986). *Coast Dune Management Guide*. Huntingdon: Institute of Terrestrial Ecology.

Rehfisch, M. (1994). Man-made lagoons – and how their attractiveness to waders might be increased by manipulating the biomass of an insect benthos. *Journal of Applied Ecology* **31**, 383–401.

Sheader, M. & Sheader, A. (1989). Coastal Saline Ponds of England and Wales: an Overview. Peterborough: Nature Conservancy Council, CSD report no. 1009.

Wilcock, F.A. & Carter, R.W.G. (1977). An environmental approach to the restoration of badly eroded sand dunes. *Biological Conservation* **11**, 279–91.

5 Rivers, canals and dykes

NIGEL T.H. HOLMES AND ROGER G. HANBURY

Introduction

Rivers, canals and dykes are probably the most accessible of all habitats to man due to their intricate, vein-like distribution throughout the land. They are also taken for granted, despite their historical and ecological interest.

The retreating glaciers of 12 000 years ago left a network of rivers which have played a major part in moulding the wildlife and landscape character of Britain. During the early settlements of man rivers were often obstacles to expanding territories; later this changed and they became the highways for population movements and trading. The deep and large rivers formed an arterial system of transportation for centuries until the end of the eighteenth century, when the first artificial canals were built to link the natural highways together. The industrial revolution was built on the traffic coursing through the waterways of Britain, a shared role between rivers and canals. Today it is a very different story, with navigation on rivers and canals being primarily by pleasure craft.

Although there are vast ecological differences between mountain rivers and lowland systems, plant and animal communities of many lowland rivers, canals and dykes may have much in common. A few plants and animals are almost exclusively found in fast-flowing rivers (e.g. stoneflies such as *Perla* spp., mosses such as *Hygrohypnum* spp.), but the majority also occur on wave-washed rocky lake shores. Dykes and canals also have few species occurring in them exclusively; however, grazing marsh dykes are known to support some invertebrates which have been recorded from no other habitats. Therefore there are extremes of habitat provided by rivers, canals and dykes, but many systems have much in common. Indeed, several rivers in their lower reaches are still navigation channels (e.g. Thames, Nene, Trent) whilst some rivers have been straightened from their original courses to become little more than drains (e.g. River Torne, Forty-foot River, King's Sedgemoor Drain). In flatlands, claimed from swamp and saltmarsh, drainage channels are normally artificial, but they can be very important open water habitats with a character all of their own.

Even when rivers are sluggish and deep, their uni-directional flow which brings nutrients, food, oxygen, a cooling influence and sometimes pollution from upstream, separates their communities from those found in canals. Where the gradient increases, the differences between rivers and canals are more pronounced. The former have obvious pool/riffle sequences and stony beds where the current is rapid; they generally have much cooler water than canals and more natural banks.

The potential of most rivers to support wildlife has, however, frequently been compromised by past management activities and degradations in water quality. In England and Wales the larger rivers are managed by the Environment Agency (EA). These rivers, totalling around 35 000 km, are called 'Main Rivers' because of the importance they have for draining of agricultural land and alleviating problems of urban flooding. Rarely are the rivers owned by the EA, but they have permissive powers to undertake maintenance and promote drainage schemes; they also have a responsibility to consent or refuse applications from third parties to undertake river engineering works, or discharge to, or extract water from, rivers. Under present legislation the EA is required to exercise its functions so as to 'further' and 'promote' conservation. Guidance on how they should fulfil these obligations is given in a statutory Code of Practice. In Scotland the situation is different with the River Purification Boards merely being advised by the Scottish Development Department to act within the 'spirit' of the English and Welsh legislation.

Dyke systems are defined here as artificial and permanent linear bodies of water not exceeding 5 m wide. Where there is flow, this is uni-directional, erratic, often artificially induced and rarely important in determining wildlife interest. They do not include the many hundreds of thousands of drainage ditches which become periodically dry and characteristically carry storm and surface water to rivers. True drainage dykes are most commonly associated with drained marshlands. These are typified by those found in Broadland of East Anglia where an area of 16 000 ha has hundreds of kilometres of dykes in the river flood-plains. Most marshlands with dyke systems were formed centuries ago by the drainage and associated embankments of estuarine mudflats, alluvial washlands and fens.

In the Gwent Levels of South Wales, it is known that the Romans were first to drain and cultivate the lowlands, but the monastic period of the twelfth to sixteenth centuries resulted in greater change. Originally the dyke systems were part of a gravity/sluice drainage system; however, in many cases the level of the drying land fell and so the water had to be pumped through dykes and over embankments. In Broadland, windmills were used as pumps in the seventeenth century with over 200 working in the nineteenth century.

Today, many interesting dyke systems still remain in intensively farmed claimed washlands. Those of the Broads, north Kent Marshes and Levels such as Pevensey, Somerset and Gwent are examples of nationally important dyke systems.

Where flat, often previously very poorly drained land is intensively farmed, drainage responsibilities are often vested in Internal Drainage Boards (IDBs). These Boards have diminished in number over the past 20 years, some being subsumed by the EA's predecessor organisations. Traditionally, the management of drainage channels by IDBs has been more intensive because of low-lying land, low gradient and richer soils. This often promotes extensive plant (weed) growth. The IDBs are also covered by the same duty to further conservation and Code of Practice in managing 'drains' as the EA.

Management of small rivers, streams and drainage ditches (outside EA and IDB control) are the responsibility of landowners and occupiers. Where the systems are natural and dynamic, nothing has been traditionally done to them save for the occasional removal of a fallen tree. In the past, man-made drains tended to be cleared of silt and reeds by hand as and when required. This was usually a small amount per year but the norm today is to use a hydraulic machine to clean whole drains, even complete systems, in one clean swoop.

Streams are regarded as semi-natural small rivers which have had little previous management. As such, the best management for them is nothing; they are thus not considered further in the chapter.

Although some canals, such as the Fossdyke, date back to Roman times, the bulk of the surviving waterway system was constructed in the late eighteenth and early nineteenth centuries for the commercial carriage of cargoes. At its peak in the mid nineteenth century the navigable waterway system extended to 6800 km of river navigation and artificial canal. It serviced the transport needs of a great swathe of England and Wales from Kendal and Ripon in the north down to the south coast and across to the valleys of South Wales. Waterways were also built in lowland Scotland. In the Highlands, the Caledonian Canal and on the west coast the Crinan provided shorter and safer passage for coastal vessels. Around 30 million tonnes of goods were carried annually at their peak. Today that figure is reduced to about 4 million tonnes predominantly on the river-based navigations of the north east.

The major use for canals is now recreational: boating, angling and towpath walking. Over the years they have also become extensively integrated into the land drainage system transferring surface water from urban and rural catchments to natural water-courses, and providing a water resource, to agriculture and industry. These functions can have a significant impact on management options.

Acts of Parliament were necessary to enable canal companies to acquire land compulsorily and build canals, and established navigation rights on condition of payment of tolls. In general terms the right to navigate remains until varied or repealed by Parliament. In many instances that right to navigation has been abolished. Boats now require a licence to use the canal. This is an important distinction from rivers where navigation is as of right. A further distinction, important in the conduct of management, relates to ownership. The banks and bed of artificial canals are owned by the successor in title to the canal builders and not the riparian owners.

Management and control of canals, navigable rivers and waterways is now vested in over 40 navigation authorities which include Trust bodies, local authorities and commercial organisations. The largest is British Waterways, responsible for just over 3200 km. The Broads Authority has 200 km and EA 800 km. The total length of navigation is about 5200 km.

The three largest authorities have, or in the case of British Waterways soon will have, the duty to 'further' conservation in carrying out their duties. This demonstrates the significant shift away from the historical priorities of transport system management for the needs of commerce to the modern need for a diverse and environmentally rich system of waterways which caters for a wide range of formal and informal recreational activities of which navigation is now only one part.

River, canal and dyke communities

A key difference between good river and good dyke systems in the open landscape is the more pronounced corridor of habitats associated with natural and semi-natural rivers. Most canals also have this characteristic encompassing within their boundary a tow path and hedgerow on at least one bank as a buffer between the waterway and adjoining land. It is the corridor which is so important to many animals associated with rivers and waterways, most notably the mammals, birds and invertebrates. For fish the important factors are diversity of underwater habitats, water quality and cover.

For animals which use the river corridor the health of the whole habitat complex is important. The Otter *Lutra lutra* is a good example of an animal which depends on so much more than just the river channel. Although it may feed on fish, Eels *Anguilla anguilla* and other animals in the water, much of its feeding is done in backwaters, tributaries, ditches and other habitats in the corridor and on the flood-plain. Cover is vital for otters, not only the resting, sleeping and breeding holts, but for safe movement up and down the river. Bankside trees, scrub and tall herbage of the corridor provide such cover. The same is true for many riverine birds and invertebrates such as dragonflies.

There are often great differences in the distribution patterns of many breeding birds closely associated with rivers. Whilst Dipper *Cinclus cinclus* and Grey Wagtail *Motacilla cinerea* are closely associated with torrent and shingle rivers, Coot *Fulica atra*, Moorhen *Gallinula chloropus* and Mute Swan *Cygnus olor* are more typically associated with sluggish rivers and canals with abundant vegetation. For the blue-arrow of our streams, the Kingfisher *Alcedo atthis*, vertical banks for nesting, and small fish and unfrozen waters in winter are its priority. Canals are often favoured feeding sites.

Fish communities are naturally very different throughout the country. Those in watercourses which discharge to the sea south of the Humber and east of the Avon are the richest because they were last to be cut off from the continental lakes at the end of the Ice Age. Sluggish rivers are characterised by Stickleback *Gasterosteus aculeatus*, Perch *Perca fluviatilis*, Dace *Leuciscus leuciscus*, Rudd *Scardinus erythrophthalmus* and Bream *Abramis brama*. Coarse gravels and faster-flowing currents favour Trout *Salmo trutta*, Salmon *Salmo salar*, Barbel *Barbus barbus* and Stone Loach *Neomacheilus barbatulus*. In canals Roach *Rutilus rutilus* is ubiquitous. In trafficked canals, other species will be Gudgeon *Gobio gobio*, and Carp *Cyprinus carpio*. Carp was widely introduced to canals in the post-war period, and although populations are not self-sustaining they are hardy and long-lived. On less disturbed canals species such as Pike *Esox lucius*, Perch, Tench *Tinca tinca* and Rudd will be present, with Chub *Leuciscus cephalus* and Dace where there is a riverine influence.

Plant communities are totally different throughout the country, being very dependent on the physical characteristics of the channel and its inherent geological richness. In upland rivers with bed-rock, the plant communities are dominated by mosses and liverworts. Where water tumbles shallowly over consolidated pebbles and gravel, beautiful white trailing crowfoot appears. As silts dominate the bed and water velocity is slow, lilies, pondweeds and reeds thrive.

Drainage dykes have a character all of their own and different parts of the country have their specialities. In the vastness of the Gwent Levels some dykes are shaded by shrubs giving the invertebrate communities shelter and providing quite a different structure to that found on other Levels. Characteristic plants of dykes are those which may be uprooted by the currents of rivers or the wave-wash of boats. Thus free-floating plants such as Frog-bit *Hydrocharis morsus-ranae* and Water Soldier *Stratiotes aloides* are characteristic alongside the diminutive Ivy-leaved Duckweed *Lemna trisulca*. Amongst them may be found Water Violet *Hottonia palustris* and fine-leaved pondweeds such as the rare Sharp-leaved Pondweed *Potamogeton acutifolius*. Reeds are also common on the banks and encroaching into the channel overwhelming the watercourse. Although this often provides good nest sites for some birds and different habitats for invertebrates, it can stifle diversity of habitat and species. It is for this reason that management of dykes is absolutely vital if varied wildlife interests are to be maintained.

Dykes often have great invertebrate interest. A wide range of dragonflies, water beetles, mayflies, water boatmen, etc., breed in the open water. The dyke, where freshwater meets saline influences on the coastal marsh systems often supports very important transition communities. Another vital area of invertebrate interest is the shallow, often trampled, vegetated margins where such specialities as the Raft Spider *Dolomedes fimbriatus* can be found, alongside hoverflies, beetles, bugs, etc., that cannot swim in deep-sided ditches. A key feature of the wildlife interest of dykes is that they often represent minute relic communities of once vast wetlands of lowland Britain, changing the buffer zone from grazing land to arable can have devastating effect. As the marshes were drained, the only refuge for the aquatic and wetland plants were the dykes. For over 300 years in some cases, the dykes have been, literally, their last *ditch* stand.

Canals can take on the characteristic of river or dyke depending on the flow regime, geomorphological context and level of boat traffic. Typically they are more disturbed than natural rivers or remote dyke systems because of the ease of access provided by towpath or boat. The best support rich plant communities, some of national import- ance. Good examples include the Basingstoke, Ashby, Chesterfield, Huddersfield Narrow and Pocklington canals, where there are well-developed communities of submerged plants such as the broad- and fine-leaved pondweeds *Potamogeton* spp., floating-leaved species such as Unbranched Bur-reed *Sparganium emersum* and Arrow- head *Sagittaria sagittifolia* and an emergent fringe of reeds *Glyceria* and *Sparganium* spp., sedges *Carex* spp., Great Water Dock *Rumex hydrolapathum* and Yellow Flag *Iris pseudacorus*. Some canals are important for nationally scarce species, for example, the Floating-leaved Water Plantain *Luronium natans* now protected under Schedule 8 of the Wildlife and Countryside Act 1981, the Grass-wrack Pondweed *Potamogeton compressus* and the Flat-stalked Pondweed *P. freisii*.

Where there is boat traffic these communities will be suppressed but evidence of each component will be present except on the most trafficked canals. The more sensitive mammalian and bird fauna will be seen only on the more remote or disused waterways or where there is close juxtaposition to a major natural river system. The Pocklington and Montgomery canals, and the River Kennet section of the Kennet and Avon Canal benefit in this way. Invertebrate communities can be rich especially where the reed fringe is well developed. Visible species such as the damsel and dragonflies can be a particular feature, but these can decline through over-management or through disturbance caused by high levels of boat traffic.

River management prescriptions

The value of a river system for wildlife frequently depends on the degree and type of management which occurs within the channel and its bank and the extent of develop- ment in the river corridor and flood-plain. Natural river features are governed by altitude, slope, geology, climate and many other factors and the range of these natural features represents a river's maximum potential for wildlife. Management practices in the past have altered these, nearly always in an adverse manner. The contrast between a pristine river and one which has been intensively managed is stark. The former is a meandering landscape feature with riffles and pools, emergent marginal reeds and cliff, and bankside trees whilst the latter is a straightened eyesore with minimal habitat diversity within the uniform channel and steep-sided trapezoidal banks.

This section is a practical guide to environmentally desirable management and assumes that works are proposed on a river for reasons other than for conservation enhancement, i.e. land drainage, flood alleviation, etc. The recommendations are therefore based on the need to ensure that the aspirations of those proposing work are not compromised, but the works are executed in a manner which will minimise impacts on, or even benefit, wildlife interests. River management in Britain is now generally undertaken in this spirit, with the needs of wildlife recognised. The same principles are to be applied where improvements to a degraded system are proposed. If the system is in equilibrium, dynamically self-sustaining, and of high ecological interest, no management is the best prescription.

For all management works it is desirable to have a pre-works ecological survey executed, preferably as part of an overall management plan. These must include data on habitats and information on vegetation. Where management is undertaken regularly, surveys may not be cost-effective, but for works undertaken infrequently it is absolutely essential. The EA now routinely undertakes river corridor surveys (RCS) prior to executing river dredging and several other management operations. The reason why dredging is regarded as a priority is that it has the greatest potential to destroy or create wildlife habitats. The habitat approach is the basic minimum that should be undertaken. If this reveals high potential for specialist or rich animal communities, surveys of these should be recommended. On the other hand there may be existing knowledge

Fig. 5.1. More interesting habitats can be re-created, even where the water course has been straightened and canalised in the past. Here a more natural looking water course with meanders, shallows and edge habitats has been created from a straight canal. (Reproduced with kind permission of M. Bjorn-Nielsen, South Jutland County Council, Denmark.)

about flora and fauna within the corridor and this should be gathered and appraised alongside the findings of the RCS.

For all river management work, a sensitive approach is to question why the work is done and whether it is necessary. If it is not, a decision is required on whether avoidance of the work is better for wildlife than some structured management which could reverse adverse impacts from previous management works.

Weed cutting

Weed cutting is still an important management activity on many lowland rivers because, during the summer, plant growth may be so great that river levels rise by almost a metre, even though the volume of water they are carrying may be less. Cutting submerged weed (typically Water Crowfoot *Ranunculus* spp.) has been traditional on chalk rivers and streams for centuries, with a close association being made with fishery interests. In sluggish rivers with clay or silt beds, weed cutting has become more prevalent in recent decades and involves both submerged and emergent plants.

The method employed on the chalk rivers Test and Itchen is particularly sensitive. The procedure is illustrated in Fig. 5.2. The whole operation is co-ordinated by the Fishing Association and the EA assists in the removal of the cut weed from the river so that this does not clog weirs and rot within the river, causing great oxygen depletion and subsequent death of biota. The benefits of this approach to aquatic weed control are enormous since there are always long stretches of uncut vegetation as refuge for animals.

In some lowland rivers and large drainage channels it is often regarded as essential to cut weeds to allow free passage of water from side channels to pumping stations or for containment of high flows within the channel. Historic practices often resulted in all weed from bank to bank being cut through entire channels. It is now becoming accepted practice to cut only a percentage of the width of channel. Where the channel is less than 5 m, the retained strip may have to be small whilst for wide channels much larger strips can be left. In the Severn Trent Region of the EA experiments leaving 20–40% of the channel uncut (depending on the river width, extent/type of growth, etc.) has not resulted in any unacceptable hydraulic performance. The same region of the EA has also experimented with cutting sinuous swathes within straight channels to promote natural establishment of habitat diversity. The same techniques have been tried and tested in Danish streams and proven to be adequate for drainage purposes.

The advantages for wildlife in adopting the strategy of retaining some weed are numerous. Some cover and food for fish and invertebrates is always maintained, as are spawning habitats for fish such as Perch, Roach, Bream and Tench which lay their eggs on vegetation. Nesting sites on marginal reeds for Little Grebe *Tachybaptus ruficollis*, Coot and Moorhen are also left or allowed to develop for the first time. An important link between the air and underwater habitats is retained for insects (e.g. dragonflies)

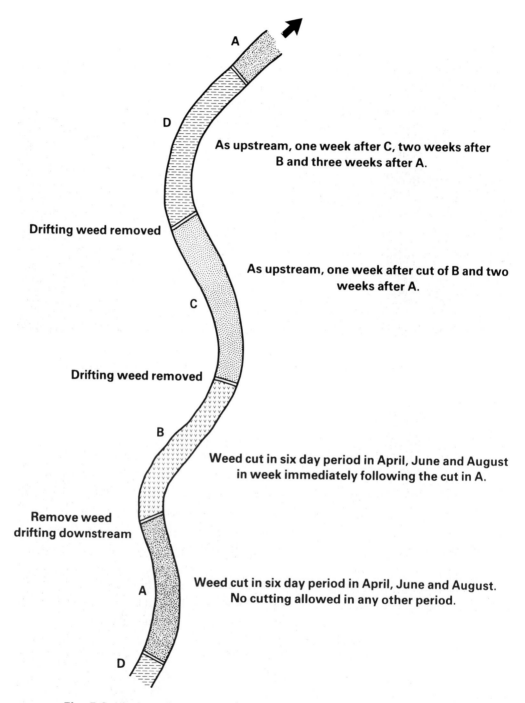

Fig. 5.2. Ideal weed cutting programme as used on the rivers Test and Itchen.

Fig. 5.3. Edge habitats, shallow ledges etc. can be maintained by lightly dredging the water course, removing silt and weed from the centre only. (N.T.H. Holmes.)

Fig. 5.4. Important edge habitats can be encouraged where there are shallow underwater margins. (N.T.H. Holmes.)

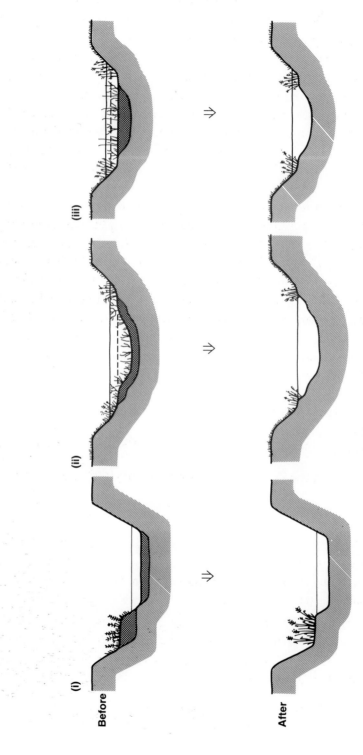

Before

(i) ⇒ (ii) ⇒ (iii) ⇒

After

Fig. 5.5. Maintaining (i and ii) and creating (iii) wetland ledge habitats.

which have aquatic larvae and flying adults. Retained margins of vegetation also concentrate river flows into a narrow central channel, thereby creating variations in velocities and subsequently creations of habitat diversity too.

Where retention of marginal strips of weed cannot be accepted, retaining wisps of plants in locations where there is slightly more capacity in the watercourse is essential. This retains habitat diversity in a system which is likely to gain habitat variety only from the physical environment which the plants themselves form. In the absence of this, leaving 1 m wide strips of submerged plants is advocated every 50 m, but this may not be advisable where vigorous emergents are present. Where pre-management surveys indicate the location of very rare plants, these too should be left if they are vulnerable to impacts of cutting.

Weed control may also involve use of herbicides and dredging. The former is only recommended where there is an abundance of vigorous weeds which are destroying the habitat for other species. Even here there is a problem since the rotting plants may cause de-oxygenation of the water and subsequent death of some animals; there is also the risk that non-target organisms may be killed. Dredging is often associated with weed control in small ditches where some silt removal is also required. Weed rakes (i.e. 'Bradshaw' buckets) are particularly good at removing vegetation and attached soft silt, leaving the firm bed intact. The use of such rakes are advocated when there is an over-encroachment of plants such as Reed Sweet-grass *Glyceria maxima* which needs to be pulled out by the roots if other plants are to grow.

Bank and floodbank maintenance

Many types of management are used on river banks. The reasons for such practices are numerous and include:

- stopping the development of woody growth which might obstruct flow during floods

- keeping banks relatively smooth to reduce the roughness which compromises a river's carrying capacity

- visual amenity purposes in urban areas

- encouragement of a thick matted root system to hold banks together to increase stability and reduce erosion

- minimise cover for burrowing animals which might damage floodbanks through tunnelling – there is a danger that when land is lower than the drainage channel, a small hole will rapidly become a major breach through the scouring action of the drainage water.

Banks provide contrasting habitats for plants and animals but the expression of interest related to the basic physical nature is fundamentally affected by the different manage-

Fig. 5.6. Spot dredging can be used to create open water in reeded or weeded sections where vegetation is otherwise completely choked. (N.T.H. Holmes.)

Fig. 5.7. If dredging or mowing are carried out with too little thought and too much enthusiasm, the resulting channel lacks practically all wildlife interest; here are two examples of how not to do it (*a*) dredging and (*b*) mowing. (N.T.H. Holmes.)

ment practices. The tightly mown banks, which may be appropriate in the urban scene, have minimal ecological interest (Fig. 5.7). A more positive approach is to change lawns into attractive features for wildlife and humans.

Banks which are mown just once a year, especially if left until the autumn, are totally different. Here herbs and grasses have an opportunity to grow to their full height, flower and set seed. Not only is the richness of species increased, but the plants are able to create greater diversity of microhabitats which provides cover for mammals, birds and invertebrates as well as a rich larder of nectar, shoots and fruits. With the greater variety of habitats comes the opportunity to support populations of small mammals which include voles, shrews, mice, etc. The presence of such animals also means that predatory birds will use a river, explaining why some river valleys remain a stronghold for the declining Barn Owl *Tyto alba*.

If annual mowing of river banks is required, and there is no scope to vary this, cutting in late autumn is ideal. This optimises botanical interest since it cuts down to size the large and invasive species but allows all species to flower and set seed. Animals can also take advantage of the available food during this period and most will be able to complete their life cycles; however, they are deprived of a rich energy source during the main time of need during late winter.

Mowing banks less frequently than once a year allows tall, often rank, vegetation to dominate and the potential for shrubs and trees to encroach and become established. Plant richness may decline but the tall vegetation provides the cover necessary for birds such as Whitethroat *Sylvia communis*, Willow Warbler *Phylloscopus trochilus*, Sedge Warbler *Acrocephalus schoenobaenus* and Blackcap *Sylvia atricapilla* to nest in tall bankside vegetation, and Reed Warbler *Acrocephalus scirpaceus* and Reed Bunting *Emberiza schoeniclus* to nest over water in reeds. The winter cover and food also attracts many birds and animals which are not found on mown banks. The protection from wind given by tall bankside vegetation to small, weak-flying insects is considerable; this in turn provides more attractive feeding grounds for hawking dragonflies, patrolling bats and insect-eating birds. Leaving some tall vegetation close to the water's edge is a great bonus to invertebrate life.

The essence of good management for river banks is sustaining and developing a good variety of vegetation stands. This can be easily achieved through the adoption of a relaxed and flexible mowing regime. Ideally some banks will be left for long periods but where annual mowing is regarded as essential, Figs 5.8, 5.9 highlight some simple approaches whereby impacts can be minimised or existing conditions improved greatly.

Fig. 5.8 shows longitudinal strips along the bottom of the bank uncut. The benefits for wildlife are greater when the lower part of the bank is left but this may impact upon the freeboard of the adjacent field. Leaving part of the top of the bank may be slightly less valuable but it should not impact hydraulic efficiency. Where strips of vegetation can be retained for several years, impressive habitats can develop.

Fig. 5.8. In order to maintain vegetational diversity and structure, the base of river or canal banks can be left uncut as shown here. (N.T.H. Holmes.)

Fig. 5.9 shows a method of selectively leaving some sections of bank uncut. This can be based on preservation of key bank communities (as for macrophytes) or for general habitat diversity only. Discharge capacity is rarely affected as the flow velocity increases due to the cut sections and effectively flattens the uncut parts. This method has the advantage that rare colonies of plants or important communities can be selectively protected. Another method routinely being used today is cutting the left bank one year and the right bank the following year. This method ensures continuity of habitats where the same range of vegetation and habitat characteristics exist on both banks. However, if this is not the case, as for instance where reedbeds are confined to a single bank, this is not appropriate and the methods shown in Fig. 5.9 should be adopted.

In most cases, tractor-mounted flays are used. If hand-cutting is employed, more precision in retaining individual plants can be achieved.

Dredging

Dredging is a relatively infrequent operation on rivers but has been responsible for massive degradations of wildlife interest in the past. Future dredging works have two options: tremendous opportunities to put life back into rivers or continue to sustain the previously adverse impacts. Dredging methods which consider wildlife retain or recreate in-stream and bankside diversity (Fig. 5.10). The pools and riffles, slacks with

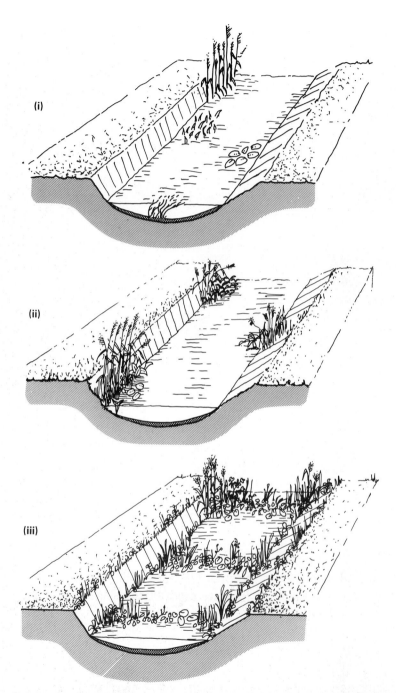

Fig. 5.9. Retention of bankside and aquatic stands of vegetation during a particularly severe dredging operation. (i) Individual species or stands retained; (ii) retention of communities; (iii) retention of strips of vegetation across the channel with intervening sections dredged.

Fig. 5.10. Partially dredging the water channel creates riffles and channel diversity in gravels, increasing the range of species which will establish themselves in different parts of the system. (N.T.H. Holmes.)

rich underwater and floating-leaved plants, marginal shallow ledges teeming with fish fry, steep cliffs, shallow banks with reeds and wetland herbs, and waterside trees are all retained or even enhanced. On the other hand dredging without consideration for wildlife often results in these features being ripped out, leaving instead barren uniformity.

It is for this reason that greatest emphasis has been placed on undertaking river corridor surveys prior to executing this type of river management to identify specially important features which still remain and identify areas where great opportunities exist for enhancement or reinstatement of lost habitats. A standard method has been in use since 1985, with an update of the system being published by the NRA (EA) in 1992.

The majority of this section dwells on approaches to be taken to minimise dredging impacts or to maximise opportunities which arise. The approaches outlined combine conserving existing interests (if they exist) with methods for reinstating and re-vitalising canalised rivers which lack interest. Both *structure* and *form* are important. The chapter deals mainly with structure since in most cases if the physical structure is right, the variations in the form of substrate will be present too. However, it is important to conserve areas of coarse and fine substrate in the bed, and on the banks of rivers. Shallow silty margins support very different communities of plants and animals from those on shingle. The latter are very, very important areas, especially in more lowland regions.

When preparing for dredging it is often necessary to fell or manage trees and shrubs. Thought should be given to the shading which many trees and shrubs give to a river, reducing luxuriant weed growth which itself will need management on a regular basis if the shade is removed. This can be done by selective cutting of mature trees (Fig. 5.11) or 'window coppicing' (not grubbing out) of shrubs so that a dredging machine can gain access to the channel, albeit having to work around some trees and shrubs. For both cases, it is necessary to be selective so that the existing tree and shrub assemblage remains. Over-mature pollarded willows should be re-pollarded at the same time.

Any tree surgery work must be preceded by a survey for possible habitation in cracks and hollows by breeding birds or bats. If Otters are present in the catchment, or may be predicted to return to it in the near future, tree root systems may be very important for them. Alternatively, holts can be built with both the brush and the logs. A design, known to work very well, is illustrated in Fig. 5.12. These should be built as frequently as possible where Otter usage may be possible; they also have the advantage of producing breeding and hibernating habitats for birds and mammals as well as excellent dead wood habitat for insects and fungi.

Where bankside bushes are causing blockages in flow, retention of single shoots to create standards is advocated. Such shoots can form standard trees which do not impede flow yet maintain some shade, create good habitat variation on the bank and help to prevent bank erosion.

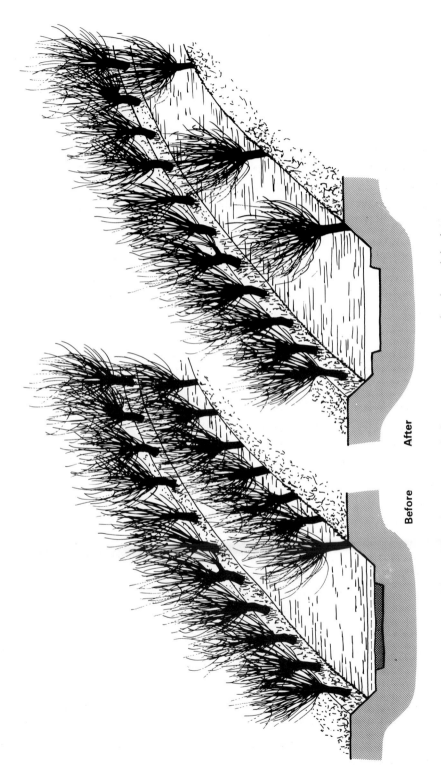

Before

After

Fig. 5.11. Selective removal of trees to gain access for channel dredging.

When dredging, retention of some channel vegetation is vital. The vegetation may have conservation value in its own right, and should be selectively maintained. In other situations it is the variety in physical habitat that it creates which is important; here it is not necessary to save individual plant species, simply retaining the communities and variety of structural forms in blocks or strips will probably suffice. Research has shown that the more diverse the community of plants in a river, the richer the invertebrate and other animal communities.

Diversity comes from the variety of physical features. Where there is a coarse substrate and faster current velocity there will be a totally different plant and animal community from that in the sluggish sections where the bed is silt. Dredging in the interests of wildlife must always aim to restore or improve both longitudinal and cross-sectional variations in profile. It must also ensure that variations in flow characteristics and bed materials are retained. In terms of bank re-profiling, shallow slopes are important for maintaining characteristic riverside plants and animals whilst cliffs should be retained as they are contrasting habitats utilised by Kingfisher and Sand Martin *Riparia riparia* and specialist solitary wasps.

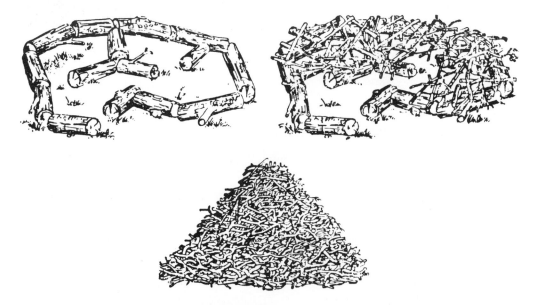

Fig. 5.12. Creation of log/stick Otter holt as recommended by the Otters and Rivers Project. The holt is constructed from logs and sticks such as may result from tree clearance. Large logs are used for the base with smaller material piled on top. The finished holt should have three to five chambers and could include as many as eight. Overall dimensions should be approximately 5 m wide, 5 m long and 2 m high. Ideal locations for the holt are at the junction between the main river and a tributary or on an island. The site should be as close to the river as possible (definitely no more than 5 m away) and above flood levels to avoid being swept away; pegging or wiring down will help to minimise this risk.

Fig. 5.13. Shallow ledges can be created on a fairly large scale with heavy machinery and can then be planted up with reeds to add interest to the system. (N.T.H. Holmes.)

There are several simple general principles which are important when undertaking dredging. The following are some of the most important:

- always work from a single bank, working over it and dumping spoil away from the edge of the river
- remove the minimal amount of bed material and vegetation from the river to achieve the drainage needs
- retain coarse materials in the bed – if this is impossible, they should be replaced at a lower level following the dredging operation
- retain variations in substrate, velocity and depth characteristics along, and across, the river channel
- never dump spoil on un-recreatable, valuable or inundation-dependent terrestrial habitats adjacent to the river
- be aware of the requirements of abutting wetlands, ensuring that the works do not lower water-tables or reduce inundation of these.

Although these are good general principles, the uniqueness of each section of river makes exceptions important. This is particularly so when the works give opportunities for reinstatement of habitats lost by previous insensitive approaches to river management. For example, where there are steep trapezoidal banks with ruderals it is desirable that these be re-profiled to create the necessary habitats for the re-establishment of typical riverside vegetation. The same is true where uniform channel features are present; additional dredging often enables habitat variations to be reintroduced (Figs 5.14, 5.15).

Dyke management

In Britain it is estimated that 5.5 million hectares of agricultural land is dependent on some form of artificial drainage; some 2 million hectares of this is claimed land from swamp or marsh needing protection from inundation from either sea or upland waters. The drainage dykes therefore perform a vital function in maintaining land in agricultural production and enabling human habitation of many areas. They are also the last remaining linear strips of wetland where the once commonplace swamp plants and animals have retreated; most of which have become rare today.

To perpetuate agriculture in the lowlands, management of dykes is necessary. In essence this is having control of both water levels in the dykes and periodicity/timing/extent of inundation on the washlands. This section is concerned with management practices of dykes which enable drainage functions to be combined with retaining and/or improving wildlife interest. It has been shown in countless cases that practices can be modified to improve wildlife interests which also achieve appropriate flood defence and

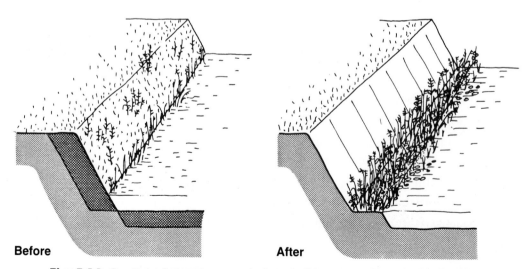

Before   **After**

Fig. 5.14. Creation of an underwater shallow shelf from re-profiling a steep, almost vertical bank of minimal interest.

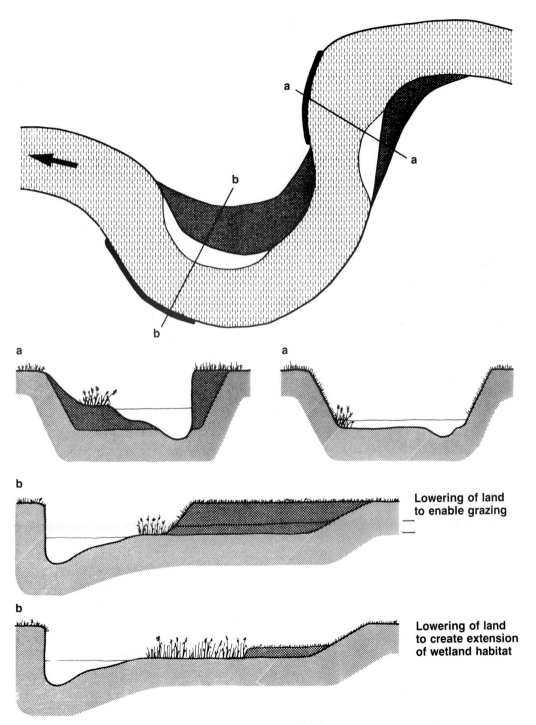

Fig. 5.15. Retaining existing good cross-sectional habitat variety, or improving poor sections through dredging.

land drainage needs. In some cases they *may* require a little land-take but in most cases changes in the timing of works and the method of re-profiling is all that is needed.

It may be surprising to many, but to maintain the wildlife value of dykes requires management too – without it many soon become over-grown and silt up. This is because they are artificial in nature, not having the natural self-maintaining mechanisms that rivers have. Without regular management, it is rarely long before a diverse range of submerged, floating and emergent plants is replaced by virtual monocultures of reeds. With the depleted range of plant species, and the habitats that each create, the animal life goes too. It is very important to check whether what appears to be very uniform does not in fact have great variety in plant species and habitats; if it has, animal life is likely to be variable too. Overgrown dykes with variety are an important habitat and should be cleared sparingly and slowly (Fig. 5.16).

Many of the general principles described for river dredging and weed control are applicable to dyke systems too. However, it is very important to remember that dykes are more regularly managed than even heavily managed rivers and there are many species associated with them that have very poor dispersal methods. For this reason *it is vital to maintain both habitat and vegetation structure in all management operations.*

Another essential element of good dyke management is maintenance of variations in water depth, amount of vegetation within the channel, bank profiles, and land–water interface influences. Where the maximum potential for the system has not been realised, efforts should be made to achieve this without any, or undue, compromise to other interests. One of the main reasons why dykes maintained their wildlife interest for centuries was because the traditional 'little and often' practices of hand weed cutting and ditching were not executed efficiently enough to remove all the vegetation or create perfectly uniform cross and longitudinal sections. Despite being regimentally straight, underwater habitats of old dykes were varied and the banks had variations in width, slope, aspect, degree of wetness and extent of cattle grazing and trampling. Management also took much longer than by hydraulic machine which can 'ditch' many hundreds of metres in a single day; previously, therefore, natural recolonisation went on in tandem with the management operations.

Although corridor surveys are rarely executed prior to dyke maintenance, rapid surveys are advocated in certain cases. For example, where rare plants are known to be present in an area (e.g. Sharp-leaved Pondweed in the Pevensey system), but their precise location is unknown, a survey could locate them and ensure they are retained. Rapid habitat surveys could also assist in promoting enhancements through different treatments for banks with blanket cover of single species of reeds or where habitats for wetland plants is limited by the steepness of banks.

The following are additional examples of watercourse management practices which are especially suitable for linear dykes over and above the ideas presented earlier for natural rivers and larger man-made channels.

Fig. 5.16. It is very important that ditches are slubbed out in rotation, and that some ditches are allowed to become 'overmature' to maintain all stages in vegetation succession right through to completely weed choked as in (*a*). (*b*) shows a Great Silver Water Beetle *Hydrophilus piceus*, Britain's largest beetle, and a Red Data Book species, now only found in overgrown sections of grazing marshes in the Thames Marshes and Somerset Levels. ((*a*) N.T.H. Holmes (*b*) R. Key.)

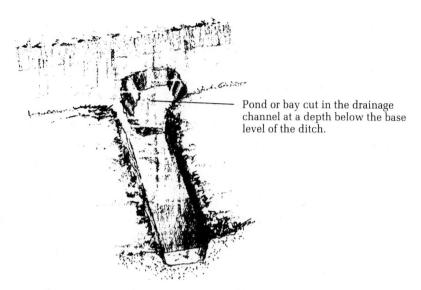

Pond or bay cut in the drainage channel at a depth below the base level of the ditch.

Fig. 5.17. Creation of deepwater pools at the junction of dykes.

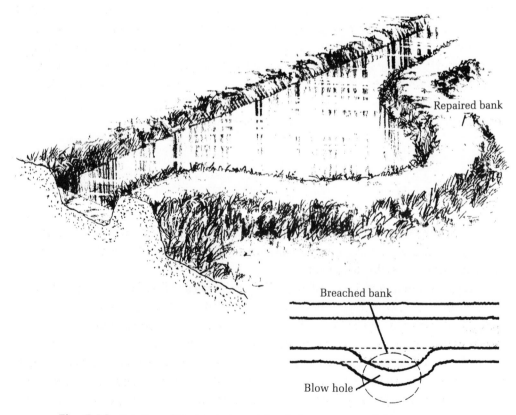

Repaired bank

Breached bank

Blow hole

Fig. 5.18. Creation of backwaters on dykes to form an area of deeper water.

The formation of deeper pools is especially valuable for small ditches where there is vulnerability to low water levels in summer or risk of drying out. Ideally they can be formed at the junction of a side channel where no extra land-take is required (Fig. 5.17). Alternatively they can be formed in the corner of fields or other locations as deep-water backwaters created as new features (Fig. 5.18). The creation of such features should be considered when repair of banks is required.

If the aquatic plant and animal communities are to remain diverse, attention to detail on bank management is also equally essential. A shallow batter which accentuates the wet-edge habitat, formed during the operation is a desirable enhancement. If this can be done where cattle trampling and grazing is possible, this will decrease the need of future management and increase the value of the 'edge' habitat for numerous specialist invertebrates and many wetland annuals. Retaining, extending, or even planting new hedges is important since communities within the dykes vary according to the amount of adjacent hedges. This is perfectly illustrated by the unique communities of invertebrates of the Gwent Levels.

Underwater dead wood is an important habitat for a number of rare invertebrates. The tendency to remove whilst 'tidying up' must be resisted. Diversity of ditch flora and fauna are so highly correlated with habitat availability that retention of even the smallest micro-habitat is important. The key to retaining existing biological interest lies in maintaining or improving upon the diversity of in-channel, bankside and edge habitat. The essential additional management issue is to undertake work in a progressive manner to retain variety in the age structures of plant communities within relatively short stretches of even individual dykes, blitzing one bank but leaving the other completely alone may not be good for retaining biological diversity.

Planting regimes and establishment

When rivers, dykes and canals have been particularly badly damaged by past engineering works, physical re-profiling of bed and banks may need to be accompanied by planting to realise maximum potential. The key to success is knowing what, where, when and how to establish vegetation.

As regards when, and how, a few guiding principles will suffice. In general, most plants establish best during the growing season. Wetland plants are best planted in situ in April, allowing them time to develop good root systems before either their shoots become excessive or dry conditions prevail. It is generally not a good idea to plant wetland plants in the autumn as the damp and cold soil can cause major losses through rotting of the rootstock. As wetland plants are established in the soils where the water table is high in April, simply digging holes large enough to take the plants is sufficient, with a need to water only once, immediately after heeling in. It should also be remembered that many plants establish well from seed on recently exposed substrates. This is particularly true of Sedge, Rushes, Marsh Marigold *Caltha palustris* and a whole host of wetland plants which often grow as annuals, e.g. Celery-leaved Buttercup *Ranunculus sceleratus*, Pink Water-speedwell *Veronica catenata* and Water-cress *Nasturtium officinale*. Simply gathering seeds from the wild and distributing them thinly in suitable locations immediately will be very successful. It should also be considered as a technique for spreading the rarer species in dykes where they only have an isolated distribution.

Aquatic plants are generally the easiest of plants to establish since no consideration has to be given to water during dry periods. Greatest care has to be given to give some protection to any plants which are being established on wave-washed shores or edges. For the vast majority of species, all that is required is to gather material, dig a small hole in the underwater substrate, and firm in. Ideally this is done during the period April to June. Seeding can also be successful, scattering recently collected fruits of pondweeds, Bur-reed *Sparganium erectum*, Flowering Rush *Butomus umbellatus* or Crowfoot *Ranunculus* spp. in shallow water.

The above recommendations are ideal for the establishment of selected species. Often it is possible to rescue plants from aquatic sites which are being managed for other purposes and utilise the material in very degraded sites. Here it is preferable to take very large clumps of material, together with the substrate, by mechanical excavator. The same excavator can dig recesses in the recipient site and firm the material in.

Management of canals

Management priorities on all waterways have shifted towards environmental enhancement and the needs of recreational use. This is sometimes a difficult fit with engineering concerns for the structural integrity of embankments and cuttings, water retention within the artificial channel, relief of flood water and navigation safety. The management approaches outlined above for rivers and dykes apply to canals in many instances. What follows highlights the constraints and special conditions applicable to canals.

Like dyke systems, canals depend on continuing management intervention to secure their survival as watercourses. Without intervention canals would eventually revert to the dry land from which they were cut. Often the best developed conservation interest is to be seen where there is a low level of boat traffic coupled with periodic management. The effect of both is to maintain habitat diversity by arresting hydroseral changes.

As on other types of watercourse before any maintenance is progressed those parts to be affected should be surveyed to identify current ecological interest so that any remedial or protective measures can be built into the work programme.

(a)

(b)

Fig. 5.19. When creating a new habitat on a river course, pay special attention to scale. Here are two examples of badly planned site creation (a) with inappropriate use of large boulders, and (b) with an inappropriate channel depth and width. (M. Perrow.)

Management of banks

Many of the general points made about mowing regimes under the heading of river management apply to canals. The distinction is that in most situations there will be a towpath on one bank bounded by a hedgerow which provides the only access to the waterway for pedestrians and bankside machinery.

For pedestrian use, regular mowing of the trodden path is usually desirable to enable access but this need not extend beyond 1.5 m in width. Depending on local use and physical constraints, mowing on either side of this path can be restricted to one or two cuts per year across a wider span to suppress woody species where they cannot be tolerated for engineering or practical reasons. For example, woody plants are unwelcome at the water's edge on the towpath side where access to the water is required, or where waterway walling or the canal lining is threatened by root penetration or woody stems. Particular care is necessary on embankments where adequate visibility of the bank surface must be provided to inspect for leaks.

Early spring cuts will favour summer flowering species such as orchids *Orchis* spp.; late summer or autumn cuts favour spring species such as Primrose *Primula vulgaris* and Cowslip *P. veris*. Where space permits, timings can be alternated along the same bank and either side of the trodden path to maximise interest, whilst also allowing adequate access.

Generally reed fringes should not be cut although cut embankments may be preferable to indiscriminate trampling and damage by anglers and boaters looking for bankside access to fish or moor.

Management of hedgerows is also important along canals. For appropriate methods see Chapter 9.

On the far or offside bank more substantial growth is more easily accommodated. Access on foot is not usually required and use of this bank by boaters is discouraged, as the land will be outside the boundary of the waterway. Engineering constraints are less: in rural situations there will be no waterway wall and where the canal is built 'side-long' (that is on the hillside with only one side embanked) the offside bank will be against the hillside and not an artificial structure. Management of vegetation will be necessary but limited to lopping branches to prevent obstruction to the navigation and lines of sight. This is the bank least disturbed by waterway use and the most important for waterway birds such as Coot and Moorhen, Sedge Warbler, Reed Warbler and Reed Bunting. Overhanging vegetation and exposed roots are important habitats for fish and invertebrates, and vital on heavily trafficked canals where aquatic vegetation is reduced. Berries and insects which fall into the water are also a significant food source for fish.

Water weed management

The physical dimensions and characteristics of canals render them almost universally desirable for plant growth. Absence of plant growth in the channel is indicative of boat traffic, polluted water or sediment, or recent dredging or weed management. Some urban waterways suffer from polluted water from highway run-off, treated sewage effluent or historically contaminated sediments. These are the exceptions and canals are generally unpolluted and well able to support plant growth, with boat traffic being the major limiting factor.

Where excessive weed growth occurs to the extent that navigation or water movement is impeded, there are usually a limited number of species involved. Algae predominate, notably *Cladophora* spp., together with the floating-leaved duckweeds *Lemna* spp. and a variety of common vascular plants such as Canadian Pondweed *Elodea canadensis*, Water-milfoil *Myriophyllum spicatum*, Rigid Hornwort *Ceratophyllum demersum*, and Curled Pondweed *Potamogeton crispus*. This type of plant community recovers in a matter of weeks after a summer cut. Nonetheless cutting, repeatedly if necessary, with a reciprocating cutter may be the only option to prevent total clogging of the channel and subsequent deoxygenation. Cut weed must be removed from the water.

Herbicide is used at a number of sites. The near static conditions typical of canals are essential. Formulations of diquat are the only realistic option for control. The gel formulation is effective in clearing fish swims and small lengths of canal, and does control some algal species as well as vascular plants. It can be applied selectively, and in slow moving water. The liquid formulation is appropriate for longer lengths but is less discriminating. All applications must be made early in the season, no later than mid June before the mass of weed has become excessive and whilst the water temperatures are below about 15 °C to minimise the risk of deoxygenation as the growth decays. However, any use of herbicide is ecologically undesirable as it may damage non-target plant and animal species and cannot be recommended until all other options have been excluded.

Grass Carp *Ctenopharyngodon idella* are another option suitable where there is adequate enclosure to prevent escapes into the wider environment. Grass Carp cannot breed in the water conditions experienced in the UK. Introductions must be licensed by MAFF and are subject to approval by the EA. Grass Carp will control the major problem weed species at stocking densities of 80 kg per hectare. Management of the stock is essential in the seasons after introduction to prevent overgrazing as the fish grow bigger. Fish will have to be removed. Compared with other methods, particularly herbicide, grass carp can be an environmentally beneficial method of management. However, the commercial supply of fish is limited, but this would improve if there was greater demand, and recognition by the environmental agencies of the relative benefits of these fish for weed management.

Another option for the future is the use of straw and other lignin-rich natural materials which on decay appear to release a substance or substances which suppress algal growth. This has been used successfully in canals and reservoirs and is the subject of current research.

Dredging

Dredging is required on all canals at some stage to remove material washed into the canal from the banks, with feed water and through the accumulation of plant debris. It is an essential and rejuvenating process which, if carried out in line with the recommendations given below, will usually lead to a substantial enrichment of ecological interest over the following 3–5 seasons.

The frequency of dredging may vary from years to decades depending on conditions. Some of the points made on river dredging apply, but there are a number of special points relevant to canals. A major point is that work may be carried out from floating, rather than land-based, plant because rights of access and opportunities for access are more restricted particularly in cuttings or on embankments. Work should be phased over several years, taking lengths of 100–200 m per year to encourage rapid recolonisation of disturbed sections.

Scope for variation in cross-sectional shape in canals is limited by land availability and the position of the clay lining. The central channel should be as deep as possible without exposing the clay lining. This minimises the disturbance of silt by boats on a navigated waterway and restricts the potential for problematic plant growth where boats are absent. Banks should grade up to shallow margins which can be usually more extensive on the offside. Existing reed fringe should be taken back to 1 m wide on the offside, 0.5 m on the towpath side, and any gaps filled with surplus material. Substantial

Fig. 5.20. A partial weed cut leaving margins and putting material back from the bank will encourage species diversity and also avoid smothering important plant species on the banks. (N.T.H. Holmes.)

renovation of reed fringe may require use of fabric reinforcement below. Invasive species such as Reed Sweet-grass *Glyceria maxima* should be substantially cleared. The short-term effect is dramatic, but without clearance regeneration is rapid, overwhelming the channel and suppressing other aquatic species.

All dredgings are controlled waste as defined in regulations under the Control of Pollution Act, 1974 and the Environmental Protection Act 1990. Under these regulations, except where specifically exempt, dredgings have to be deposited in a licensed site. However, as a specific exemption dredging spoil may be placed on the bank as work progresses without licence so long as the material is harmless. Demonstration of this may require chemical analysis. Any dredgings deposited in this way should be placed *well behind, rather than on*, the immediate reed fringe or marsh vegetation. In some circumstances this will involve placing the material on the trodden path where it will drain and consolidate as top soil. The tow path may have to be closed or diverted for public safety. This temporary disruption is preferable to permanent loss of the reed fringe which is the single most important vegetation type in the canal cross section. Towpath vegetation will usually recover in 1–2 seasons, although dominated by opportunists characteristic of disturbed ground for a number of seasons thereafter.

Bank protection

Protection of banks is vital for the integrity of the waterway, to prevent erosion where there is navigation and prevent leaks whether boats are present or not.

Original bank protection was vegetation, 'dry' stone walling or pitching and in urban areas masonry. Much of this has decayed and collapsed over the years particularly since the introduction of powered boating. In critical areas, for example on embankments, various forms of piling have been used to renew bank structure, including timber, sheet steel, concrete, asbestos, as well as some new masonry and concrete walls. The most widespread has been steel trench sheets (light weight piles) galvanised for longevity and held back with a galvanised walling and tie bars anchored into the bank. Despite its lack of visual appeal it has been widely used as a relatively cheap and cost-effective method for securing the long term survival of waterways. Ecologically it is undesirable on waterways with heavy traffic. It presents a hard reflective surface to boat generated wash which scours out the margins and eventually destabilises the reed fringe leaving the critical marginal area barren (Fig. 5.21). Its use today is limited to a few critical sites where there is no practical alternative. However the method is acceptable on canals with light traffic where, for example, control of leakage is required through piling placed within any existing reed fringe (Fig. 5.22) so that the marginal plant and animal communities are left undisturbed.

Wherever possible, banks should be protected with natural vegetation. Reeds can be introduced to the eroding bank using a composite system of fabric and vegetation. This method has been used on heavily trafficked canals. One arrangement is shown in Fig.

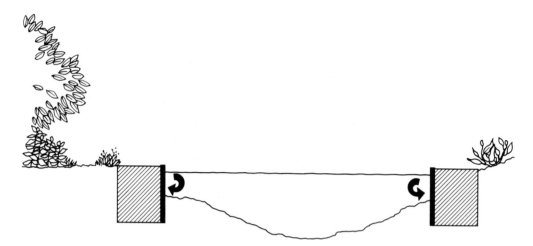

Fig. 5.21. Hard bank protection. Retention of water within the artificial canal structure is essential for the survival of the waterway. In the past, piling has been used to provide bank reinforcement and waterproofing. The disadvantage on trafficked canals is that the hard surface is highly reflective to boat wash. The resultant scouring leads to loss of vegetation and deepening of margins. In this condition, the bank is inimical to the regeneration of plant life.

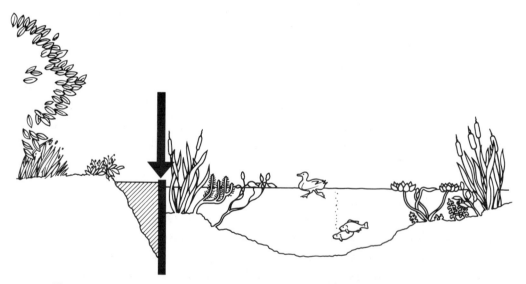

Fig. 5.22. Piles inserted in reed fringe. On trafficked canals, the effects of piling on the marginal flora can be mitigated if piles are inserted within, rather than in front of any pre-existing reed fringe (which can be reinforced with additional surplus reed bed vegetation collected locally). The retained fringe will absorb rather than reflect wave action. Positioning of the piles must however be compatible with the position of the clay lining. To retain water, the piles must be driven into the clay to provide a good seal. If backfilling of the new piled bank is required, spoil should be dredged from the centre channel, *not* the margins of the waterway.

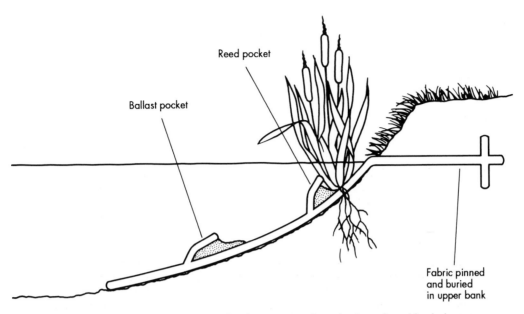

Ballast pocket

Reed pocket

Fabric pinned
and buried
in upper bank

Fig. 5.23. Reinforced vegetative bank protection. Introduction of reed bed plants to an eroding canal bank is difficult without reinforcement. The system illustrated stabilises the structure of the bank and provides support to plants while root development takes place. Clumps of plant material should be cut to give a tight fit in the pockets to minimise movement from wave action. Ballast pockets can be filled with stone or dredged material from the channel. Bank slope will depend on local ground conditions, but should not be steeper than 1:1.5.

5.23 and described below. Other commercial systems are now appearing on the market and should not be overlooked. The fabric component essentially provides short-term protection of the bank and anchoring for the plants while they become established. It may not be necessary on waterways without boats, but should be used where there is significant wind-generated wave movement.

The bank is reshaped using dredged material and graded to a slope of about 1:1.5 with a terrace formed just above the water-line. The pocket fabric is laid out on the bank in lengths of 25–50 m, and the ballast pocket weighted. The fabric is launched down the bank and pegged to the terrace so that the reed pocket is at water line. Clumps, about 20 × 20 cm, of locally obtained reed or sedge species, such as Greater Pond-sedge *Carex riparia* suitable for low traffic, or the high traffic-tolerant Reed Sweet-grass are placed in the pockets and the bank regraded above water. Other species such as Branched Bur-reed, Reed Mace *Typha latifolia* and Common Reed *Phragmites australis* can be used but they should be selected from local sources, and the mix of species should include sufficient proportion of those with a dense growth habit to establish a continuous fringe. Gaps in the fringe will be the focus for erosion. If planting is completed in early summer an established reed fringe is obtained by autumn.

Critical points for success are:

- ensure pockets are positioned so that the plants have their roots and no more than 10 cm of stem in the water
- fabric above waterline must be buried in the upper bank, and not left superficial to the bank as it will not readily revegetate
- if Common Reed is used, plants must be large enough for the shoots to be above the waterline when planting.

Boat traffic

Balancing the needs of navigation and ecology is an important management consideration. In research conducted at the University of Liverpool, modest levels of traffic up to 600 boat movements a year have been shown to be positively beneficial in maintaining open water, habitat diversity and conservation interest. At higher levels of traffic up to 3000 movements a year, all but the most sensitive plant species are retained although total plant biomass may be reduced, creating a clear central channel with vegetated shallow water margins and reed fringes. Above 3000 movements per year vegetation is progressively lost as the waterway becomes ecologically impoverished and less interesting, although this does not appear to deter use.

In some circumstances there can be a major clash of interest between the interests of navigation and ecology, particularly where restoration of a disused waterway is proposed. This is a matter on which there must be involvement with relevant local and national conservation organisations and a careful appraisal of the impact of any action including the 'do nothing' option. This may be threatening to the long-term interests of wildlife if the waterway is left unmanaged. Space does not permit a detailed consideration of restoration but three general points are worth making. First, restoration programmes proceed slowly in time-scales measured in decades rather than years, some will never reach fruition as the obstacles are too great and costly to rectify. This time can be used to explore the opportunities to enhance ecological interest within the restoration programme. Second, redundant waterways are usually out of use for good reasons such as inadequate engineering design or water supply. Both are formidable problems. The latter may be a fundamental control on the level of boat traffic that may be achieved. Third, restoration can be beneficial for wildlife interests. The key factors are how the work is done and the level (if any) of boat traffic expected thereafter. Low levels of traffic can hold the hydrosere at its maximum diversity, but high levels are damaging. Much depends on the attitude of the restorers and their remit and, as is so often the case, maintenance of a constructive dialogue between the interested parties.

Waterbodies

6

JOHN ANDREWS

Introduction

Britain contains a wide variety of natural and artificial waterbodies and differences in each site's character and location are important in deciding what, if any, management may be desirable. The first section of this chapter looks at the characteristics of the main types of waters in the lowlands and uplands. The second section considers the main factors relevant to wildlife management and describes how to modify them; these are the amount of nutrients present in the water, depth and fluctuation in water level, the effects of wind and waves, the context of the waterbody relative to other habitats and the effects of human disturbance. Other factors such as geographical location are also

Fig. 6.1. Large waters – Loch Neagh in Ulster is the biggest waterbody in the British Isles – need an integrated management system which should ideally extend to the whole catchment. The lack of mechanisms to achieve this is a very serious defect in conservation powers. Many big sites are of national or international importance for wildlife. Most of those which formerly had valuable commercial fisheries have now lost them due to mis-management of the water or to overfishing. (J. Andrews.)

important influences on wildlife but are not considered because they cannot be altered.

Brief management prescriptions for plants, invertebrates, fish, amphibians, birds and Otters are set out in the third section. The main emphasis is on vegetation as this is fundamental to the fauna. The detail of management methods is summarised. Much of it is a matter of common sense or the application of techniques which are familiar to most land managers and experienced estate workers. However, some aspects are complex and specialised; for these, it is sensible to take further site-specific advice before attempting the work. Any particular pitfalls are identified.

Not all sites need intervention. In some cases, the best course will be to let well alone but other waters may need considerable attention if they are to achieve or maintain their potential. Where rare species are known to be present, take advice from one of the conservation bodies before making any change to past practice.

Waterbody communities

Regardless of their size or origins, waterbodies are normally classified according to the amounts of nutrients available to support growth of plants and their dependent fauna. As this classification helps in assessing a waterbody's existing and potential wildlife importance, it is an important guide in making management decisions.

The abundance of dissolved nitrogen and phosphorus largely determines the amount of plant growth, including free-floating unicellular algae (phytoplankton) and other aquatic plants (macrophytes). Nitrogen is rarely inadequate for prolific growth but phosphorus is often limiting.

The availability of plant nutrients changes with the seasons due to a variety of influences. A more constant measure of productivity is provided by alkalinity, indicated by the amount of calcium carbonate dissolved in the water. Applying this, waterbodies are classed as eutrophic, in which calcium carbonate and productivity levels are highest, mesotrophic, oligotrophic and, the lowest level, dystrophic. Marl lakes form a fifth class in which there are very high levels of alkalinity but phosphate is bound into the sediment and not available to fuel growth by phytoplankton. The different types support different wildlife assemblages.

Waterbodies in the lowlands

Most of the waterbodies of lowland Britain are eutrophic. The water contains mineral salts derived from the catchment and is thus moderately enriched with natural nutrients. Today, most sites are additionally enriched by the run-off of agricultural fertilisers and, in some sites, by inputs of river water which contain or largely consists of treated sewage effluent. Productivity of algae and macrophytes is high and the former

Fig. 6.2. Occupying naturally fertile catchments, most lowland waters develop prolific submerged and emergent plant growth which supports a varied and abundant fauna. Most sites are now affected by nutrients derived from agriculture or treated sewage effluent so that increased algal abundance means reduced light penetration and a decline in macrophyte and faunal populations. (J. Andrews.)

are normally so abundant in the water column that, by reducing light penetration, they limit the depth to which rooted plants can grow. The potential diversity of rooted aquatic plants is high, precise composition varying with substrate, depth and exposure to wave action. In small sites and still-water zones, there are likely to be water-lilies or floating mats of duckweeds *Lemna* spp.

Overall, the marginal zone is likely to have the most diverse and often most abundant invertebrate fauna of any standing water. The larvae of many insects live in the sediments, including non-biting midges *Chironomidae*, craneflies *Tipulidae* and hoverflies *Syrphidae*. Both submerged and emergent plants support many other insects and molluscs are well-represented as there is abundant calcium for shell-building. Predators include waterbeetles and dragonflies. The fish fauna is dominated by *Cyprinidae* with powers of smell and touch which equip them to feed in turbid conditions; often its composition has been much modified by angling interests.

The wide range of food resources supports an abundant bird fauna including species which feed on plants, molluscs, insects and fish (see Table 6.1) and others which depend on the physical structure of marginal vegetation to provide nest sites.

Waterbodies supplied from catchments of soluble limestone and chalk may contain

Table 6.1. *Feeding zones and food of selected waterfowl*

Species	Main Feeding zone	Feeding depth[a]	Age[b]	Status[c]	Diet[d]						
					Aquatic plants	Marginal plants	Seeds	Crops/ grass	Insects, c'stacea	Molluscs	Fish
Little Grebe	Inshore	<2 m	Adult	B W					+++	+++	+
			Young						+++		
GC Grebe	Offshore	<5 m		B W					++		+++
Cormorant	Offshore	<3 m		S W							+++
Mute Swan	Inshore	0–1 m	Adult	B W	+++			+			
			Young		+++	+++					
Shelduck	Margins	0–0.4 m		B					+++		
Wigeon	Ashore	0–0.3 m		W	++	+++		+++	+++	+++	
Gadwall	Margins	0–0.35 m	Adult	B	+++	+++	++	+++			
			Young						+		
Teal	Margins	0–0.2		W	+	+	+++		+++		
Mallard	Margins	0–0.35 m	Adult	B W	++	++	+++	+	+		
			Young						+++	+	
Shoveler	Margins	0–0.3 m	Adult	B W	+		+++		+++	+++	
			Young						+++	+	
Pochard	Inshore	<2.5 m		W	+++				+	+	
Tufted Duck	Inshore	<5 m	Adult	B W			+		++	+++	+
			Young						+++		
Goldeneye	Inshore	<4 m		W			+		+++	+++	+
Goosander	Offshore	<4 m		W						+++	+++
Coot	Inshore/ ashore	<2 m	Adult	B W	+++		++	+	+	+	
			Young						+++		

Note: [a] <depth, feeds mainly by diving; 0–depth, feeds mainly by dabbling.
[b] For breeding species, diet of young is shown if different from diet of adult.
[c] B, breeding; S, summering; W, wintering.
[d] +++, major diet; ++, important; +, minor part of diet.

very high levels of calcium carbonate, resulting in phosphates being precipitated and unavailable to support algal growth. In these so-called 'marl lakes', water clarity is such that rooted plants may grow to a depth of 6 m or more. Stoneworts *Chara* spp. and molluscs are extremely abundant, the former supporting large wintering flocks of Pochard *Aythya ferina* and the latter Tufted Duck *Aythya fuligula* and other species. Because the shores are usually fairly silt-free they may support invertebrates characteristic of nutrient-poor upland lakes such as stoneflies and mayflies. The native crayfish *Austropotamobius pallipes* occurs in southern marl lakes where it is at risk of displacement by several non-native species now spreading widely due to inept control (Holdich, 1991).

The great majority of eutrophic waterbodies and marl lakes are man-made and include reservoirs, mineral-extraction pits, ornamental lakes and farm ponds. At least in the summer months, Fenland dykes and abandoned canals have more in common with eutrophic standing waters than with rivers. Most lowland standing waters are subject to management or use such as active recreation unconnected with and sometimes inimical to wildlife conservation goals. There is often considerable potential to modify habitats within them to benefit scarcer species and minimise the impact of human disturbance, but integration with these other uses always requires careful planning.

Waterbodies sited at the upland fringes or on the Tertiary sands of southern England, notably in a heathland context, may have less calcium carbonate and a lower nutrient status than most lowland waters. In such mesotrophic sites slight algal discolouration of the water may occur in summer but light penetration is generally good. They can support a wider range of aquatic plants including several pondweeds *Potamogeton* spp. and, where marginal sediments are deposited, varied stands of emergents or poor fen communities develop. These waterbodies may hold uncommon species not found in eutrophic sites. For example, some of the southern sites on heaths contain soft-water molluscs which are otherwise confined to northern waters. A main management need is to protect these sites from eutrophication, which is most likely to arise through application of inorganic fertiliser to adjacent land. If it occurs, scarce species will be replaced by others which are common and widespread in the lowlands.

Waterbodies in the uplands

Most of the natural waterbodies in Britain are in the uplands, occupying sites formed by glacial action, and are fed by base-poor catchments today used predominantly for grazing, forestry and field sports. The majority of these bodies are oligotrophic, with clear water but plant growth limited by lack of nutrients. Phosphates are unavailable for algal growth, being bound into deep water sediments during summer, and there is a restricted range of rooted plants. Shorelines are mostly wave-beaten and comprise coarse inorganic materials but on small sites and sheltered shores where organic

(a)

(b)

Fig. 6.3. (*a*) Most upland waters receive drainage from inherently infertile catchments. They are relatively unproductive and infertile. (J. Andrews.) (*b*) Black-throated Divers *Gavia arctica* breed on freshwater lochs in Scotland. The nests are susceptible to flooding and disturbance from fishermen. Providing turf-covered rafts has greatly improved their breeding success. (W.J. Sutherland.)

sediments can accumulate there may be a fringe of emergent plants including Water Horsetail *Equisetum fluviatile*, Common Club-rush *Schoenoplectus lacustris* or Bottle Sedge *Carex rostrata*. Stable stony shores support mayflies and stoneflies but most invertebrate fauna is confined to the sheltered shallows where there is vegetation.

The fish fauna of these waters is often of conservation and economic importance being largely salmonids. A few sites contain Powan *Coregonus lavaretus* or Vendace *Coregonus albula*, while others hold Charr *Salvelinus alpinus* (Maitland & Lyle, 1993).

Eutrophication can occur due to run-off by artificial fertiliser or discharge of sewage, and may damage the wildlife interest. Some sites may be subject to acidification, proximately due to forestry, and several measures may be taken to reduce this, including removal of conifer cover from the banks of feeder streams. Application of calcium carbonate is still at the experimental stage and should not be attempted without first taking advice from the appropriate forest authority and water authority as it may have harmful effects on terrestrial vegetation and is in any case only a temporary palliative.

Dystrophic waterbodies are either on or receiving drainage from peat bogs so, by definition, they are rain-fed and do not receive water which contains mineral salts dissolved from the underlying rock (see Chapter 7 on bogs). The water is peat-stained and productivity is very low due to lack of nutrients and poor light penetration. The plant assemblage is limited with bog-mosses *Sphagnum* spp., Bladderwort *Utricularia minor* and Bogbean *Menyanthes trifoliata* often the most obvious elements. Many dystrophic pools are naturally too acid to support fish, while waterfowl numbers are low in summer and negligible in winter so a wide range of dragonflies, water bugs and midges flourish in the virtual absence of these predators.

Most dystrophic waters are in the north of Britain and many are the result of the hand-cutting of peats. There are also a small number of bogs in the south and there are differences in the wildlife assemblages in different regions. Normally, management is neither necessary nor desirable. The main risk to these sites is drainage associated with commercial peat extraction. If pools are drying out or becoming overshaded, for example by scrub development, this is a symptom of a serious problem which will affect the whole bog system and needs to be addressed.

In summary, few upland waters need management but all require protection from changes which affect their nutrient status or hydrology.

Factors influencing wildlife

Nutrient availability

Too much nutrient input is harmful. When phosphate levels in particular become too high, the abundance of phytoplankton discolours the water and cuts off light from the

Fig. 6.4. 'Dubh lochans' on blanket peat may hold great numbers of damselflies including uncommon species. Those in Shetland and other coastal areas of northern Scotland are often used for breeding by Red-throated Divers *Gavia stellata* which feed on the sea. A main management need may be to avoid breeding season disturbance. (Unst, Shetland; J. Andrews.)

macrophytes. In extreme cases, all submerged rooted plants may die as a result. This affects all other life using the water.

In some instances, the site may appear healthy and then suddenly and cata-strophically switch to an algal-dominated community. Factors which may be implicated include: unnatural or extreme fluctuation in water levels which may expose and destroy macrophytes – a situation typical of water supply reservoirs; stocking with fish species that consume large numbers of the zooplankton such as *Daphnia* which graze phy-toplankton and so limit their abundance, and long periods of calm, hot weather. Once the ecosystem has switched, recovery of macrophytes does not necessarily occur in the short term even if nutrient inputs cease, so prevention is most important.

It may be possible to protect artificial waters fed by pumping, such as lowland reservoirs and off-stream ponds, by monitoring the quality of the supply and minimis-ing recharge at times when the water has high nutrient loadings. Monitoring requires chemical analysis and may be impracticable for a private landowner. In this case, take water only when flows are high, so that any nutrient loading is diluted. Bear in mind that for the first few days after rain on farmland the increased flow is likely to carry higher levels of nutrients than it will later when they have been flushed through.

Where a feeder stream brings nutrients and cannot be diverted at high-risk times, consider the possible benefits of establishing a bed of emergent plants such as Reed *Phragmites australis* or Bulrushes *Typha* spp. on the inflow as this will take up some of the nutrients in the summer months while the plants are growing. However, the process may be less effective in winter so the method is probably only applicable where winter flows are high and retention times in the waterbody are low.

In small ponds, early spring application of barley straw at 1 bale per ha, assuming 1 m depth, is reported by the Aquatic Weeds Research Unit (Barrett & Newman, 1991) to inhibit growth of algae. With sites such as farm ponds, it may be possible to withhold fertiliser use on the land which drains into the waterbody. In garden ponds, refilling from a water-butt holding rainwater is much better than using tap-water.

If a small site has become so enriched that it has lost most or all of its macrophytes and their dependent wildlife, the only solution may be to dredge or pump out the bottom sediments during the winter months. At this season most of the phosphate will be in the mud and, though the disturbance of the sediments will cause some release of nutrients, much will be removed and recovery of the macrophytes should occur progressively over two or three years provided further enrichment is prevented.

For large sites, restoration methods have been devised to take account of nutrient inputs and internal loadings and which use equipment which achieves efficient removal of sediments and, where appropriate, beds of emergent vegetation (Bjork, 1992).

Fish can inhibit recovery. Some disturb bottom sediments and others feed on *Daphnia* and other zooplankton which eat the phytoplankton. Removal of fish (e.g. by netting or electrofishing) may aid recovery. The fish may be sold to stock other waters.

A licence additional to the normal fishing licence is required to remove fish in these ways and advice on techniques may be obtained from the appropriate water authority.

In new waterbodies fed by eutrophic waters – ranging from lowland reservoirs to garden ponds – it is common to have algal discolouration and abundant growth by filamentous algae (blanket weed) for the first few seasons. In small sites, it is a temptation to pull out the blanket weed and try to treat the phytoplankton chemically. This is unwise. The filamentous algae provide good cover for invertebrates and amphibian tadpoles. Chemical treatment will have short-lived effects. The water will only clear effectively when macrophytes are growing strongly and progressively take up the available nutrients.

Depth and fluctuation

The greatest variety of wildlife is usually found in shallow waters. As depth increases, light penetration decreases and so plant growth is reduced. In lowland waters with normal algal populations, most plant growth is confined within about 2 m depth. In marl lakes and upland waters which have greater clarity, the depth of growth may be greater.

For amphibians and many invertebrates, relatively high spring and summer water temperatures are essential and only develop in shallow, sheltered waters. Birds too require or prefer shallow waters. Waders feed in depths less than 0.15 m, dabbling ducks can reach down 0.35 m or less when upended and many diving species such as Pochard tend to select feeding sites with depths of less than about 2.5 m (see Table 6.1).

Natural seasonal variation in water depth is significant. On soft shores not colonised by emergent vegetation, the drop in level exposes ground on which annual plants such as Celery-leaved Buttercup *Ranunculus sceleratus* can germinate, grow and set seed free from competition. When levels rise in autumn, their decay adds to the organic resources of the waterbody and often their seeds are an important food for wildfowl such as Teal *Anas crecca* and Shoveler *Anas clypeata*. Some other plants, including Shoreweed *Litorella uniflora* and the Slender Spike-rush *Eleocharis acicularis* persist under shallow water as green plants but flower only when levels fall and expose them.

In small sites such as farm ponds and those on village greens, the drop in summer level means better penetration of light and warmth, benefiting plants and those invertebrates and amphibians which require high temperatures for growth and reproduction.

Some waterbodies are ephemeral and dry out entirely in summer, at least in dry years. A surprising number of aquatic animals can cope with or even benefit from this. Some molluscs become dormant, sealing off the shell opening to resist desiccation. Some crustaceans, such as the Fairy Shrimp *Chirocephalus diaphanus* and the Tadpole Shrimp *Triops cranciformis*, survive as eggs in the mud. They die out rapidly in permanent pools because of predation by fish and invertebrates. Similarly, most amphibians do best in sites which dry out occasionally and so do not build up substantial predator populations. Many of the invertebrates of temporary pools are now very rare as most such sites in the lowlands have been drained or made into permanent ponds.

In new waterbodies such as large ponds, gravel pits and reservoirs, design to create extensive shallows with about half their area less than 0.5 m deep and the rest up to 2 m deep. Make allowance for seasonal change in level and the effect this will have on the extent of available habitat. With good restoration, as the upper shore is exposed and no longer suitable for feeding, areas lower down the shore become accessible.

In small ponds a maximum depth of 1 m is acceptable. Aim to have some areas which are so shallow that they will dry out in summer. Do not artificially top up the natural drop in levels.

Bear in mind that sheltered waters less than about 1 m deep over muds or sands may be rapidly colonised by emergent plants such as Bur-reeds *Sparganium* spp. and Bulrush *Typha latifolia* unless they are isolated by deep-water trenches which the plants' rhizome systems cannot cross. Depending on the wildlife goals, it may be necessary to control their spread (see 'Wildlife Management' below).

Variations in water level which exceed the natural range or which occur at the wrong time of year may occur on reservoirs. Often, draw-down by several metres must take place in summer and recharge may not be possible until the late winter. This results in the loss of macrophytes due to exposure and desiccation. Any new growth at the lower level is likely to be lost after recharge as the water depth and algal density may be too great to permit adequate light penetration. Each time the cycle is repeated, further plant recolonisation is destroyed. This has significant effects on the composition of invertebrate and vertebrate fauna. A limited solution would be to create isolated bunded zones where a natural regime can be mimicked, as has been done at the upper end of Rutland Water, Leicestershire, for instance. Unless such an area has an adequate water supply, it may be necessary to recharge the zone annually in late winter to make up any shortfall in precipitation.

Wind and waves

Wind blowing over the water surface generates waves, the size and effect of which depend on both windspeed and fetch (distance across the water). Wave height and length increase with fetch. The water within a wave has a roughly circular motion, the depth of this being roughly half the wave length. As large waves run into shallowing water, they progressively interact with the bed. On inland waters this effect rarely extends below 2 m. If the bed is soft, the water will become turbid, plants may be uprooted and shorelines eroded. Few invertebrates can survive either in substrates or on plants.

Wave action can be a serious problem in the management of sites such as lowland reservoirs and gravel pits where natural erosive processes cannot be permitted to run to their conclusion. From a wildlife standpoint, the worst solution is to dump stone or rock on the shore itself or to install piling or other 'hard' protection. These solutions prevent plant growth and deprive birds, invertebrates and other fauna of the natural shoreline. Geotextiles should be considered as they will resist erosion but allow plant growth. However, where wave action is considerable, such growth may still be impossible. Perhaps the best solution is to create a reef of an inert hard rock such as granite dumped at least two metres from the shore to allow plants to grow in the calm waters inside it. The reef itself will become a valuable habitat for fish fry and invertebrates. A less effective but useful alternative is a floating timber boom chained to robust uprights set in the bed.

For smaller waters, a long term answer may be to plant a shelterbelt comprising three or four lines of trees sited to intercept the prevailing wind. If possible, the belt should be set back from the shore so that the area of greatest shelter remains sunlit. Poplars and conifers are often used as they are quick growing and conifers remain equally effective throughout the year. Belts work through the flexing of twigs and branches which thus absorb wind energy and reduce its speed. Correctly designed shelterbelts have a useful effect on wind speed for 20 to 30 times their height downwind (Caborn, 1965).

Wind shelter is also important for small pools, not because of wave action which is negligible but because wind cools water and air temperatures and disrupts the activity of flying insects. Small-scale shelter with shrubs or even stands of tall emergent plants is invaluable.

Fig. 6.5. Trees are an important element of bank vegetation and some aquatic invertebrates spend part of their life cycles in amongst their foliage. However, dense and continuous tree cover can result in the loss of other plant communities and of variation in vegetation structure such as tall herb and scrub cover. The optimum is a mix of tree-lined and open banks. (J. Andrews.)

In new sites, wind should be taken into account from the design stage. Do not create waterbodies with a long fetch aligned with the prevailing wind (i.e. from the south west in most of Britain), locate features which may erode, such as islands and spits, at the upwind end where waves will be small and install essential protection for exposed shores in the form of offshore reefs. Such reefs can be made into, or used to protect, islands surfaced for nesting by waterfowl (see Wildlife Management below).

Context

The context of a waterbody is highly significant at every level from the range of land uses which affect its whole catchment to the vegetation and habitats in its immediate proximity. Often the uses of adjacent land are outside the control of the site manager but where possible resist those which involve the application of artificial fertiliser or may result in plant, invertebrate or fish kills due to agricultural spray drift. Small ponds on farmland are particularly vulnerable but may be protected by a screen of scrub or a dense hedge at least 1.5 m tall.

Within 10 m of ponds and at least 50 m of larger waters, aim to maintain or create a variety of semi-natural vegetation types including scrub, tall herbage and permanent grass. These are important nesting or feeding places for waterfowl and as foraging and roosting areas for the adult stages of many aquatic insects including dragonflies.

Disturbance

Human disturbance can be a major influence on waterfowl and Otters *Lutra lutra*. In the lowlands, the main effect is likely to be on wintering waterfowl; in the uplands few birds are present in winter but there may be rare breeding species such as Black-throated Diver *Gavia arctica*.

The response to disturbance is to restore a 'safe distance'. This varies according to species, activity and habitation. If there is no cover into which they can retreat or if they are species which do not use cover (such as wintering Goldeneye *Bucephala clangula*), a lone dog walker or birdwatcher may put all the birds off a site which is too small to provide safe withdrawal distances. Even on large sites, if there is disturbance on banks and the waterspace simultaneously, birds may be 'squeezed' between the two and forced to leave.

It is not possible to lay down quantified guidelines on the integration of recreational use with wildlife as much depends on site conditions but some rules apply to most situations (Ward & Andrews, 1993). In general, wildlife reacts most to activities such as water-skiing and board sailing in which the craft move fast and unpredictably, less to dingy sailing on a fixed course and least to static activities such as fishing. Seasons when birds are breeding or moulting (during the latter ducks and geese become flightless) are the most sensitive and any activities which impinge on potential or actual nest sites are problematic.

Important areas of shoreline such as sheltered bays or islands where birds feed, roost or breed can be screened from landward activities by dense plantings with an evergreen component which is effective year round. Hides are an often unrecognised cause of disturbance and are best set back 50 m from the water's edge. Islands and stands of tall emergents can provide screening from water-based activity, which may also be zoned in space and time. Islands themselves may hold breeding birds, so close mooring or landing should be prohibited in the nesting season.

Consider using disturbance as a tool, for instance, by opening up areas to disturbance in winter so as to reduce depletion of food resources required by birds when coming into breeding condition.

Wildlife management

Plants

Aquatic plants are of great importance as the primary food resource and the physical habitat of much of the fauna. They can also play an important role in binding substrates and preventing erosion. Table 6.2 lists common wetland plants, their growth depths and some of their values for other wildlife.

As well as depth, eutrophication and wave action, all of which are considered in the previous section, several other factors may impair plant establishment and growth. For example, propeller craft may break up stands of submerged plants and anglers may wish to cut open 'swims' into which they can cast. On sites where submerged plant growth is limited, for instance because there are few shallow or wave-sheltered areas, it is important to limit or exclude potentially damaging uses. However, in smaller sites with ample sheltered, shallow water the creation of swims may help to diversify the interest.

Fish can have a significant effect on plant growth. Studies have shown (Giles, 1992) that bottom-feeders including Bream *Abramis brama* uproot plants and create turbidity. When these species are removed, plant growth may become luxuriant, with consequent effects on the rest of the fauna.

Most of the flora of standing waters has a remarkable ability to colonise suitable new sites. In principle, there is no need to introduce species and good arguments for allowing natural processes, not least that they cost nothing and provide plants which will flourish in the conditions. In practice, where particular species-related goals are set, introduction of relevant food-plants or other particular requirements may be necessary. For instance, a site in which it is desired to encourage wintering wildfowl will benefit from plants like Bur-reed which produce nutritious seeds.

Wherever possible, use material of local provenance preferably obtained from nearby waters of similar size and nutrient status. Possible sources include river dredgings and nature reserves where plants are being removed as a part of management. Plant material

Table 6.2 *Common wetland plants*

Species	Depth	Comments
Submerged plants		
Stoneworts *(Chara* spp.*)*	To 2 m	Good invertebrate habitat. Main food of Pochard. Requires high quality water
Water Crowfoot *Ranunculus aquatilis*	To 1 m	Seeds eaten by wildfowl
Spiked Water-milfoil *Myriophyllum spicatum*	0.7 m (2 m)	Favours lime-rich waters. Excellent invertebrate habitat
Mare's-tail *Hippuris vulgaris*	To 1 m	Favours lime. Good invertebrate habitat. Seeds eaten by wildfowl. Cover for duckling broods
Common Water Startwort *Callitriche stagnalis*	To 1 m	Good invertebrate habitat
Canadian Pondweed *Elodea canadensis*	0.5 m (3 m)	Introduced but widely naturalised. Excellent invertebrate habitat
Curled Pondweed *Potamogeton crispus*	To 2 m	Seeds eaten by wildfowl
Fennel Pondweed *Potamogeton pectinatus*	0.5–2.5 m	Tolerates heavily polluted and turbid water. Seeds and tubers eaten by wildfowl
Horned Pondweed *Zannichellia palustris*	To 2 m	Seeds eaten by wildfowl
Floating-leaved plants		
Yellow Waterlily *Nuphar lutea*	1–5 m	Duckling broods forage among leaves
Amphibious Bistort *Polygonum amphibium*	Damp ground To 2 m	Good invertebrate habitat. Good seed production taken by wildfowl. Good duckling foraging habitat
Broad-leaved Pondweed *Potamogeton natans*	To 3 m	Good seed production
Duckweeds *Lemna* spp.	Free-floating	Plants eaten by wildfowl
Emergent/marginal plants		
Marsh Marigold *Caltha palustris*	Damp ground To 0.3 m	
Watercress *Rorippa nasturtium-aquaticum*	To 0.3 m	Favours chalk and limestone
Fool's Watercress *Apium nodiflorum*	To 0.3 m	Favours chalk and limestone
Great Water Dock *Rumex hydrolapathum*	Damp margins To 0.5 m	Seeds eaten by wildfowl
Water Forget-me-not *Myosotis scorpioides*	Damp ground To 0.2 m	Submerged leaves used for oviposition by newts.

Table 6.2 (*cont.*)

Species	Depth	Comments
Brooklime *Veronica beccabunga*	Damp ground, shallows	
Water Mint *Mentha aquatica*	Damp ground, shallows	Good butterfly nectar plant
Water-plantain *Alisma plantago-aquatica*	To 0.75 m	
Arrow-head *Sagittaria sagittifolia*	0.3 m	Plant eaten by wildfowl. Brood foraging habitat
Rushes *Juncus* spp.	Damp ground	Seeds eaten by wildfowl
Yellow Flag *Iris pseudacorus*	Damp ground	Seeds eaten by wildfowl
Bur-reeds *Sparganium* spp.	To 1 m	Good invertebrate habitat and seed production
Great Bulrush *Typha latifolia*	To 1 m	Winter cover for wildfowl
Common Spike-rush *Eleocharis palustris*	Damp ground	Very good seed producer. Used by duckling broods
Sedges *Carex* spp.	Damp ground Some to 0.5 m	Very good seed producers. Nesting cover for wildfowl
Reed *Phragmites australis*	To 1.5 m	Prime habitat of Reed Warbler. Winter cover for wildfowl
Reed Sweet-grass *Glyceria maxima*	To 0.5 m	Highly invasive and limited value
Reed Canary-grass *Phalaris arundinacea*	Damp ground To 1 m	Favoured by Harvest Mouse
Purple Loosestrife *Lythrum salicaria*	Damp ground	Butterfly nectar plant
Meadowsweet *Filipendula ulmaria*	Damp ground	
Great Willowherb *Epilobium hirsutum*	Damp ground	
Water Figwort *Scrophularia aquatica*	Damp ground	Good nectar and seed producer

Note: Unless otherwise stated, the plants are suitable for any site with a medium to rich nutrient status, medium to fine substrate, moderate to high pH.

All submerged plants and most floating and marginal species are used by a variety of invertebrates and may be assumed to be of general value. Only those which are known to be particularly favoured are indicated as such.

Where appropriate, the average preferred water depth is given with maximum depths in brackets.

Source: Reproduced from Andrews & Kinsman (1990) by kind permission of the RSPB.

may not be taken from the wild without the landowner's consent. For small waters, avoid the most vigorous species such as Great Bulrush *Typha latifolia* or Canadian Pondweed *Elodea canadensis*, now an ineradicable honorary native.

It may be best to plant single-species stands to minimise the risk of vigorous species overwhelming those which establish more slowly. For submerged aquatics on large sites, areas free from wave action and with an unconsolidated sand or muddy bed may be planted with single-species stands in stretches about 5–10 m long down to 1 m depth. Planting at about 0.5 m spacings is probably adequate as spread can be rapid. Rooted material is tied to weights (stones will do) and thrown in. In small sites it may be possible to plant at greater density and this can be advisable because the sooner macrophytes are established, the sooner algal blooms will cease. Often one plant species will be dominant initially but variety develops as nutrient levels fall.

In larger waterbodies, marginal and emergent plants should also be planted in single-species stands, taking account of the depth of water which they will be able to colonise. To minimise future invasion and management need, do not plant emergent species next to extensive shallows important for dabbling ducks, for example. In new site design, it may be possible to restrict the invasion of emergents by surrounding designated open water shallows with trenches 2 m deep and 2 m wide which the plants cannot cross because their rhizome systems cannot grow at that depth.

Most larger emergent and marginal species are best moved during the winter months as clumps of root or rhizome and soil. Do not allow them to dry out. Under no circumstances use topsoil or fertiliser as they are unnecessary and may promote algal growth. Set the plants in moist ground, not in water, with individual clumps at about 1 m spacings as they will quickly spread. In small waters, only one or two clumps will be required. If necessary protect from grazing by livestock, geese or coot by fencing or netting at least until the plants are firmly rooted and growing strongly.

Emergent plants rapidly invade shallows on soft substrates in sheltered waters less than 1.5 m deep. They can choke a small site in a few seasons. However, eradication is undesirable as many insects and birds depend on them for food, shelter or other functions. Where rare plants or invertebrates are known to be present, specialist advice should be sought from one of the conservation bodies before initiating any clearance.

As a rule of thumb, allow emergents to cover no less than a quarter and no more than half the total area of shallows. Aim for a varied structure and composition of emergent and fringing vegetation. In large sites, natural variation and processes such as erosion or low-intensity grazing may do the job without any need for management. On small sites, grazing needs to be watched carefully as a few days' inattention can allow damage through overgrazing and trampling. In small, sensitive sites it may be best to exclude stock and devise a summer cutting regime. Where these options or constraints do not apply, undertake periodic clearance in winter by machinery (e.g. back-acter or dredger bucket) on big sites, by hand digging on small ponds or by herbicide use where these methods are not practicable.

Where possible clear the advancing front face of some beds of emergents, so maintaining stable conditions on the landward interface, and cut others in bands from water to land, so putting back the full process of succession. Different invertebrate species will live in the different conditions created. Do not eradicate any plant. For example, though Great Bulrush *Typha latifolia* can pose a recurring management need, it has a number of dependent species.

Herbicides have disadvantages compared with cutting or dredging. Notably, the target vegetation may take weeks to die and its decay releases nutrients which may cause algal bloom, whereas cut material can be removed. Any form of control creates a vacant niche for other plants to colonise, so both mechanical clearance and herbicide use may become a long-term commitment.

There are several herbicide formulations specifically developed for aquatic application. As the range of available formulations and brands changes over time and new regulations on use continue to be introduced, it is essential to refer to up-to-date guidance. At present, the key text is 'Guidelines for the use of herbicides on weeds in or

Fig. 6.6. Emergent plants have a distinctive invertebrate fauna and provide cover for Otters and waterfowl, but they rapidly invade shallows which are themselves of importance. Rotational clearance is essential. Where the water level can be controlled, vegetation management is greatly facilitated. (J. Andrews.)

BOX 6.1 Pond creation

Site
Select a site with no existing wildlife interest. Do not destroy a marsh to create a pond as the marsh probably holds rarer species than will colonise the pond.

Context
Ideally close to an existing semi-natural habitat and with wind shelter from scrub or a good hedge. Avoid heavily shaded sites.

Water supply
Ensure supply is likely to be adequate throughout the year, accepting 20–30 cm drop in summer as normal and desirable. To take water from a bore hole or a watercourse, an abstraction licence may be required; check with the water supply authority for your area. Check that the water quality will be satisfactory: for example, drainage from permanent pasture will probably be fine but that from intensive arable will not be.

Bed
Simplest way is to excavate to below the normal summer water table but many sites will not be suitable. Traditional method of sealing is puddled clay but this is very laborious. Otherwise install a heavy-duty black polythene or butyl liner which are more expensive and longer lasting. Suppliers are in 'Yellow Pages'. Remember that liners may be punctured by livestock or machinery and may be unsuitable for any site where this is a risk.

Size
As big as possible but frogs, newts and dragonflies will thrive in ponds as small as 1 m diameter.

Depth
Depends partly on size because it is important to have plenty of water less than 0.4 m deep. If the pond is big enough, slope to a central area 1–1.5 m deep.

Margins
Irregular as possible. Artificial liners cannot be stretched and this limits the shape compared with a natural bed.

Substrate
Liners should be buried under at least 15 cm of nutrient-poor, stone-free subsoil. At all costs do not use farm or garden topsoil which will cause algal blooms. No additional substrate is needed in unlined ponds.

Stocking

Not essential to plant but if desired select species which occur locally from the list at Table 6.2. Avoid vigorous emergents which will have to be cleared annually. Do not introduce invertebrates. Do not introduce fish or encourage ducks if the pond is for amphibians or invertebrates.

Aftercare

There may be some algal blooms in the first years until submerged rooted plants are well established. Do not remove filamentous algae. Though unsightly, it is good cover for invertebrates.

If the pond is on farmland, keep artificial fertiliser and spray applications well away. Control livestock access to ensure that they do not destroy marginal cover or over-enrich the pond with dung.

Manage bankside vegetation to have about half tall herbage allowed to stand over winter and half an open shore with short vegetation and some bare mud; this may require controlled grazing, mowing or hand cutting depending on the site.

near watercourses and lakes' published by the Ministry of Agriculture Fisheries and Food. In all cases, ensure that the plant species to be controlled has been correctly identified and choose the appropriate formulation. Check to ensure that no rare plants are at risk. Apply only in the growth period and under the weather and water conditions recommended.

In small waters, herbicides may affect more than the target unless applied with the utmost care. Physical removal by hand or by back-acter bucket may be the best solution, taking care to minimise disturbance to sediment and avoid damage to the bed – farm ponds often have a puddled clay liner which must not be broken. Spoil and plant material must be disposed of away from the water in areas without wildlife importance, so that nutrients released by decay will not wash back during rainfall.

Control of submerged plants and floating plants may be necessary in ponds where injudicious introduction of vigorous species such as White Waterlily *Nymphaea alba* can result in the loss of other vegetation but on larger waterbodies there are no circumstances in which it is desirable to control the growth of submerged or floating-leaved plants unless they are vigorous non-natives, such as *Crassula helmsii*, which endanger the ecology of the site. If there is no choice in the matter because of overriding navigation or recreational interests, the minimum necessary area should be cleared.

Stands of tall herbage, scrub and trees adjacent to the water are invaluable for wind shelter and many invertebrates with aquatic larval stages require them for feeding or shelter when adult. Overhanging foliage is used by some for oviposition. The rain of leaves and invertebrates helps fuel aquatic food chains. The shading effect influences water temperatures and plant growth. A few insects require shaded ponds; a more

Fig. 6.7. In small ponds, frequent small-scale intervention may be required to maintain the balance of plant cover. Here at Upwood Meadows nature reserve, Cambridgeshire, the use of the pond by cattle is retarding invasion by Common Reedmace, but some additional clearance would be beneficial. Up to a point, poaching is good, creating bare ground and plant regeneration niches as well as basking sites for dragonflies. The nearby stand of scrub shelters the site from wind but does not shade the water surface. (J. Andrews.)

diverse fauna and flora develops in sunlit sites. With ponds, the general rule is to keep trees and tall scrub to the north side so that they do not cause shading. However, for wind shelter, planting on the south side may be necessary, in which case it should be set back from the shore far enough to minimise shade. On large sites, some parts of the shore and shallows may beneficially be shaded but avoid the development of a continuous band of willows or other trees with no open banks, as this will reduce the development of marginal plants and deprive waterfowl of roosting and feeding areas.

Invertebrates

Many aquatic invertebrates are rare and all are important in the food chains of fish and birds. Individual species may have complex habitat requirements. For example, dragonflies need appropriate larval habitat which, depending on species, may be submerged plants, soft sediments or a gravelly bed, suitable emergent plants for the transition to the adult stage, hunting habitat and shelter which may be tall herbs close to water for many damselflies and woodland glades and margins for some of the big hawkers, *Aeshna* spp. They also need correct oviposition sites – some use bare mud, others need different plant species. Water temperatures affect larval development, while air temperatures, wind speed and rain all affect the ability of adults to breed successfully; thus water depth and adjacent vegetation are most influential.

Although the precise requirements of most invertebrates are unknown, it is possible to manage existing sites and create new ones to provide the broad range of conditions for a diverse assemblage or, in some instances, for particular species groups. The most important main habitat types are submerged aquatic plants, emergent and marginal vegetation, shingle, bare mud and sand, vertical earth banks, and tall herbs, scrub and trees near the water (Kirby, 1992). Do not alter hydrological regimes from natural seasonal fluctuations. Do not introduce fish which may increase predation or alter plant communities or turbidity. Exclude fish entirely from small invertebrate ponds or remove them if practicable.

Many invertebrates depend on particular plant species. Some species, such as molluscs, may have very limited powers of dispersal so that if major management results in their extinction at the site, they may be unable to recolonise. Therefore, management

Fig. 6.8. Where stock have unrestricted access to shorelines as here at Loch Nacnean Lower in Fermanagh, emergent and marginal vegetation may be destroyed completely. Creation of small enclosures will greatly increase wildlife value with negligible loss of grazing. (J. Andrews.)

must not completely remove any existing type of vegetation structure or plant species as this is likely to result in loss of invertebrates species. Flourishing populations may need quite large, continuous blocks of habitat. This means that diversification of small sites by introducing more plants or altering the actual physical form of the waterbody is hazardous but it is least risky with large waters provided that at least one and preferably two examples of all existing habitats are retained with a minimum extent of 0.1 ha or 100 m of shoreline length.

For most aquatic invertebrates, submerged plant growth cannot be too dense though some water beetles appear to need open water and other species require areas of exposed bed in which they bury themselves. It is important that such areas are not taken over by dense stands of emergents though these too are necessary.

Bare margins of soft, organic muds are important for the larvae of many species of flies, while more consolidated mud or wet sand will hold burrowing organisms. Where cattle or ponies have access to areas of the shoreline, their grazing, trampling and dunging will maintain these conditions by controlling plant colonisation. Otherwise, rotational cutting or dredging of open shores will be necessary.

Unvegetated shingle is a hot, sunlit habitat with abundant crevices for shelter and hunting. Stabilised, sparsely vegetated shingle has a different fauna and its warmth and free-draining character make it particularly valuable in north and west Britain where summer temperatures are lower and rainfall is greater. Prevent colonisation by tall or dense cover which will destroy its microclimate. Ensure that necessary wave disturbance is not prevented by shore protection, jetty construction or tree shelterbelts.

Vertical earth banks composed of consolidated but soft soils, especially if they are south-facing, will be used by burrowing wasps and bees. Where possible, tolerate erosion which creates and renews such faces and prevents the development of plant cover which will shade the soil surface. Otherwise, faces can be cut by hand or machine.

Retain or encourage stands of herbs, scrub and trees near but not shading the water for their value as food plants, breeding sites and shelter.

Fish

There is an extensive literature on fishery management for angling but little which treats fish as a wildlife resource. Care should be taken to safeguard populations of uncommon species. A key consideration is protection of spawning habitat and the most frequent problem occurs in the uplands where redds in headwaters may be destroyed by siltation due to ploughing in the catchment for afforestation or pasture reseeding.

In many sites, the fish community will have been altered by introductions, 'weeding' of undesired species and alterations to age classes due to removal by anglers. Some large upland lakes formerly contained populations of commercial importance.

Fish are important influences on the aquatic ecosystem, both as food resources for piscivorous birds and otters and as competitors for invertebrate food. They may also be implicated in the initiation and continuance of algal-dominated conditions. Thus, introduction of new species to any water may bring about significant effects not only on other fish but on the food-chains sustaining other wildlife and should be avoided. Further information on their effects is given at other points in this chapter. If removal or introduction appears to be appropriate, consent from the National Rivers Authority or, in Scotland, the River Purification Board, is required under the Salmon and Fresh-water Fisheries Act.

Amphibians

Amphibians require small, sunlit waters with no or little current or wave action, so most breeding colonies are in ponds or, occasionally, well-sheltered bays on larger waters. Common Toads *Bufo bufo* lay their egg-strings in water up to 2 m deep or more, twining them round plants or structures such as submerged branches. Frogs *Rana temporaria* spawn in water which may be only 15 cm deep, often over bare substrates. Newts fold individual eggs in the leaves of aquatic plants. The context of an amphibian site is important because the animals spend much of the year away from the water. Margins with damp ground and a mix of tussocks and bare surface will aid dispersal of young. Adults feed in grassland and herbage containing ample invertebrate food so sites with a natural plant community are likely to be better than managed swards. Frogs may overwinter in the water but newts and toads overwinter under stones or in the soil, the former choosing moist conditions and the latter often in dryer sites. Many populations thrive in garden ponds because of the diversity of cover, feeding opportunities and hibernation sites.

Tadpoles of all species are predated by some invertebrates, fish, including Stickle-back *Gasterosteus aculeatus*, and wildfowl. Occasional late-summer drying kills fish and so is beneficial even if the year's amphibian production is lost. Do not introduce fish to small pools.

Birds

There is a rapidly growing body of knowledge of the precise requirements of many waterfowl so it is possible to manage for selected species. As most species are quick to find suitable sites within their range, on sites where existing bird interest is low and there are no overriding priorities for other groups, create conditions suitable for species which are nationally rare or whose populations are of international importance; details of their status and habitats are given in Red Data Birds in Britain (Batten *et al.*, 1990). The following section gives general requirements of waterfowl and management principles. Specialist advice should be sought on major schemes.

Waders feed mainly on invertebrates on soft, open shores in water depths to 0.15 m or less. Rich, organic muds are ideal. Terrestrial margins may be fringed by emergents or herbage but most species dislike close tree cover. Dabbling ducks can feed to depths of up to 0.4 m depending on species. In the breeding season all are largely dependent on invertebrates taken from the water surface, bare substrates and aquatic vegetation; at other times, seeds and green vegetation are important for some species. Margins with good cover of emergents are favoured by some species. Swans take aquatic vegetation in depths to 1.0 m and also graze short swards as do geese and Wigeon *Anas penelope*. Diving waterfowl favour depths of 0.5–2.5 m though some mollusc and fish feeders habitually go deeper. Deep water is used by dabbling and diving species for roosting; there is a strong preference for sheltered waters so on large sites birds mainly use bays and lee shores.

Waders nest in tall grass, short vegetation or on shingle, depending on species. For Little Ringed Plover *Charadrius dubius* and Ringed Plover *C. hiaticula*, retain un-disturbed shingle beaches (or create shingle islands) over 0.2 ha. To prevent vegetation invading the shingle, place a 10 cm layer over a plastic membrane – old fertiliser bags will do – and bury the edge deeply or weigh it down with rocks to prevent waves from washing it free. For Redshank *Tringa totanus* and Snipe *Gallinago gallinago* maintain required feeding conditions by ensuring that adjoining grasslands remain surface damp (and not trampled by livestock or cut for hay) until July, or provide as much water as possible less than 3 cm deep as chick feeding areas.

Breeding wildfowl favour islands which give extra safety from ground predators. On new sites, create these as far offshore as possible, consistent with possible erosion problems. If possible, space at least 20 m apart. Any size will do but smaller islands have relatively more shoreline which is important for feeding. Gently shelving margins ease access for ducklings.

On existing sites, install rafts anchored as far offshore as is consistent with shelter from waves. There are many raft designs but the key considerations are to provide an edge which will prevent the surfacing material from washing off but with external ramps to allow birds to gain access from the water. A practical size would be 3 m × 3 m. For terns, surface with shingle over a solid bed. For wildfowl create a bed of rustproof netting which will retain soil bound by the roots but allow the vegetation to draw moisture from the water beneath. Install buoyed mooring lines so that the raft can be moved inshore for maintenance.

On islands, wildfowl rafts and on shorelines, retain dense cover at least 0.5 m tall in blocks as large as possible to miminise predation. Do not clear stands of dead herbage in winter as these are important cover for birds such as Mallard *Anas platyrhynchos* nesting in early spring. Do not plant trees or allow regeneration in or near potential wildfowl and wader nesting areas or they will be used by corvids as lookouts and this may increase predation of eggs.

On existing sites, maximise the extent of sunlit shallows by clearing dense stands of emergents and overhanging trees. Retain some fringing emergents as nest sites for grebes and trees which reduce wind effects without shading shorelines. On new sites,

(a)

(b)

Fig. 6.9. Extensive shallows are a valuable feature which can be modelled from scratch in new reservoirs and extraction pits. Most sites are too deep to be of use to dabbling ducks and waders, but parts of this example (*a*) a gravel pit at Great Linford in Milton Keynes have been designed for optimum value to waterfowl such as great crested grebes (*b*). The low-profile islands allow easy access by wildfowl for loafing or nesting. (D.A. Hill.)

margins should slope at less than 1:10 and preferably over 1:50. Allow for variations in water level; a 1:100 slope gives a 40-m wide feeding zone for dabbling duck but this gradient must extend 80 m offshore to maintain the same amount of feeding if summer levels fall by only 0.4 m. Where major fluctuations occur, as on reservoirs, create bunded bays in which a natural hydrological regime can be simulated.

Pools with hydrological control can be deliberately drawn down to expose fresh feeding for breeding or passage waders, for example. On sites where shallows cannot be created, convolute shorelines as much as possible to increase the length of feeding habitat.

If soft shore features and shallows cannot be given adequate wind shelter, they must be protected from severe wave action by reefs or other means (see page 132).

Fish are an important food for some grebes and other waterfowl. They are also major competitors with waterfowl (Giles, 1990). For example, Bream and other species consume the bulk of the chironomid production on which many duckling species depend in their first days of life and consequently affect their survival. Fish also reduce aquatic plant growth, affecting Gadwall *Anas strepera* and other wildfowl both directly and through the effect on other invertebrate populations. Endeavour to exclude or net out fish from smaller waters where wildfowl breeding success is a goal.

Mammals

A small number of mammals use waterbodies but positive management is likely to be considered only in relation to Otter which, after a catastrophic decline in much of lowland and parts of upland Britain, is now slowly recolonising former sites. In addition to adequate food resources in the form of fish, there are two main needs: freedom from disturbance and suitable resting and breeding sites.

So far as disturbance is concerned, apply the general guidelines. For cover, retain extensive stands (over 0.1 ha) of reeds, tall herbs, bramble or dense scrub at the water's edge. Breeding holts are often in cavities, particularly where water has undercut the root systems of bankside trees; willows *Salix* spp. and Alder *Alnus glutinosa* have fibrous roots which do not lend themselves to cavity-forming but many other native trees are very suitable. If natural sites are not available, artificial holts can be created by constructing piles of large timber or rocks, leaving a central cavity about 1 m in diameter. Elaborate variants include a chamber of unmortared bricks roofed with slabs, with 20 cm drainpipes for access, the whole buried with soil. It is a good idea to create several possible holts, located in the least disturbed parts of the site in the hope that at least one will prove attractive (see also chapter 5).

References

Andrews, J. & Kinsman, D. (1990). *Gravel Pit Restoration for Wildlife: a Practical Manual*. Sandy: RSPB.

Barrett, P.R.F. & Newman, J.R. (1991). *Aquatic Weeds Research Unit Progress Report*. Bristol: Long Ashton Research Station.

Batten, L.A., Bibby, C.J., Clement, P., Elliott, G.D. & Porter, R.F. (1990). *Red Data Birds in Britain*. London: Poyser.

Biggs, J., Corfield, A., Walker, D., Whitfield, M. & Williams, P. (1994) New approaches to the management of ponds. *British Wildlife*, **5**, 273–287.

Bjork, S. (1992). Principles and methods for the restoration of shallow lakes and wetlands. In: *Integrated Management and Conservation of Wetlands in Agricultural and Forested Landscapes*, ed. M. Finlayson. Slimbridge: IWRB.

Caborn, J.M. (1965). *Shelterbelts and windbreaks*. London: Faber and Faber.

Giles, N. (1992) *Wildlife after Gravel*. Fordingbridge: Game Conservancy/ARC.

Holdich, D. (1991). The native crayfish and threats to its existence. *British Wildlife* **2**, 141–51.

Kirby, P. (1992). *The Invertebrate Management Handbook*. Peterborough: JNCC.

MAFF (1985). *Guidelines for the use of herbicides on weeds in or near watercourses and lakes*. London: HMSO.

Maitland, P.S. & Lyle, A.A. (1993). Freshwater fish conservation in the British Isles. *British Wildlife* **5**, 8–15.

Ward, D. & Andrews, J. (1993). Waterfowl and recreational disturbance on inland waters. *British Wildlife* **4**, 221–9.

Reedbeds, fens and acid bogs

7

NEIL BURGESS, DIANA WARD, RICHARD HOBBS
AND DAVID BELLAMY

Introduction

Reedbed, fen and acid bog habitats are generally part of the succession from open water to woodland, differences between them being due to water chemistry, hydrology, geology and climate. Naturally developing successions of all these habitats have considerable conservation significance. However, in many British reedbeds and fens, lack of management has allowed succession to proceed which has produced large areas of woodland, with relatively small areas of early succession open vegetation communities remaining. Hence conservation management in these habitats now tends to focus on the early successional stages, and generally aims to slow or halt the succession. In contrast, many of the open acid bogs remaining in Britain are relatively stable and often require low levels of management, or even none at all. They are, however, threatened by peat extraction, drainage and invasion or afforestation by exotic conifers.

The three habitats vary widely in the vegetation communities and the associated animals that they support. Management experience for reedbeds, fens and acid bogs is summarised in this chapter.

Reedbed, fens and acid bog communities

Reedbeds

Reedbeds are dominated by Common Reed *Phragmites australis*. Wheeler (1992) defined a pure reedbed as a stand where the *Phragmites* cover exceeded 90% and an impure reedbed where the *Phragmites* cover exceeded 75% but was less than 90%. He considered that any stand of vegetation containing *Phragmites* in lesser proportion than 75% cover constituted a fen community.

Reedbeds may be brackish and tidal, but the majority are freshwater and either riverine or in waterlogged depressions. The principal sites are located from Dorset eastwards to the Humber, with the greatest concentration in East Anglia. Other important beds are found in Lancashire, Wales (principally Anglesey) and Scotland where the Tay reedbeds are of particular interest. Surveys undertaken in 1979 and 1980 (Bibby & Lunn, 1982) indicated that there were only 109 reedbeds in England and Wales that exceeded 2 ha in extent and only 33 over 20 ha. This size (20 ha) is considered important as it is the minimum size of reedbed supporting breeding populations of some of the rare reedbed birds.

In the absence of management, the period of dominance of Common Reed at a particular site may be limited, except where the bed is regularly inundated by tidal or freshwater flooding. Where this is not the case, dead vegetation or litter builds up and the reedbed becomes drier. This then allows colonisation by other vegetation types, culminating in woodland.

Many reedbeds have been, and some still are, managed by cutting dead reed stems during the winter to produce a valuable crop of thatching reed (Fig. 7.1). In many instances this management is the reason for the survival of the reedbed and hence any consequent value for wildlife.

Fens

Fens can be distinguished from other types of mire because they receive their nutrients and water from the catchment as well as from rainfall (Minerotrophic). The hydrology of an individual fen has a fundamental influence upon the type of vegetation which is present, although this may have been modified by management.

In the past, fens produced many useful materials. Peat was cut from deep pits or shallow strip cuttings to provide fuel while components of the vegetation were also gathered and utilised. Rush *Juncus* spp. and 'Sedge' *Cladium mariscus* were used for thatching, making flooring materials and a wide variety of other uses. Mixed vegetation was cut for fodder or bedding and aftermath grazing took place on many fens in the summer months. In some parts of the country Willow *Salix* spp. and Alder *Alnus glutinosa* within the fen were also managed and cut for human use.

Owing to economic changes, most fen management had ceased by 1940 and many

Fig. 7.1. Bundling cut reeds on the Tay reedbed in Scotland. A traditional craft that maintains the reedbed through the removal of standing plant material and hence slows the rate of build-up of the bed. (D. Ward.)

fens were subsequently drained or partially drained (e.g. Rowell, 1986). Undrained fens are now of little economic significance and management has largely ceased except for conservation, or for the shooting of Pheasants *Phasianus colchicus* and ducks. Lack of management, coupled with drying of many fens has led to large areas of former fen being taken over by the later successional stages dominated by shrubs and trees, and now early stages of fen succession are poorly represented in Britain. The remaining areas of open fen are threatened almost everywhere by lack of water, declining water quality and lack of management.

The fine classification of fen types is complex. Two main criteria are used, the first utilising the method of formation and hydrology of the fen, and the second the vegetation type according to the National Vegetation Classification system (Rodwell, 1984). The major fen categories are: Flood-plain, basin, 'Schwingmoor', open water transition, spring, valley, ladder within an oligotrophic mire, or soakaway within an oligotrophic mire. Within these categories, fens can also be divided into poor and rich fen, based upon the chemical properties of the water. Poor fens have a base-deficient water supply and typically contain *Sphagnum* spp. and sedges; they mostly occur in the north and west of Britain. Rich fens have a base-rich water supply and support vegetation types ranging from open water communities through to woodland. A useful summary of the major categories of fen, and details of all NVC fen vegetation types, with a brief description of the vegetation, its required habitat conditions and appropriate management techniques, is given by Fojt (1989).

Some of the main categories of fen are outlined below:

Basin mires

Typically these consist of a floating mat of *Sphagnum* species often pioneered and supported by the rhizomes of Bogbean *Menyanthes trifoliata* and Marsh Cinquefoil *Potentilla palustris*. The vegetation mat is often floating on a nutrient-poor waterbody.

Valley mires

These develop where nutrient-poor water creates conditions suitable for the development of poor fen. Valley mires develop alongside watercourses in small valleys. In broad valleys, and especially broad valley heads, the system may take on some of the characteristics of a raised bog.

Springs and flushes

These are located in upland areas where relatively nutrient-rich water comes to the surface and runs down a slope.

Mixed fens

These are generally characterised by vegetation containing Common Reed and Bottle Sedge *Carex rostrata*. Mixed fens may also include vegetation dominated by Great Sedge *Cladium mariscus* and Reed Canary-grass *Phalaris arundinacea* as their management usually falls within that of the reed-containing mixed fens.

Fen meadows

The most important fen meadow communities contain one of the following main species: Blunt-flowered Rush *Juncus subnodulosus*, Hard Rush/Soft Rush *J.* × *diffusus*, Sharp-flowered Rush *J. acutiflorus* and Purple Moor-grass *Molinia caerulea*. These have developed in response to long-established regimes involving cutting and grazing, and may contain a number of rare plants.

Fen woodlands (carr)

These are dominated by willow, birch or alder and in some stages of development, a combination of these. These have often developed in drier areas or where management has ceased allowing succession to occur.

More detail on the vegetation of British Fens can be found in Fojt (1989), Wheeler (1980a,b, 1984, 1988) and Wheeler & Giller (1982).

Acid bogs

The surface vegetation of these systems derives its water and mineral supply solely from the rain which falls directly on its surface (Ombrotrophic mire), and the water chemistry is oligotrophic (acidic and nutrient poor). In Britain, these mires have generally been termed bogs, in contrast to fens, and their vegetation is characterised by acidophilous plant communities in which the genus *Sphagnum* usually is a conspicuous component. The main types of acid bog are raised bogs and blanket bogs. Conservation of the ombrotrophic communities must be tackled along with the conservation of the whole system (Kulczynski, 1949; Mackenzie, 1992; Schimming & Blume, 1987).

Biologically the acid bogs of Britain are of global significance. For example, Great Britain is one of the main world locations for blanket bog, holding 10–15% of the total global area. Also, whilst raised bogs are a relatively widespread peatland type in the northern hemisphere, the remaining 10 000 ha of this habitat-type in Britain is important structurally and floristically as an extreme oceanic type.

Bogs are characterised by short vegetation, particularly *Sphagna*, but including other mosses, sedges and dwarf shrubs. Other vegetation, including trees, is almost always the result of disturbance, for example drainage, although some areas in the north-east of Scotland can support Scots Pine *Pinus sylvestris*. In other parts of Europe trees may naturally occur over a larger proportion of bogs.

There are two main ways of classifying these systems, the first based on hydro-morphology and the second on vegetation. In terms of vegetation the acid bog or ombrotrophic mire communities of Europe are all members of two classes (*sensu* Braun Blanquet: see Reiley & Page, 1990) of vegetation, namely the *Scheuchzerietea* and *Oxycocco–Sphagnetea*.

The two broad hydromorphological types of acid bog are outlined below.

Raised bogs

Raised bogs are usually formed in the flood-plains of rivers where they have become independent from ground-water influence. Here they represent the terminal acidification stages of the classic hydrosere: swamp – marsh, fen, poor fen, transition mire – ombrotrophic mire, the latter occupying the dome (or cupola) of acid peat which may be hundreds of hectares in extent. However large the cupola, in the long term its continued existence depends on the hydrological integrity of the whole complex: dome, rand (the sloping sides) and the other mire communities which flank and surround the dome, which are usually referred to as the lagg.

Drainage of the lagg will have an effect on the hydrology of the cupola. This will eventually lead to more rapid drying of the predominantly *Sphagnum* covered surface permitting the colonisation by trees and leading to the loss of the open communities of hummocks and hollows, pools and ridges: the patterned mire typical of actively growing bogs.

Blanket bogs

Blanket bog refers to the enormous complexes of ombrotrophic mire communities which once covered thousands of square kilometres in the wettest mountain and especially oceanic regions of Europe. True blanket mire peat can form directly on acid nutrient-poor podsols and even over bare leached rock. The depth of the peat is related to the degree of waterlogging and stagnation and can be up to several metres deep.

There is also much evidence to suggest that slash and burn neolithic and early iron age agriculture helped tip the balance towards the development of blanket peat, so some may be regarded as man-made communities.

Wherever embrotrophic communities exist, local disturbance which produces dust, for example from unsealed roads, ploughing, tree felling, and the use of lime, fertilisers and pesticides by farmers, foresters, gardeners or even conservation management, must be minimised. If these residues blow onto the mire surface they can have disastrous effects on the vegetation.

Wildlife communities

Reedbeds

Reedbeds are often considered to be a uniform monoculture of *Phragmites* with little habitat variation and supporting few species. In fact, they contain a number of features used by different plants and animals. It should be noted that more than one feature may be required to maintain a species on a site.

Botanical interest

The botanical interest of near monodominant reed stands is relatively low. However, where the proportion of reed is lower, a range of other species may be present. For example, the Reed-Milk Parsley vegetation community may support national rarities including Marsh Pea *Lathyrus palustris*, Greater Water-parsnip *Sium latifolium*, and Milk Parsley *Peucedanum palustre*. Fojt & Foster (1992) provide more detail on the plant communities of reed-dominated vegetation.

Vertebrates

Reedbeds are important for birds. There are five nationally scarce breeding species (Batten *et al.*, 1990), Bittern *Botaurus stellaris*, Bearded Tit *Panurus biarmicus*, Marsh Harrier *Circus aeruginosus*, Cetti's Warbler *Cettia cetti* and Savi's Warbler *Locustella luscinioides*. Habitat requirements of these species are summarised in Table 7.1 and Fig. 7.2. Other bird species common in reedbeds include Water Rail *Rallus aquaticus*, Reed

Table 7.1. *Main habitat and management requirements of reedbed birds*

Habitat and management features	Use (% of site)	Bittern (20 ha)[a]		Marsh Harrier (100 ha)		Bearded Tit (20 ha)		Cetti's Warbler (<1 ha)		Savi's Warbler (<1 ha)		Reed Warbler (<1 ha)		Sedge Warbler (<1 ha)		Waterfowl (various)	
		N	F	N	F	N	F	N	F	N	F	N	F	N	F	N	F
Open water 0.05–2.5 m deep		–	–	–	/	–	–	–	–	–	–	–	–	–	–	–	*
Islands	10–15%	–	–	–	/	–	–	–	–	–	–	–	–	–	–	*	/
Dykes		–	/	–	/	–	–	–	–	–	–	–	–	–	–	–	/
Reed/water interface	maximise	–	*	*	–	–	*	–	–	–	–	*	*	–	*	*	–
Wet reedbed 0.1–0.3 m deep	60%	*	*	*	*	–	*	–	–	–	*	*	*	–	*	/	/
Dry reedbed	15–20%	–	–	*	*	*	*	–	–	/	*	/	–	/	*	–	–
Reed/fen herbage mix		–	–	–	/	/	/	–	–	*	–	–	/	*	*	/	–
Reeds cut annually	<50%	/	/	/	/	/	/	–	–	/	/	/	/	/	/	/	/
2–3 yearly	>40%	/	/	/	/	–	–	–	–	/	/	*	*	/	/	/	/
5–10 yearly	>10%	/	/	/	/	*	*	–	–	/	/	*	*	/	/	–	–
Reed stubbles wet		–	/	–	/	–	*	–	–	–	–	–	–	–	–	*	*
dry		–	–	–	/	–	*	–	–	–	–	–	–	–	–	/	/
Patchy scrub		–	–	–	–	–	–	*	–	–	–	–	/	*	–	–	–
Carr/reed interface	5%	–	–	–	/	–	–	–	*	–	–	–	/	–	/	–	–
Carr		–	–	–	–	–	–	–	*	–	–	–	/	–	–	/	*

Note: N, nest site; F, feeding; *, primary requirement; /, regular use; –, no significance.
[a]Areas in parentheses refer to site size.
Source: Adapted from Andrews and Ward, 1991.

Warbler *Acrocephalus scirpaceus* and Sedge Warbler *A. schoenobaenus*. Reedbeds are important pre-migration roost sites for hirundines and act as winter roost sites for Pied Wagtails *Motacilla alba*, Starlings *Sturnus vulgaris*, and Hen Harriers *Circus cyaneus*, as well as many other species. Bibby & Lunn (1982) and Tyler (1992) provide further information on reedbed bird species.

Reedbeds also support a similar diversity of mammal species to other British habitats. Some species also reach high levels of abundance, for example Water Shrew *Neomys fodiens* in shallowly flooded sites (Jowitt & Perrow, in press). In contrast, Harvest Mice *Micromys minutus* are found in greatest abundance in reed growing in water where their winter diet of *Phragmites* seed is most abundant (Perrow & Jowitt, in press).

Invertebrates

All stages of the development and disappearance of a reedbed support important invertebrate communities (Kirby, 1992). For example, in East Anglia over 700 species of invertebrates have been recorded from reedbeds and 23 of these are listed in British

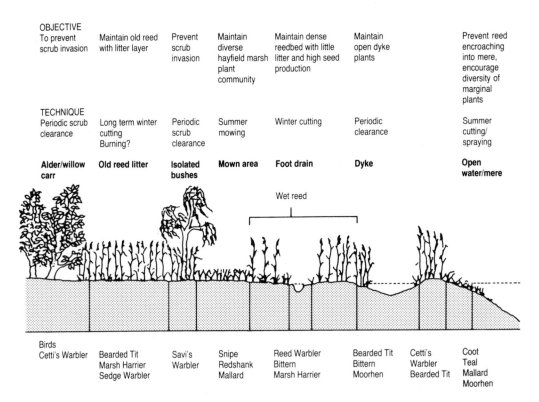

Fig. 7.2. Habitat features of significance to reedbed birds. (From Andrews & Ward, 1991.)

Invertebrate Red Data books (Shirt, 1987; Bratton, 1991). Forty species in four orders, Hemiptera (bugs), Lepidoptera (only moths), Coleoptera (beetles) and Diptera (flies) are known to feed only on reed, with a further 24 insects feeding partly on reed during their life cycle. Of the moths, there are four species which are sufficiently rare to be listed in the Insect Red Data Book: Reed Leopard *Phragmataecia castaneae* (Hubner), Flame Wainscot *Senta flammea* Curtis, White-mantled Wainscot *Archanara neurica* (Hubner) and Fenn's Wainscot *Photedes brevilinea* (Fenn). A wide range of invertebrates are also associated with reed, even if they do not feed directly on it. These include predators (mainly beetles and spiders) and parasites of the reed-feeding invertebrate species which live in the stems, including gall-forming flies and solitary wasps. The least interesting parts of a reedbed for invertebrates are those areas of standing open water (Kirby, 1992).

Fens

Plants

A great diversity of plants can be found in the various types of fen, including a number of rare and threatened species, particularly in the lowland fens of East Anglia and the mountain flushes and springs (covered in Chapter 11). Nationally rare species found in fens include Fen Orchid *Liparis loeselii*. On a more local level they often contain many attractive and relatively scarce species such as Grass-of-Parnassus *Parnassia palustris*, Marsh Helleborine *Epipactis palustris*, Round-leaved Wintergreen *Pyrola rotundifolia*, Marsh Fern *Thelypteris palustris* and Fibrous Tussock Sedge *Carex approprinquata*. A more comprehensive treatment of those plant species typical of fens and the rarities associated with this habitat is presented in Wheeler (1988).

Vertebrates

A wide range of mammals will use fens as part of their territory but no species are specific to this habitat. In the East Anglian tall-herb fens, Water Vole *Arvicola terrestris* and Short-tailed Field Vole *Microtus agrestis* can occur in relatively high numbers, together with Harvest Mice, Common Shrew *Sorex araneus*, Pygmy Shrew *Sorex minutus* and Water Shrew.

Taller mixed fens provide breeding habitat for birds such as Reed Warblers, Sedge Warblers and Reed Buntings *Emberiza schoeniclus*. Grasshopper Warbler *Locustella naevia* and Cetti's Warbler also nest in drier mixed fens with associated carr woodland. Marsh Harrier, which is mainly associated with reedbeds, will sometimes breed in fens dominated by *Cladium mariscus*. Recently cut fens can provide nesting places for Snipe *Gallinago gallinago* and fen meadows can, with suitable management, provide breeding habitat for Black-tailed Godwit *Limosa limosa*, Snipe and Redshank *Tringa totanus*.

Invertebrates

Fens support a very wide range of species. Many species are confined to a small number of vegetation types at a handful of sites. For example, the only true fenland species of butterflies in Britain, the Swallowtail *Papilio machaon* and the Large Copper *Lycaena dispar*, are both rare and the Large Copper has been reintroduced from the Netherlands. Another example is the Fen Raft Spider *Dolomedes plantarius*, which is known from only a handful of sites and is one of a number of rare spiders confined to fens.

A useful overview of management to favour invertebrate species in fens and other habitats is provided by Kirby (1992), although the precise management required for individual species is often not well known.

Acid bogs

Plants

Bog vegetation is characterised by a dominance of acidophilous plant communities in which the genus *Sphagnum* usually is, or has been, a conspicuous component. The distinctive bryophyte communities found on bogs are an internationally important feature of these habitats, and the floristic composition of blanket bog and associated wet heath in Britain is unique in the world, and demonstrates the highly Atlantic influence on plant distribution and vegetational development found in this country.

Growing within the carpet of mosses are a low diversity of vascular plants. However, a number of these species are scarce, including Great Sundew *Drosera anglica*, Bog-rush *Schoenus nigricans*, Wavy St. John's Wort *Hypericum undulatum*, Pale Butterwort *Pinguicula lusitanica*, Brown Beak-sedge *Rhynchospora fusca*, White Beak-sedge *R. alba*, Rannock Rush *Scheuchzeria palustris*, Crested Bucker-fern *Dryopteris cristata*, and in some upland sites Bog Orchid *Hammarbya paludosa*. A full list of rare and scarce vascular plant species found in acid bogs is presented in Rodwell (1991).

Vertebrates

The upland blanket mires form an important part of the habitat of many vertebrates. Short-tailed Field Vole is most widespread and abundant on drier ground where there is most vascular plant cover, the Polecat *Mustela putorius* frequents the raised bogs of central Wales, and Red Deer *Cervus elaphus* and Mountain Hare *Lepus timidus* are commonly found in Scotland.

Uncommon bird species nesting on blanket mires include Golden Plover *Pluvialis apricaria*, Curlew *Numenius arquata*, Dunlin *Calidris alpina*, Greenshank *Tringa nebularia*, Merlin *Falco columbarius*, Hen Harrier and Short-eared Owl *Asio flammeus*.

Raised bogs have a relatively limited breeding avifauna, occasionally including Nightjar *Caprimulgus europaeus* where conditions are suitable.

Invertebrates

Butterflies and dragonflies are the most conspicuous invertebrates on acid bogs, although a wide range of species may be found. Typical butterflies are the Large Heath *Coenonympha tullia*, in the north, and the Silver-studded Blue *Plebejus argus* in the south. Of the dragonflies, the widespread *Aeshna juncaea* and the rare and local *A. caerulea* frequently lay their eggs in *Sphagnum* peat pools.

Information on other groups of invertebrates on peatlands has been collected primarily in the past 40 years. There are large numbers of scarce species for Britain, and it has been shown that the invertebrates of blanket bogs typically show close affinities with the fauna of northern Scandinavia (Coulson, 1992).

The invertebrate fauna of lowland bogs differs from those on peat soils at higher altitudes; the number of species is impoverished, and they contain characteristic species which are replaced at higher altitudes.

Fig. 7.3. Structured vegetation in addition to *Sphagnum* bogs is required for the endangered Bog Raft Spider *Dolomedes fimbriatus*. (R. Key.)

Management prescriptions

Setting objectives

As with other habitats, it is important to ascertain what species are currently present and recently used to occur at a site before decisions are made over management regimes. A particular site may already have major biological significance and this needs to be understood and taken into consideration. Indeed, the decision not to manage a site at all may be correct, for example on raised acid bogs. Another site may have little biological value and more radical management ideas can be considered.

A table of desired species, habitat, their requirements, and the likelihood and desirability of attracting them is a good starting point when defining biological objectives. As with other habitats, it is much easier to succeed in conserving a species or habitat at a site where it is still present, or has been present in the recent past, than at sites where the species has never been present. It is generally better conservation practice to try and manage for rare or uncommon species than to try and attract a wide diversity of common species.

Management constraints

If one type of management has been followed at a site and it remains important for wildlife, there needs to be a compelling wildlife reason to change this management. Furthermore despite having identified what the desired wildlife objectives might be, there may be reasons why these are not achievable. Practical constraints on management at a particular site may include limited availability of water, poor water quality, severe scrub encroachment, land-ownership, access problems, lack of funds and/or manpower. Given the species or habitat type for which it is desired to manage the site, these will be of varying significance, but they should all be considered.

Of critical significance to most radical management ideas is the availability of water, and its chemistry (nutrient-poor, nutrient-rich). For example, if it is planned to make a large reedbed wet enough for Bittern, then the quantity and quality of water available must be checked. Similarly, it should be known whether the site can hold water, without it all draining away or evaporating during the summer. For most other species and habitats which might form the targets of management, similar questions will have to be asked and answered before progressing further.

Once these considerations have been made, then it is time to decide management prescriptions to provide conditions for species of particular interest, or to solve management problems.

Hydrological control

Reedbeds and fens

The hydrology of these wetlands must be considered carefully before any management activities are undertaken and it may be necessary to seek professional advice. Ward & Robinson (1989) provide a good introduction to hydrology. In particular, information will be needed on the quantities and positions of the inflow into a site, the outflow, the soil type and ground levels. Once these data are available, they can be used to assess whether water can be held on a site and whether it is feasible to construct the appropriate structures to manipulate water levels on the land to support the desired wildlife objectives. Where sites are raised above the surrounding drained land (as for example with remnant fens) or where the soil is particularly porous, it may not be feasible to seal the entire boundary to prevent water loss and water availability will determine whether or not management of the water levels can be achieved.

It is important to assess how much water is available within the catchment in each season as it may be necessary to compensate for high evapo-transpirative losses in summer, especially in the south and east of England. In some cases abstraction licences may need to be sought from the Environment Agency or other appropriate authority.

Water may be pumped on to a site or allowed to flow on by gravity and dams and bunds are the normal ways of holding water on a site (Figure 7.4). It should be noted that an impoundment licence is required for any impoundment, and that if more than 25 000 cubic metres of water is held above ground level the structure may then fall within the constraints of the Reservoirs Act 1975. Water movement should take place through a reedbed and fen and this can be ensured through a system of ditches and foot-drains. Fine control of water levels can be provided by sluices installed at appropriate points within the ditch system. Examples are shown in Figure 7.5; Brookes (1981), Burgess & Hirons (1990), Rowell (1988) and Scott (1982) all provide useful designs of water control structures. Although there is temptation to install complex arrangements of dams and sluices, simple systems are easier to manage and this is especially important when resources are limited.

It is helpful if the system of ditches and sluices allows the site to be divided into hydrologically discrete units which can be managed independently to allow greater diversity of management and/or rotational management. On large sites (>25 ha) four management blocks may be appropriate, with fewer on smaller sites.

Ideally water feeding a site should be of high quality and this is particularly important on fen systems where vegetation communities can show rapid changes. Saline incursions may adversely affect reedbeds. Although bunding and dams can keep out polluted water, there is likely to be a trade-off between acquiring adequate water supply and preserving water quality, and the decision to allow water of less than desired quality to

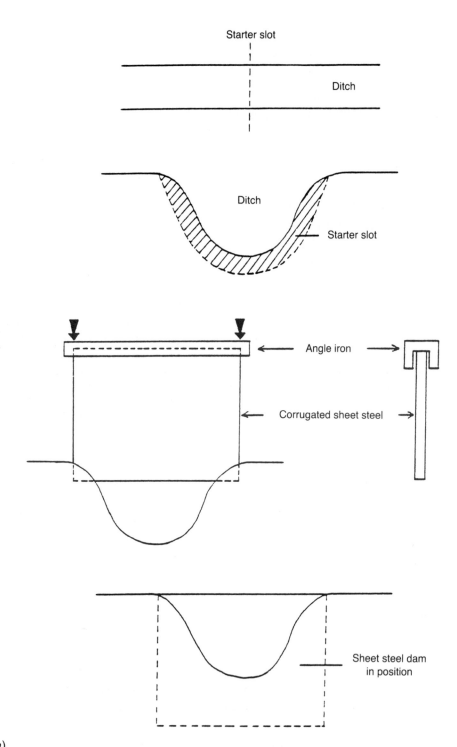

Starter slot

Ditch

Ditch

Starter slot

Angle iron

Corrugated sheet steel

Sheet steel dam
in position

(a)

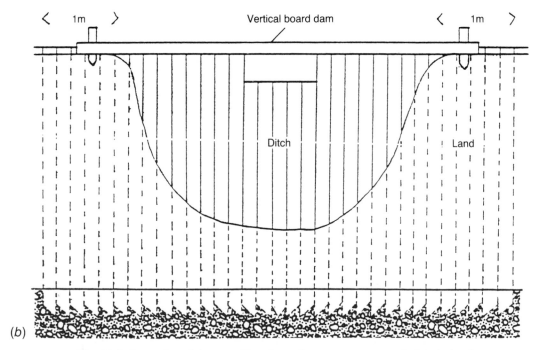

Fig. 7.4. Examples of dams and bunds used on nature reserves in Britain for the purposes of holding back water. (a) *Constructed dams for silt drains of up to 0.5 m across.* Simple dams consisting of a single sheet of some impervious material, such as steel sheets with a plastic coating on both sides are suitable for narrow ditches. For increased longevity, the edges of the sheet can be painted with bitumastic paint. A starter slot should be cut into the vegetation and the sheet inserted. A spillway can be cut in the top of the dam if required, although a simple dent from a sledge-hammer will often suffice. (b) *Constructed dams for large ditches.* Vertical board dam. This should be built using untreated timber, preferably oak or elm, with planks being hammered into the ground. The dam should extend at least 1 m into the bank on either side. The upper stringer should be laid at ground level and secured by stakes, or excavated a small way into the ground. (Modified from Brookes, 1981 and Rowell, 1988.)

enter a site may be necessary. A lagoon system at the inflow may assist in improving water quality.

In all cases any interference with the hydrology of the site should be done on a properly monitored experimental basis.

Acid bogs

Bogs may be divided into two layers: the surface layer or acrotelm (20–30 cm deep), and the lower catatelm. In an undisturbed bog the catatelm always remains completely saturated with water, while the acrotelm is the principal zone of water transfer with the environment.

Any modification of the bog structure, stature, surface roughness or phenology will

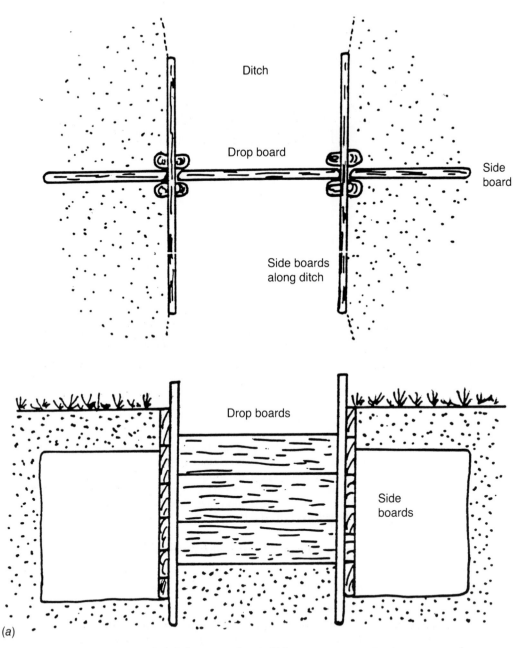

Fig. 7.5. Examples of sluices used on British nature reserves for the purpose of controlling water levels. (a) *Drop-board sluice*. Wooden sluices should be built of hardwood such as oak, elm or chestnut. Side boards should be set at least 1 m into the soil to prevent seepage and stop animals burrowing around the sluice. The height of the water outflow from the sluice can be controlled by adding or removing drop boards. More robust sluices can be made with concrete sides and base, using wood for the drop boards.

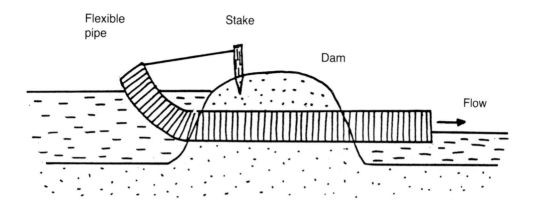

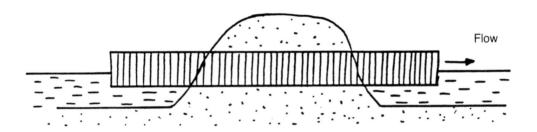

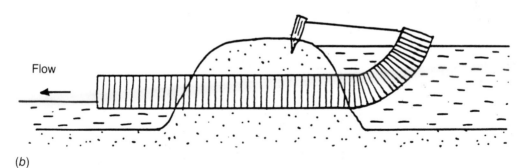

(b)

Fig. 7.5 cont. (b) *Adjustable flexible pipe sluice.* Large bore pipes (350–500 mm diameter) can be used to move large volumes of water rapidly from one location to another and also to finely control water levels. The pipes are flexible, light and can be cut to length with a hacksaw. They are also much cheaper than concrete pipes, have similar strength and are much lighter. Pipes are set into a standard earth dam when it is constructed, and corrugations on the outside of the pipe help to slow water seepage. The height of the inflow or outflow is controlled by varying the position of the flexible end to the pipe by means of two tensioning ropes attached to stakes (Burgess & Hirons, 1990).

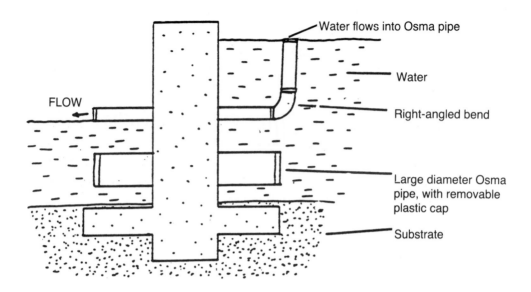

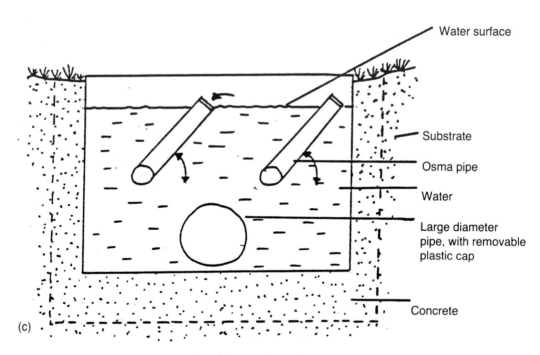

Fig. 7.5 cont. (c) *Adjustable solid pipe sluices.* Solid 110–200 mm diameter plastic pipes are inserted through concrete dams. A rotatable right-angle bend is fitted to these pipes, and a fine degree of hydrological control can be achieved by adjusting these. Larger volumes of water can be moved through such a dam, by having a larger volume pipe towards the base. This is normally capped off. Non-return valves can also be fitted to any of these pipes to prevent unexpected back-flow into a hydrological system. (Burgess & Hirons, 1990.)

affect its water balance and, indirectly, its soil structure. If the lower layers become de-watered, their physical structure is disrupted by shrinkage and cracking which weakens support for the upper layers: this can result in 'slumping' of the bog, particularly in the region of drains, and the loss of water which allows suppression of mosses by dwarf shrubs and disruption of the peat production processes.

Management of acid bog systems is based around maintaining an appropriate hydrological regime, by preventing any negative influences on the bog hydrology from human activities in the surrounding area (total catchment management) and minimising unnecessary disturbance to the bog, including conservation management activities. If any hydrological management is being considered, efforts should be made to gain water table data using a system of levelled dipwells before any management decisions are made. Any management must be monitored.

In an undisturbed bog the water is nutrient-poor. Every effort should be made to maintain this. However, it is recognised that activities in the catchment to the bog may adversely affect water quality and it may be impossible to prevent deterioration of water quality. It is thought to be more important to maintain the quantity of water in the bog, rather than worry unduly over the water quality. However, if the water quality is considered to be poor then expert advice should be sought from the statutory conservation agency.

To restore the hydrology of a damaged site all drains from it should be dammed and any cracks in the peat sealed to prevent drainage. Peat dams can be effective, but the peat used must be 'black peat' from the lower levels. Material should not be taken from the undisturbed bog. It must be compacted (with the shovel of a digger or by driving the digger over the dam repeatedly), and must not be allowed to dry out. Such dams should be built in the autumn and will start to vegetate in the following spring. An alternative method of creating dams is to use plastic sheets which can be pushed into place in a soft bog using the back of a digger's shovel. The number of dams on a site will be determined by the degree of slope, but up to 1000 have been used on the 180 ha Roudsea NNR in Britain.

In general it is only the large conservation agencies which will have the resources to manage large areas of blanket bog whereas some of the remaining areas of raised mire can be tackled by smaller organisations.

Hydrological regimes

Reedbeds
Reed can grow in water tables which are from 1 m below to 1 m above the ground (Haslam, 1972a) but in general these extremes are not recommended for management purposes. Reedbeds must hold sufficient water to prevent them drying out during the summer and allowing scrub to encroach, but further consideration must be given to the

depths and flooding regimes required by the wildlife. For example, Bittern can feed in water up to approximately 20 cm deep and require this throughout the year, while Harvest Mice require dry areas of reedbeds for nesting. Summer flooding is considered to be particularly harmful to reedbed invertebrate interest (Kirby, 1992), but it should be noted that this is normal practice on many sites. Seasonal water level variations may be tolerated by the reed plant provided that they can be replicated from year to year. Sudden changes in regime appear to be particularly deleterious. It appears to make little difference to reed survival if summer flooding or winter flooding is the chosen regime, although this has major implications for wildlife.

A number of different hydrological regimes can be devised for reedbeds. For example:

1. Water maintained high all year
2. Water above ground level until after the bird breeding season, then gradually dropped to just below soil surface and raised again in the autumn
3. Water held above soil level over the summer, gradually dropped to below soil level in late autumn and raised again in the early spring.

Additionally, regimes can be devised which allows the reedbed to remain moist with little or no flooding. Although this may be beneficial for some invertebrates, it allows scrub invasion to commence and may facilitate the dereliction of the reedbed.

Each of the above water regimes will fulfil different management objectives and favour different types of wildlife. Regime number one will favour those wildlife which depend on wet conditions, such as Bittern. Those which depend on the development of scrub, such as Cetti's Warbler, will not benefit. This management will also preclude reed-cutting and thus possibly benefit overwintering moth larvae. However, some other invertebrates may be adversely affected.

Regime number two creates damp soil conditions in the late spring and summer. This will allow summer mowing to occur and favour a diverse plant and invertebrate assemblage. However, reedbeds that dry out unduly in the summer risk becoming impoverished of elements of their invertebrate fauna so water levels need to maintain moist/wet conditions. Low summer water levels may also help maintain some species, for example the rare Milk Parsley, the food-plant of the Swallowtail Butterfly *Papilio machaon* in Britain can die back under summer flooding. Dropping the water level can also facilitate oxidation of the litter and thus help prevent the reedbed becoming drier with consequent scrub encroachment.

Regime number three has winter drying to allow a reed crop to be taken. This also has the benefit of reducing the build up of litter and thus successional change. Summer flooding (10–60 cm deep) may also reduce the need for scrub control on sites where this has become a problem. This regime is used where harvesting of reed takes place and is the traditional regime on many reedbeds.

Fens

Much of what has been said above for reedbeds, can be applied to fens. However, fens are much more complicated systems than reedbeds as they possess a much higher species diversity of plants and animals. On fens, finer-scale variations in hydrological regimes (depth, timing, cycles) will potentially have a major effect on the fen vegetation and the animal (especially invertebrate) communities. In general, the rarer fen species tend to be confined to the wetter sites (Wheeler & Shaw, 1987). However, the same study showed there was no significant correlation with species diversity although it appears that communities with low minimum water levels have low species density figures. There appears to be little information on the relative merits of different water regimes although if mowing or grazing is to take place, summer drying will be required.

Acid bogs

Managing water regimes in acid bogs is not well understood. Current thinking suggests that the optimum water table height is at or a few centimetres above the bog surface during the winter and not more than 10 cm below the bog surface during the summer (preferably close to the surface).

Dyke management

Reedbeds and fens

Dykes are important wildlife habitat in fens and reedbeds in their own right, but are also essential to maintain a flow of water through the site, thus maintaining the habitat. They can also be useful for access.

Many reedbeds and fens have grips (grupps or footdrains) across them. These are small spade-sized channels which allow water to flow off and on blocks of the reedbed or fen and are connected to the main water system. These should be maintained, providing that water levels have not dropped so much in the main channel that the flow is exclusively off the site. If this is so, they should be blocked to prevent them draining the site further and the level in the main drain raised as an urgent priority.

Clearing dykes out on regular rotations is a standard management technique, and will help maintain certain types of invertebrate and plant. The length of time between each clearance depends on the speed with which sediment accumulates and the degree of choking by plants. Signs which are normally taken to indicate the dyke should be cleared out are: choked with rank vegetation of one species, no open water remaining, or drying out during the summer. Regular clearance regimes work well for those species, particularly invertebrates, which need open water and the vegetation succession leading to a choked dyke. However, it is not often recognised that there is a substantial invertebrate fauna which is only found in ditches which have been choked for many years. Some of these ditches (which are not essential for water transport) are better left overgrown, in both open situations and beneath carr woodland.

Dredging is the usual method of clearing dykes, and can be undertaken on rotation along short sections of dyke to maintain their biological interest. Dredged material should not be dumped adjacent to the banks as this restricts the movement of water into the bed and also leads to drier areas along the dykes which allows the invasion of ruderals and subsequently scrub.

An alternative method of clearing dykes is mud pumping, where material is sucked from the dyke and sprayed onto surrounding land. This method has several problems; it is expensive, long sections of dyke have to be cleared at one time to justify the cost and thus it is difficult to use on rotation, it is not possible to profile dyke sides during the clearing process, there is a considerable loss of invertebrates, and the semi-liquid spoil needs to be disposed of. The spoil is particularly troublesome; care must be taken to keep it from running out of the desired repository area and this repository should be of low wildlife value, ideally arable land. An often used alternative is to pump material into areas of carr woodland. This is not wholly recommended as the mud may kill the trees and reduces the considerable wildlife importance of these habitats.

Fig. 7.6. Different management regimes create different wildlife habitats. At Upton Fen, the maintained ditches provide habitat for aquatic species, the summer-mown margins to the ditches support a rich fen community and the winter-cut reedbeds support the typical reedbed species. Areas that receive little or no management have reverted to carr woodland. Hence it is possible within a small area of a single site to use management to create a wide variety of wetland habitats. (J. Burgess.)

Dykes which are cleared out by dredging can be profiled to minimise the need for clearing. Ideally, the centre of a dyke should approach 1.5 m in depth to prevent both rapid silting and invasion by reed and other larger plants, although local conditions such as the depth and types of soil may preclude this ideal. A side sloped at approximately 1:10 will allow reed to grow down into the water and prevent it forming floating mats (hover) which can shade out the channel (Andrews & Ward, 1991). This batter also has the advantage that, whatever the water level, there will be a suitable margin for birds to use for feeding. Where an increase in the amount of light reaching the dyke surface is desired, perhaps for the benefit of dyke flora or dragonflies, then the marginal vegetation may need cutting back. Brushcutters are often used, but spraying with approved herbicides may be an option. Cutting in mid summer reduces the vigour of *Phragmites* and the management frequency may then be reduced. In order to maintain cover for secretive species only limited areas should be summer cut. Cutting is thus best undertaken in sections of the dyke margin, preferably on a rotation. The length of the rotation will depend on the width of the dyke, its aspect and the height of the bankside vegetation.

Open water

Reedbeds

Open water adds habitat diversity and may be important in attracting certain rare species. The ornithological analysis of Bibby & Lunn (1982) suggested that reedbeds with open water features were more likely to support the rare bird species, and up to 15 per cent of open water is considered optimal by site managers. However, waterbodies should never be created in areas of already high conservation significance and biological surveys must be undertaken during the planning process. Large shallow waterbodies provide a wider range of interest although even small pools have value for invertebrates such as dragonflies and corixids. The value of waterbodies is increased if they have a convoluted shape, to increase the length relative to surface area. Their depth is important as deep waterbodies have less wildlife value and thus they should contain as much shallow water as possible, 0.05–0.3 m depth below the summer water level being optimal for waders and wildfowl (Andrews & Ward, 1991). Islands and rafts (either natural or comprised of vegetation mats such as Yellow Flag, *Iris pseudacorus*) will provide loafing and breeding sites for wildfowl and may be used by terns. Some island designs are provided in Brookes (1981) and Burgess & Hirons (1992).

Most new waterbodies will need to be excavated and the paramount consideration is the disposal of spoil. This usually needs to be taken off site involving extensive and expensive use of heavy machinery. The need for spoil removal can partly be avoided if it is bulldozed into the middle of the new waterbody to create islands. Other considerations when creating waterbodies may include access for machinery and vehicles, and minimising disturbance.

Encroachment of larger plants, particularly Reed, Bulrush *Typha* spp. and Reed Sweet-Grass *Glyceria maxima* will occur in shallow waterbodies and must be controlled if the open water is to be maintained. This may be done either by scythe, brushcutter (cutting late summer – July/August), or by spraying with herbicides approved for use in and adjacent to water, such as glyphosate. Only part of the margin of an open water body should be managed at any one time in order to maintain some areas of old reed adjacent to open water which act as refugia. Where the reed has been cut or removed, margins may be used by waders and various wetland plants, e.g. *Equisetum* may appear. These areas are very good as invertebrate habitat.

Fens

Many of the rarest fen communities are the earliest stages of succession, either from open water or bare peat. A technique used in East Anglia to set back succession in small areas and favour these scarce communities is to create peat diggings. Often known as turf ponds, these shallow diggings can soon become colonised and develop into important fen communities. Before this operation is carried out it is useful to carry out peat corings on the site as they work best on deeper peats where the influence of marine clays or whatever is underneath is less pronounced. Such ponds should not be located in the last damp places on a site, as this may be where scarce invertebrate species are concentrated.

Acid bogs

Pools on acid bogs support a wide variety of plants and animals and the transition from open water to the drier surface of the bog is important. On undisturbed bogs, artificial pools should not be created.

Mowing vegetation

Winter cutting of reedbeds

In the past, most reedbeds were cut during the winter to produce thatching reed. Cutting of conservation reedbeds also takes place to increase the dominance of *Phragmites* and arrest the build-up of litter (Gryseels, 1989a; Haslam, 1972a, b; Wheeler & Giller, 1982). The major difference is that in commercial beds a larger proportion of the standing reed will be cut on annual or biennial cutting regimes. In conservation managed reedbeds longer rotations may be used in some parts of the bed to provide cover for secretive species such as Bittern, Marsh Harrier and Reed Warbler, and maintain overwinter habitat for scarce invertebrates (Burgess & Evans, 1989; Haslam, 1973; Ward, 1992a, b).

Whilst the life histories of many invertebrates are not known in detail, some general points in relation to winter cutting can be defined. Various insects, including the larvae of some of the rarer moths, feed in the stems of reed and there is thus no totally 'safe' period for cutting reed. There is also a substantial invertebrate fauna of reed-litter, such that small areas of a reed bed should be managed on long rotations to allow this to develop. Equally, other elements of the fauna thrive where regular cutting exposes bare wet peat. In essence, varied management will encourage a varied invertebrate fauna, and it is essential that annual cutting does *not* take place over the whole of the reedbed. For many invertebrate species it is probably only necessary to modify the approach in small areas of the reedbed, and along rides or other reedbed edges. It is most important that consistency of management is maintained in order to allow invertebrate communities to persist.

All winter-cutting should take place after the reed leaves have fallen and the stems have dried out, normally after the start of December. The water level must be dropped to allow the ground to dry out and allow machinery onto the site without causing compression of the ground which damages dormant reed shoots and rhizomes. After the bed is cut, it should be exposed to one or two moderate frosts before reflooding, which will increase the density of reed in the following season (Haslam, 1972a, b). Flooding protects the reed shoots from many hard frosts and prevents competing vegetation becoming established.

Large scale cutting of reed requires specialised cutting machines such as the Olympia Reed Harvester or Iseki KC450, although on smaller areas brushcutters or scythes can be used. The Olympia Reed Harvester both cuts and bundles the reed, but it is a heavy machine and requires fairly hard ground. Conversely, the Iseki is a small walk-behind machine and can be adapted to work on soft ground. Cutting machinery must also cut close to the ground as in thatching terms the toughest and thickest part of the reed is that closest to the soil.

Cut material must be disposed of. It cannot be left lying on the ground as it will impede the flow of water, increase shading, increase the nutrient status and speed the accumulation of litter and the build-up of the reedbed. Commercially harvested material is removed automatically, but non-commercially cut material will either have to be removed from the site, burnt, or piled up on site. Removal is very expensive on most sites and thus seldom used. Burning leads to an input of nutrients onto the site and may be contrary to SSSI regulations. There is debate as to the value of burning small piles of reed as this may spread nutrients more widely. Piling material without burning is usually only possible to a small extent. Such piles should not be located on the edges of carr woodland as this removes the best invertebrate habitat and they should not be used to fill wet depressions and shallow ditches as these are also very important for invertebrates. One advantage of some piles is that they provide refuges for invertebrates, particularly Coleoptera, basking and hibernation sites for Grass Snakes *Natrix*

Fig. 7.7. Pressure tracks in the Tay reedbeds in the spring following reedcutting.
Regeneration of the reeds appears to be inhibited in this area. Note the presence of
uncut dead reed stems in the distance, and the area of carr woodland at the margins
of the bed. (D. Ward.)

natrix, and loafing and breeding places for wildfowl, particularly where they are adjacent to water. The decision over whether to burn cut material, or pile it up to rot will have to be taken individually for each site.

Typical cutting regimes are annual (single wale) or biennial (double wale). Longer cutting rotations also take place, generally of 3–15 year duration, and normally for purely conservation reasons.

Annual cutting (single wale): Annual cutting produces a high yield of reed by increasing stem density and hastening spring emergence, although annual cutting may reduce the long-term yield (Haslam, 1972b, 1973). All the stems harvested are from the previous season's growth and where this system is continued tussocks do not develop and few weed species are present. However, the litter layer is poorly developed. This regime may damage invertebrate populations such as overwintering moths (Ditlhago *et al.*, 1993; van der Toorn & Mook, 1982), and reduce numbers of breeding reedbed birds such as Bittern, Bearded Tit and Reed Warbler. Even Coot will not find suitable nest sites in very young reed (Nilsson *et al.*, 1988). However, annual cutting produces good seed heads which are of benefit to winter feeding Bearded Tits and Harvest Mice. Cut areas are also used as winter feeding grounds by waders and wildfowl, and certain invertebrates prefer exposed bare peat.

Biennial cutting (double wale): Reedbeds cut every other year produce up to 50–75% more reed than those cut every year, but contain more old stems and waste matter. There are some advantages over annual cutting. Many specialised invertebrates, such as the Reed Leopard Moth, can overwinter in dead stems, there are old stems for nesting Reed Warblers, reed litter is available for use by invertebrates, and the higher levels of cover are beneficial to secretive birds such as Water Rail and Bittern.

Longer rotations: These can vary from 3 to 15 years (or longer) depending on the wildlife objectives and the speed of build-up of litter. Rotations of at least 4 years, preferably with some areas subject to longer rotations, have been recommended to provide habitat for scarce invertebrates (Kirby, 1992). The build-up of litter also provides ideal nesting locations for Bearded Tits. Because of the importance of areas of reed managed on long rotations for the conservation of scarce invertebrates, then conservation managed sites should always retain some of this habitat. The precise area will depend on the manpower available to rehabilitate the reedbed once it starts to dry out and scrub takes over.

Summer cutting

Reedbeds

Cutting reed during the summer reduces its competitive advantage by weakening the reed and allowing light to reach the ground, thus increasing plant species diversity and altering the vegetation towards a fen community with associated plants and animals

(Gryseels, 1989b; Wheeler & Giller, 1982). Reed can be eliminated altogether if it is cut twice in a growing season and flooded all summer over a two to three year period. As with winter cutting, the timing and cutting regime should depend on management objectives and only parts of a site should be treated in this way, preferably under rotation. For example Milk Parsley needs to be managed on a four year rotation if it is to obtain the height preferred by the Swallowtail Butterfly. Summer cutting should be undertaken in late July and August, after the main bird breeding season.

Fens

Annual summer mowing can be a very useful management technique but is also a potentially dangerous one. The dominant species of plant can be changed quite easily, and invertebrate assemblages can be radically altered. Thus on a SSSI advice should be sought on the appropriate regime. If the bulk of the fen is on an annual late summer cut, cutting patches only every 3 years with some not cut at all where small areas of scrub are allowed to develop, helps to benefit species intolerant of regular mowing, such as many specialised invertebrates (Kirby, 1992).

   Mowing is particularly appropriate where the fen has been traditionally managed in this way, or where falling water tables make it necessary to stop the area being invaded by scrub. In this case consideration also ought to be given to restoring appropriate

Fig. 7.8. Tussock sedges left in an area of fen at Upton Fen which has been cut.
Sensitive management allows the retention of notable features such as these sedges.
(J. Burgess.)

Fig. 7.9. Mowing fen vegetation with a traditional scythe at Ranworth Fen in the Norfolk Broads. This is a slow, but sensitive way of cutting summer vegetation to maintain the species diversity and structure of the fen vegetation. (Norfolk Naturalists' Trust.)

levels. Mixed fens with a low *Phragmites* content, often containing *Schoenus nigricans* and a variety of sedge species can be mown annually in late summer, although it is important that some areas are managed on a longer rotation.

When the fen is not in its desired state then the mowing regime can be adjusted to bring it back under controlled management. If a fen has a high content of *Phragmites*, *Filipendula* or *Epilobium*, an early summer cut can reduce their dominance, as they will have a relatively high proportion of their stored nutrients above ground. All three of these species can be effectively controlled in a three year mowing programme. Once they are reduced to the required proportions then the mowing can take place rather later in the year.

Where controlling scrub invasion is the main aim of mowing then again it can take place in late summer, and the frequency can be dropped to say once every three years. It is also valuable to allow some areas to progress to more mature scrub carr to benefit the invertebrate and bird fauna.

Mowing is also an appropriate technique for communities dominated by *Cladium mariscus*, but in this case the cutting must be carried out during the summer to allow adequate growth to occur before water levels start to rise. This plant can be killed if it is flooded too soon after a late cut. Cutting on a three or four year cycle is normal (Rowell, 1988).

On fen meadows it is often best to follow closely the traditional cutting regime, as many species will be present because of this regime. If there is no traditional regime, or the site has been subject to a period of neglect, then a good general regime for the majority of a fen is cutting for hay during mid July (although this will vary depending on the part of the country), with other areas being managed under a more varied regime, including some longer rotations. If necessary the mowing can be followed up by aftermath grazing, normally with cattle and then topping before the winter. Particularly where *Juncus* species are encroaching into a fen, a late cut can be useful as the new shoots are frost sensitive and many will be killed during the winter.

It is important to think carefully about the machinery to be used on a fen. Reciprocating cutters are the best machines for cutting large areas of fen. It is important that tussocky formations and the peat surface are not damaged and double or cage wheels should be used to reduce ground pressure. Independent steering on the wheels will save dragging and the exposing of bare peat. A machine needs to be well balanced with the weight not on the cutter bar. The machine should have at least 5 inches (12.5 cm) ground clearance under the engine. A 1 m cutter bar is ideal, and a 2 m cutter bar should be avoided as it can cause serious damage to tussocks. Scrub cutters are good for mowing small areas of fen, particularly where the fen vegetation is mixed with scrub, along dyke sides, or is particularly tall and thick. The ideal machine is two handled with a harness and an engine of about 35 cc. For cutting non-woody material a four-tipped steel blade is best.

All-terrain motor cycles ('quad-bikes') can also be useful for access and especially when it comes to removing cut material. The types that can be used in 4-wheel-drive at all times are preferable and a ground pressure of not more than 2 lb per square inch (0.5 kg/cm²) is recommended.

Cut material must be removed from the fen and it is unlikely that the mowings will have any commercial value. The material can be stacked off the fen, or burnt. Wherever possible carr edge and mature carr woodland should be avoided as locations for stacking materials, as they possess important invertebrate faunas. Also, never burn the litter where it has fallen but always take it to one fire site. Cut material should be burnt on sheets of corrugated iron and the ashes removed from the site. Removing the ashes prevents them from increasing the nutrient status of an area, which can be damaging, and the corrugated iron ensures that no piece of ground gets so hot that large patches of vegetation are burnt. It is important to remember, particularly if mowing fens for the first time, that dealing with cut material will take approximately 10 times more time and labour than cutting.

Grazing

Fens

Grazing animals can effectively manage many types of fen. Cattle are usually used and some of the lighter traditional breeds may be preferable. Single suckler beef cows are also suitable. Cattle grazing usually provides a variable sward when stocked at low to medium densities, while sheep provide a more uniform vegetation structure. Except for some of the 'rare breed' sheep, the damp ground may cause worm and foot problems in sheep and graziers may be reluctant to allow animals onto a fen. Horses crop certain areas very closely and leave others, principally latrine areas, untouched allowing rank vegetation to develop. In the light of experience on neutral grasslands and heaths it is possible that a combination of horses and cattle may be most suitable for fens. However, if this regime is chosen its effects should be monitored.

Light and late grazing is recommended, with starting dates usually sometime in July, and finishing dates in the autumn when the fen becomes wet, generally October. Fens are prone to poaching and while some poaching may be beneficial to open up the sward and create regeneration niches an extremely wet spell may lead to the grazing animals causing severe damage.

Ideally there should be flexibility in the deployment of grazing stock and as a rule of thumb with the equivalent of one cow to the acre (*c.* 2 cows/ha), but remember that it is much easier to start low and increase than the other way round. Any grazing licence agreement with a farmer should include dates, numbers of stock allowed together with conditions prohibiting the use of fertilisers and herbicides.

Fens that are being grazed will rarely have hedges or fences but are more likely to be bounded by dykes. These can provide an effective stockproof barrier although this is not guaranteed, and as animals may become stuck regular supervision is essential. Temporary electric fencing can be used but by far the most important factor is to get the right stock. Young adventurous cattle should be avoided.

Valley mires have been grazed in the past by sheep, ponies and, of course, native deer. At high densities cattle and horses can do much mechanical damage to the mire, and eutrophication from both dung and urine can alter invertebrate and plant communities. However, grazing of the New Forest valley mires, largely by ponies, has been shown to be essential to maintain the species-richness of these communities. Current thinking suggests that grazing levels on these mires should not exceed the equivalent of 0.3 ponies per hectare.

Acid bogs

Light grazing has probably been a natural feature of many acid bogs for thousands of years, e.g. Red Deer. It is not harmful on most sites and may be beneficial, especially on sites where rank vegetation or trees are invading. However, for most peatland

Fig. 7.10. Cattle are one of the most useful methods of managing fen vegetation. The species composition and structure of the resultant vegetation can be controlled to some extent by the numbers and types of cattle used in the management. The cattle are prevented from wandering by the man-maintained and water-filled ditches. These may also support a highly notable assemblage of wetland and aquatic plants and animals. (D. Ward.)

PLATE 1.

Oxlip *Primula elatior* is almost entirely restricted to ancient woodlands as it is a poor disperser, unable to cross even a hundred metres of open countryside. (J. Geeson.)

PLATE 2.

Heath Fritillaries *Mellicta athalia* are poor dispersers dependent upon recently cut coppice. Their survival is thus dependent upon continual management within sites. (W. J. Sutherland.)

PLATE 3.

Rocky cliffs such as this at (*a*) St Finan's Bay in South-west Ireland and (*b*) The Lizard, South-west England are of major importance for their diversity of plants. They are often diverse because they are inaccessible to people and grazing animals. The best way to conserve these habitats is through non-intervention. (*a*, D. A. Hill, *b*, W. J. Sutherland.)

PLATE 4.

Water quality is very important in rivers and streams. The Beautiful Demoiselle *Calopteryx virgo* disappears very rapidly with pollution or enrichment of the water. (R. Key.)

PLATE 5.

The natural drop in water level in summer exposes ground where waders can feed and ruderals such as Celery-leaved Crowfoot *Ranunculus sceleratus* can germinate and set seed. They in turn are a food resource for waterfowl and fish. However, unseasonable or extreme variation, typical of reservoirs, can be very detrimental to wildlife. A good feature here at Grafham Water in Cambridgeshire, is the retention of tussocky grass over winter as this provides hibernation sites for invertebrates and potential nesting sites for wildfowl in early spring. (J. Andrews.)

PLATE 6.

The creation of open water features in the reedbed at Strumpshaw Fen RSPB reserve in the Norfolk Broads. These open water features encourage a wider variety of species to use the reedbed area, and the provision of islands enhances use of the area by breeding ducks. (D. A. Hill.)

PLATE 7. (above)

Unimproved hay meadows typically contain a diverse mix of grasses and bulky herbs. (D. A. Hill.)

PLATE 8. (right)

Traditionally managed damp lowland hay meadows provide an important habitat for Fritillaries *Fritillaria meleagris*. (Mickfield Meadow, Suffolk Wildlife Trust Reserve, Suffolk; M. Ausden.)

PLATE 9.

Sulphur Beetles *Ctenopus sulphureus* need very short, warm, dry grasslands. (R. Key.)

PLATE 10.

A conservation headland in full bloom (*a*) benefiting butterflies, many other arthropods and partridge chicks which depend on invertebrates during the first two weeks of life. (*b*) One of the rare arable weeds, Pheasant's Eye *Adonis annua* which has been encouraged back to cereal crops from the seed bank. (*a*, N. W. Sotherton; *b*, D. A. Hill.)

PLATE 11.

(*a*) Dorset heathland in pristine managed condition and (*b*) showing destruction caused by housing. Note in (*b*) that Pine, *Pinus sylvestis*, and Birch, *Betula pendula*, have begun to establish, further reducing the value of the remaining heathland. (*a*, D. A. Hill, *b*, J. Andrews.)

PLATE 12.

Bogs such as this with actively growing Bog-moss (*Sphagnum*), or with open water pools should not be burned even when conditions are sufficiently dry to permit burning.
(A. J. MacDonald.)

PLATE 13.

Flushes add to the habitat diversity of moorlands and are important sources of insects for many moorland bird species and their young. This flush in the central Highlands is irrigated by lime-enriched water and characteristically has the conspicuous Yellow Mountain Saxifrage *Saxifraga aizoides*. (A. J. MacDonald.)

PLATE 14.

Singled coppice in Easton Hornstocks, Soke of Peterborough. Coppicing had lapsed for several decades, but the regrowth has been thinned to retain good-quality stems grown from seed and the strongest stems grown from stools. Although most of the stems in this picture are Silver Birch *Betula pendula*, in other parts of the stand, more Oak *Quercus* sp., Ash *Fraxinus excelsior* and Lime *Tilia cordata* were retained. (G. F. Peterken.)

PLATE 15.

This high-rise housing scheme in Delft, the Netherlands has a naturalistic landscape rich in wildlife and popular with local residents too. (C. Baines.)

PLATE 16.

Woodland wildflowers such as Foxglove *Digitalis purpurea* can be introduced beneath the canopy of trees in town parks to help rebuild the popular support for woodland habitat. (C. Baines.)

communities there is little information available on the optimum grazing regimes which should be employed to meet defined objectives. Guidelines are mainly available for *Calluna*-dominated blanket bog, as there is an economic value in maintaining this vegetation for Red Grouse (Hudson, 1992) and species associated with this bird.

For blanket bogs, current thinking suggests that very light grazing (probably less than 0.37 sheep/ha) will be beneficial to the blanket bog vegetation. This is more probable if the site has a long history of grazing. However, Coulson (1992) suggests that the greater nutrient inputs onto bogs caused by grazing sheep, and the local effects of trampling, may be disadvantageous. An increase in the intensity of grazing observed on many blanket bogs, caused by a combination of increased sheep numbers and the practice of winter grazing, can have a disproportionate effect on certain key vegetation types. Summer grazing alone is preferable: on blanket bog in the northern English uplands, the traditional pattern of grazing management was to drive the sheep onto the high moors during the late spring and move them onto lower pastures at the end of October or the start of November.

Tree and scrub control

Reedbeds and fens

Many reedbeds and fens are subject to scrub encroachment and ultimately conversion to carr woodland. Scrub can add to the diversity of the site as it supports many insects and birds that are not found in open reedbed and fen communities (e.g. Cetti's Warbler), but total domination of a site by scrub will ensure the extinction of most specialist reedbed and fen plants and animals. In general scrub cover in the region of 15–20% is considered valuable and if possible it should be around the margins with only scattered bushes in the site.

Scrub can be cleared from fens to good effect and the resulting herbaceous vegetation can be similar to that present before the scrub invaded. This is particularly true if there is a viable seed bank of fen plants in the peat or if remnants of the original vegetation are still present. The most cost-effective way of controlling or removing scrub is to manage the advancing face rather than tackle large blocks.

Scrub clearance is best carried out by hand, even if it is possible to get a tractor with cage wheels and a heavy swipe on to the site. Extensive damage to the turf should be avoided if possible, although smaller-scale disturbance may not be harmful and may benefit some species such as the Fen Violet *Viola persicifolia* which require disturbed ground to germinate. Cut stumps should not be winched out of a fen, but instead should be painted with an approved herbicide at the recommended concentrations (e.g. Garlon), or flooded. Where there are large numbers of small stumps inclusion of some dye with the herbicide will identify those stumps which have been treated. It may take two or three years to kill all the stumps. When a small tree or large bush growing in a wet peaty fen is cut down, then the stump can rise up from the fen surface and may need a second cut the following year. This is particularly important if the follow-up management is to be mowing.

Flooding of stumps which have been cut as low as possible during the growing season may achieve a much more successful kill than herbicide treatment. However, the often negative effect of prolonged summer flooding on other plants and animals, particularly soil invertebrates must be considered carefully before undertaking such management.

An alternative to clearance for some areas of carr woodland, is coppicing. Coppicing on rotation has the advantage of benefiting the invertebrate and bird communities in addition to keeping the carr under control. A 5–10 year rotation of small blocks (*c.* 0.2–1.0 ha depending on the size of the site) of carr woodland is recommended. Cut wood should be burnt to prevent regrowth and this should be done at special burning sites to avoid scattering the nutrients widely.

Acid bogs

There is little or no evidence that active peat-forming *Sphagnum* vegetation is invaded by trees, it is always the other way round. Almost all cases of tree invasion are caused by a reduction in the water balance of the bog, either through drainage (most), or natural reduction in rainfall (historical examples only). Pines can, however, be planted onto drained bogs and even onto the hummocks of undrained bogs, although they cannot naturally invade an undrained bog.

Planting of coniferous trees on ombrotrophic peat will eventually destroy the plant communities, mainly because their high transpiration rate combined with their interception of rainfall will eventually lower the water table. However, bog communities may persist beneath the trees for up to 20 years, thus restoration is possible within this period. Although some damage may be caused to the bog surface *Sphagnum* fragments should establish rapidly.

Trees can be removed from a site with minimal damage by cutting them down by hand and removing the trunks and branches using sky-hook logging systems. Regenerating seedlings can be removed by hand for the next 3–4 years, after which the bog should be incapable of invasion by trees (provided the hydrology has been restored).

Small areas of trees can be taken out by hand and burnt on site, although this is very labour-intensive. Material should be burnt on sheets of corrugated iron raised above the bog surface by bricks. In this way the heat of the fire does not reach the bog surface which prevents scorching and ashes can be removed off site which prevents nutrient enrichment.

Rhododendron *Rhododendron ponticum* can also be successfully removed from acid bogs by spraying with glyphosate and a suitable wetting agent, followed by hand-pulling of seedlings.

Habitat restoration and creation

Reedbeds

Reedbeds that have been left unmanaged for some years may contain high levels of litter, be dry, have sparse reed and be invaded by carr.

Burning

Where the reed is sparse with large tussocks and a considerable depth of litter, burning may be used to try and bring the reedbed back into a condition where it can be cut. Burning disposes of the standing crop cheaply and quickly, and provides a nutrient input into the site, after which seed production can be extremely high (Cowie *et al.*, 1993).

Derelict reed is burnt when it is completely dry, usually in late winter and when the water table is close to the surface. Opinion differs as to the merits of burning up or down wind, the former burning deeper as the fire moves slowly (thereby risking burning the rhizomes and dormant shoots), and the latter burning faster, leaving a long stubble. Care must be taken to control the fire and adequate firebreaks (at least 3 m wide) should be cut.

Burnt plots should be less than 2 ha in size on a large site and smaller on a small site. In experiments in Norfolk, there was no evidence that burning had any greater effect on the invertebrate community than cutting (Cowie *et al.*, 1993; Ditlhago *et al.*, 1993). The burn was also shown to be beneficial to plant species such as Marsh Pea and Milk Parsley, and may also reduce abundance of two invasive species of plant, Goat Willow *Salix caprea* and the alien Orange Balsam *Impatiens capensis*. Nevertheless, burning should be used with caution and only to bring the reedbed back under normal management regimes. Burning a dry reed bed or one in summer is likely to have greater impacts.

Removing surface layer

A potentially expensive option which can become necessary where it is impossible to exercise any form of hydrological control (including pumping water) and drying out has occurred. Excavation up to a depth of approximately 1 m should enable reed to regrow from rhizomes located deep in the soil, although in dry beds the water table may be at greater than 1 m depth in the summer. However, excavation to this depth may result in water levels on the renovated bed that are too deep for some species and the depth of excavation should therefore be considered carefully before work starts. Further excavation will allow meres to be created within the framework of the new bed. Spoil must be removed from the site in order to allow free flow of water across the reedbed and this can be extremely difficult and expensive.

Fens

Fens that have been unmanaged will be dominated by tall rank vegetation and may be invaded with carr. It may be impossible to undertake standard management practices on such a site and restoration may be required before any further activities are possible.

Mowing

Mowing can be used to bring areas of a fen back into an annual mowing regime. The cutting regimes for a fen being restored should probably be started at an earlier cutting date than would be normal. This helps to remove nutrients from the site and get the vegetation back under control which permits more normal management regimes to be introduced. However, in terms of invertebrates it is important not to manage the whole site in one go using the same regime, as species will be lost. As far as possible a fen should be restored bit by bit.

Carr woodland removal

Removing carr woodland is a possible option for fen restoration, but is extremely expensive, and unless the site contains remnant populations of a rare species, or could otherwise be justified, then it should be thought about very carefully. Methods for removing carr have been considered earlier in this chapter.

Burning

The restoration of fen vegetation is also possible by burning. Species-rich vegetation communities can be associated with burning regimes.

Creation of new sites

Before attempting to create a reedbed, fen or acid bog, biological surveys should be carried out to determine that nothing of importance is present at the proposed site, and careful consideration should be given to the likely biological value of the restored site in relation to the often extremely considerable cost of creation. Detailed plans of the biological objectives, management opportunities and constraints, and the likely cost of the venture should be drawn up and considered carefully before progressing further.

Reedbeds

Before creating a new reedbed, the manager must ensure that the land to be converted to reeds has lower conservation value than the likely conservation value of the new reedbed, and that soil and hydrological conditions are suitable. The most valuable sites are those exceeding 20 ha, ensuring that the site is large enough for all the rarer reedbed species and allowing the manager to incorporate different management regimes to cater for many specialised species. Reedbeds established adjacent or near to existing sites are likely to be colonised more rapidly than those some distance away and consideration should be given to establishing beds adjacent to other wetland habitats.

Reedbeds should be designed in at least four hydrological blocks with an adequate system of dykes and foot drains for the transport of water and control of levels. Although reedbeds are traditionally flat or sloping, a ridge and furrow land form may also be considered as it provides a range of water depths of benefit to different species and may help reduce maintenance requirements. For example, 4 m wide dykes set 50 m apart with a slope between the dykes of 1:100, the top height being just above the summer water level, will provide suitable depths for feeding Bittern, and the top drier areas will develop a litter base suitable for Bearded Tit nests. Meres can be incorporated into the design to provide open water and these should also have sloping margins with convolutions to extend the reed interface. Drawdown will provide large areas of exposed mud and shallow water over the whole site (Andrews & Ward, 1991).

Reeds may be planted as pot grown plants or as rhizomes and establishment trials are being undertaken using seeds; however, reed establishment over large areas is still in its infancy and no foolproof method can be given. Ekstam *et al.* (1992) provide a useful summary of experience in Europe. The best current options for creating a reedbed from scratch are planting rhizomes or pot-grown plants. Unfortunately, rhizomes are difficult to obtain in bulk, expensive to transport and success is not guaranteed. They must also be kept moist in transport and should be planted in moist ground with part of the rhizome in contact with air if they are not to suffocate and rot. They should not be planted underwater, but as they establish, water should be available to flood the site gradually, following the growth of the plants, and suppressing potential competitors. Pot-grown plants are expensive, but seem to be more reliable. A 0.5 m spacing between plants seems to be optimum for establishing a bed quickly. If reed is already present on the site, then shoots can be pegged down to the damp substrate, which increases their rate of spread as the reed will root at the internodes.

'Regeneration' of ombrotrophic mire

In many parts of Europe where the vast majority of ombrotrophic mires have been destroyed, regeneration is the only way ahead in mire conservation (Dierssen, 1992).

There are no completed examples of the regeneration of any whole ombrotrophic mire complex, and some counsel that this is impossible. However, a number of experiments are underway, which range from cessation of burning and grazing to blocking all drains which have been cut in the complex. All that can be concluded to date, however, is that the time span for the regeneration of whole bog complexes is very long (Vasander *et al.*, 1992; Joosten, 1992).

There are, however, many examples of the natural regenerative capacity of ombrotrophic vegetation communities (Smart *et al.*, 1986). For example, when the drains of old peat diggings or cutaways which ring many raised bogs and margins of blanket bogs have been blocked, ombrotrophic mire plants begin to grow within a few years. Figure 7.11 shows a range of such natural regeneration. This offers enormous potential for the regeneration of acid bogs. There is always the problem of invasion by weed species such as *Juncus effusus* and *Molinia caerulea*, but stabilisation and manipulation of the water table can limit their spread (Rogers & Bellamy, 1972).

Modern methods of peat extraction using large scale drainage and 'milling machines' which leave vast expanses of featureless cut away peat prove a much greater problem. However, if they can be wetted sufficiently, then ombrotrophic plants may be able to start growing again.

In some locations in Western Ireland, where the peat has been removed down to the impermeable clay base, natural regeneration of mixed and transition mire communities have quickly developed in wet expanses and pools which receive run-off from the surrounding peat. In time these may act as foci for mire regeneration. There is however one problem, such pools attract water birds and wildfowl – even the endangered Greenland White-fronted Goose *Anser albifrons flavirostris* – causing eutrophication of the area and allowing invasion by weed species.

Rehabilitation of raised bogs

The best examples of rehabilitated raised bogs, at least in their early stages, are in Lower Saxony in Germany, where 30 000 hectares of exploited bog are scheduled for rehabilitation for nature conservation by the process of 'rewetting' (Egglesman, 1988), although there are also many examples of restoration work being undertaken in the UK, for example Thorne Moors, Westhay and Fenns and Whixall Mosses.

The critical factor for such regeneration is the ability to kickstart the regrowth of *Sphagnum*. This can be achieved through the creation of permanently wet conditions, across the cutaway or milled surface (Egglesman, 1988). Permanently wet conditions allow the system to change from aerobic (oxygen-rich), where plant matter is oxidised to carbon dioxide, to anaerobic (oxygen-poor) where plant material tends to accumulate and where the production of methane may be highly significant to the process of bog formation (Brown *et al.*, 1989). This chemical switch is absolutely essential for the regeneration of acid bog communities and once it has been achieved then invasion by *Sphagnum* species can be rapid. To achieve this very large quantities of water may be required, and it may be necessary to obtain abstraction licences. It should be noted that when drainage for peat extraction takes place, pumping rates of up to 8000 litres/minute may be used.

The Lower Saxony experiments

Rules now applied in Lower Saxony state that at least a 0.5 m thick layer of undisturbed peat must be left *in situ* at the end of any current peat-winning operation, together with at least 0.3 m of 'top spit'. Drains are then blocked and the rewetting begun. In the absence of a thick layer of living *Sphagnum*, the peat surface will tend to dry out during periods of drought. This can be compensated for with sprinklers or some other form of pumped irrigation, although this is costly and the source of the water must be 'ombrotrophic' (best taken from drains on adjacent peat cuttings).

Where the water storage capacity of the cut peat surface is not sufficient to allow the growth of a deep layer of living plants which prevent drying in periods of drought, reservoirs of open water are dug. From hydrological calculations, Beets (1992) recommends that the reservoirs should be no more than 10 m apart in slightly decayed peat and no more than 5 m apart in moderately to strongly humified peat. They should have vertical sides and their depth should be 0.5 to 0.6 m deep, a fact which could cause problems if the depth of *in situ* peat is only 0.5 m as these drains may enter a permeable substrate beneath the peat. If this has occurred reservoirs must be adequately sealed, and PVC sheeting has been shown to work well. The reservoir pools should be no more than 20 m long to prevent erosion by waves. Bog species should be transferred to these pools to start the process of bog regeneration. *Sphagnum cuspidatum*, *S. subsecundum*, *S. recurvum* and *S. papillosum* have been shown to transplant well. The use of peat-cutting machinery to prepare the mire for the rewetting process will greatly cut the cost and hopefully speed the process (Rogers & Bellamy, 1972; Beets, 1992).

The only proof that these methods will eventually work is the plethora of examples of accidental regeneration of vegetation communities rich in ombrotrophic plant species (e.g. Table 7.2).

One cautionary note is that although it may be possible to re-create passable vegetation communities, unless the re-created areas are close to existing mature bog systems restoration of the invertebrate fauna will not be possible. Hence we must be cautious in stating that bogs can be re-created if a large component of the bog fauna cannot be transplanted or restored. Moreover, the examples of bog regeneration given here should never be used or accepted as an excuse for the continued destruction of pristine peat sites, or the exploitation of others.

Conclusions

This chapter has concentrated on identifying the principles behind managing reedbeds, fens and acid bogs for wildlife but recognises that the mechanics of undertaking such management may be site-specific and vary widely.

Reedbeds
Before attempting any management the wildlife value of the existing site should be assessed and decisions made over which the target species or communities are and a

Table 7.2. *Species composition of main types of acid bog in Britain, with examples of regenerating and re-created sites*

	1	2	3	4	5	6	7	8	9	10	11	12	13	14
	P	P	R	R	R	R	R	P	R	R	P	R	P	R
Pristine/Regenerating														
Andromeda polifolia	1/75	2/80	0.1/80	2/20	2/45	–	–	0.2/30	0.1/10	0.5/90	1/90	0.5/40	–	–
Calluna vulgaris	–	4/80	–	0.2/5	–	–	11/10	15/75	0.2/20	–	11/95	15/80	0.5/50	1/55
Carex limosa	0.3/20	–	2/80	–	–	–	–	+	–	–	0.2/20	–	0.6/10	–
C. panicea	–	–	–	2/5	–	–	–	0.6/15	3/25	0.7/7	–	2/40	–	–
C. pauciflora	–	0.5/50	–	–	–	–	–	0.6/15	–	–	–	–	–	–
C. rostrata	–	–	2/40	–	0.6/60	3/65	1/30	–	15/55	1/7	–	–	–	–
Drosera anglica	0.1/10	–	–	–	–	–	–	0.1/5	–	–	0.2/20	0.5/20	0.2/40	+
D. intermedia	–	–	–	0.1/10	–	–	–	–	–	–	–	–	0.7/40	0.6/50
D. rotundifolia	1/80	0.5/40	0.7/70	0.2/55	–	–	–	0.3/30	+	0.5/40	0.7/65	0.7/10	0.9/45	0.5/40
Erica tetralix	–	–	–	–	6/100	2/65	0.7/15	+	15/20	6/65	6/90	1/80	0.6/50	1/65
Eriophorum angustifolium	–	–	0.4/30	0.2/20	10.90	1/60	4/30	3/80	35/80	2/60	1/60	1/50	0.2/15	0.6/55
E. vaginatum	3/65	10/90	–	–	+	–	–	8/65	–	–	0.2/15	1/40	–	–
Menyanthes trifoliata	–	–	–	+	–	–	–	0.1/5	1/5	–	–	10/60	0.3/25	0.3/30
Molinia caerulea	–	–	–	0.8/20	5/70	+	16/40	0.2/5	5/30	1/20	–	–	6/95	6/75
Myrica gale	–	–	–	–	–	1/10	15/40	0.7/20	0.5/15	–	–	–	0.2/15	0.3/30
Narthecium ossifragum	–	–	–	–	3/80	–	2/15	0.5/5	1/15	3/65	1/75	1/5	0.7/65	0.4/45
Oxycoccus palustris	–	1/80	0.5/45	4/80	0.5/30	–	–	–	–	–	0.1/5	0.1/5	–	–
Rhynchospora alba	10/80	0.2/15	6/85	16/30	–	0.1/20	1/20	–	–	–	2/60	3/30	12/90	22/100
Scheuchzeria palustris	3/85	0.4/15	0.4/40	–	–	–	–	–	–	–	–	–	–	–

Species	1	2	3	4	5	6	7	8	9	10	11	12	13	14
Schoenus nigricans	–	–	–	–	–	+	+	–	0.5/5	–	–	–	12/90	3/60
Sphagnum capillifolium	+	10/55	+	–	–	–	–	9/55	1	20/40	28/20	4/50	3/30	3/20
S. cuspidatum	80/100	19/30	71/90	4/30	3/25	–	9/20	24/50	–	2/8	9/20	13/40	4/50	18/55
S. fuscum	2/20	+	+	–	–	–	–	–	–	–	1/20	+	–	–
S. imbricatum	–	–	–	15/35	13/34	–	–	0.5/5	–	–	9/15	+	3/5	+
S. magellanicum	0.5/10	25/70	–	40/80	40/90	40/45	–	5/20	20/25	–	10/15	8/50	10/50	7/70
S. papillosum	4/15	2/12	12/45	20/65	20/60	45/50	–	26/75	25/30	40/75	10/40	15/80	2/15	22/80
S. recurvum	1/10	25/70	–	–	–	20/30	–	–	–	–	–	–	–	–
S. submitens	–	–	–	–	1/10	–	–	–	10/20	–	–	–	–	–
S. subsecundum	–	–	–	–	–	–	–	+	15/20	20/60	–	–	12/50	11/60
S. tenellum	–	–	–	0.4/8	0.5/10	–	–	3/30	–	+	+	+	0.5/10	–
Trichophorum cespitosum	–	–	–	–	–	–	–	0.1/10	–	3/27	4/60	8/70	+	0.1/5

Note: All vegetational analysis based on at least 10 randomly placed 1 m square samples, the first figure an estimate of species cover, the second figure the percentage of the sample in which the species was found.

Column 1 Central pool complex of raised bog cupolas in southern Germany. Column 2 Main mire plane of raised bog cupolas in southern Germany. Column 3 Ombrotrophic mire-communities affected by slight eutrophication, southern Germany. Column 4 Regeneration in 40-year-old cutaways, southern Germany. Column 5 Ombrotrophic mire communities Thursley Bog National Nature Reserve in southern England, regeneration time 20 years (after destruction by tank training during the Second World War). Column 6 Basin mire developed behind an earth dam Bagshot, Surrey, England, regeneration time *c.* 40 years. Column 7 Basin mire developed behind a brick dam Bagshot, Surrey, regeneration time *c.* 30 years. Column 8 Central Scotland Blanket Mire, pristine site although some grazing by deer and sheep. Column 9 Central Scotland, regeneration in cutaways, over 10–60 years. Column 10 Ireland, western side of central plain, regeneration where peat has been removed down to underlying clay. Column 11 Central Irish Plain, pristine raised bogs. Column 12 Central Irish Plain, regeneration in cutaways 10–60 years. Column 13 Western Ireland, Blanket mires. Pristine mire site, though affected by grazing. Column 14 Western Ireland. Blanket mires. Regeneration in cutaways after 10–60 years.

The pristine bog sites have been chosen to show the main range of change in the makeup of ombrotrophic peats, passing from the central continent to the atlantic seaboard.

Source: All data are from Bellamy (1971) and Bellamy & Bellamy (1967).

(a)

(b)

Fig. 7.11. The Leegmoor (Lower Saxony) comprises some 300 ha of the once extensive raised bogs of the North German Plain. The site has been extensively cut for peat, leaving a depth of only 50 cm of well-humidified peat in places. Operations have started in parts of the site to try and encourage the growth of bog species by raising the water table by damming ditches and creating lagoons to divide up the large areas of bare peat, using low walls of peat to retain the water. (a) Shows how in some areas, the surface of the peat has been ploughed in an experimental attempt to simulate the 'hummock–hollow' microtopography of typical raised mires. (b) Shows the initial recolonisation of a re-wetted area by such species as *Sphagnum cuspidatum, Eriophorum angustifolium* and *E. vaginatum*. (From S.C. Shaw, University of Sheffield.)

plan developed for appropriate management regimes to suit them. For many species these are not known, hence a variety of regimes in a site provides a useful safeguard, and allows results to be evaluated scientifically.

Within a reedbed, all the habitat components from open water through all the successional stages to carr woodland have value for different species of wildlife. In order to maintain the full range of wildlife, it is important that a diversity of management is employed on the site to maintain each stage in the succession.

Of principal importance is the maintenance of the reedbed proper, and appropriate water management techniques must be employed to prevent the rapid encroachment of weeds and the subsequent development of carr. The maintenance of an adequate ditch system, both for its intrinsic wildlife value and for water transport, managed on rotation will be required. Open water bodies increase the biological value of reedbeds and the margins of both these and ditches will require periodic management to prevent reed encroachment reducing the area of open water. The optimum area of open water on a site is believed to be approximately 15%.

In order to maintain a reedbed and prevent excessive litter build-up, appropriate cutting regimes should be employed. Although some areas can be managed on long rotation, thus developing substantial areas of litter, more regular cutting will need to be employed over much of the site. Some areas may be annually cut and some biennially cut but the same regimes must not be employed over the whole site. It is important to remove cut material from the site. Commercial sale will reduce the cost of managing the site.

Carr is a valuable habitat, but can rapidly take over a site. Management by cutting and treating stumps should take place to maintain it on 15–20% of the site, principally around the margins.

Burning old reed, and lowering the soil level of derelict reedbeds may help to renovate the site. These techniques should be used judiciously.

Creation of substantial reedbeds is in its infancy but basic guidelines can be given. Ground-forming is likely to be critical if management resources are limited. New beds should not be created on land with existing high conservation interest and must be sited where hydrological conditions are suitable.

Fens

Fens are grouped into two types, poor and rich, by their hydrology, water chemistry and mode of formation. Vegetation communities vary between the two types, the former typically holding *Sphagnum* communities and the latter a more diverse range of communities. Management options will depend on the communities in question.

Appropriate water management techniques must be employed, if fens, particularly poor fens, are not to dry out. Some communities depend on high stable water regimes, although others require drier conditions. Water passage is important to fen communities and dykes and other waterbodies are important in addition to sub-surface flow. Dyke and open waterbody management is similar to that for reedbeds above.

Vegetation management is required to help restrict invasion of woody plants and succession to scrub. Grazing or mowing may be used. Grazing is used on more sites and may create a more diverse sward. Cattle are principally used, although sheep and horses are alternatives. Numbers should be low initially until their poaching effects have been assessed. Mowing in summer will also be valuable, but material should not be allowed to remain on the ground. It is best to follow traditional regimes where possible and cutting can be followed by aftermath grazing.

Like in reedbeds, carr forms a valuable habitat but can take over a site. Management is similar to reedbeds.

More research is required into the effects of burning on fens, but it is possible that this may be a useful restoration tool.

Small-scale peat cutting may provide areas of open water which can undergo the natural colonisation process as part of the hydrosere.

Acid bogs

Acid bogs fall into two main types, raised and blanket, based on their hydrology and mode of formation. Vegetation at the surface layer receives its total water supply and nutrient input from rainfall. Vegetation communities are typically dominated by *Sphagnum*.

Management options will depend mainly on whether the site has been disturbed (drained, planted with trees, or subjected to high grazing pressure), or whether it remains undisturbed. Most undisturbed sites require little management and should be left completely alone as far as possible. In comparison, disturbed sites may require considerable management effort to restore their hydrology and vegetation type to that of acid bog.

Acid bogs should have water tables at or just above the surface during the winter and no more than 10 cm below the surface during the summer. The water should be acidic and nutrient poor. Ensuring that the sites remain a complete hydrological unit and that drainage does not occur should maintain the hydrological condition.

Vegetation management may help maintain certain communities, particularly those containing Heather *Calluna vulgaris* on blanket bogs. Light grazing by sheep or ponies is the normal practice, and this grazing should follow traditional summer-grazing regimes.

Invasion of acid bogs by trees is normally a symptom of disturbance to the bog. Removal of these trees will help to restore the height of the water table, but preventing water from escaping from the bog by blocking drains and preventing further disturbance will probably be required as well. Trees can be removed from the surface with little disturbance, but special methods have to be employed and several years of follow-up work such as hand-pulling of seedlings will be required to eradicate all trees.

Drained sites can be restored to bog. Firstly the peat must be waterlogged and the chemical processes in the peat switched from aerobic to anaerobic. This normally requires that all drains from the site are blocked, all cracks in the peat are filled, and that substantial quantities of water are pumped onto the site. Once achieved, invasion by *Sphagnum* species can be rapid and the bog can start to regenerate. Digging pools into the peat may also provide a location where *Sphagnum* species can colonise and start to reclaim drier areas of peat. In all cases, although it may be possible to re-create vegetation communities of acid bogs, re-creating the invertebrate communities may be impossible. The fact that it may be possible to recreate bog vegetation should never be used as justification for the drainage or destruction of a pristine site.

More research is required into many aspects of acid bog management and restoration.

References

Andrews, J. & Ward, D. (1991). The management and creation of reedbeds, especially for rare birds. *British Wildlife* 3, 81–91.

Batten, L.A., Bibby, C.J., Clement, P., Elliott, G.D. & Porter, R.F. (1990). *Red Data Birds in Britain*. London: Poyser.

Beets, C.P. (1992). The relation between the area of open water in bog remnants and storage capacity with resulting guidelines for bog restoration. In *Peatland Ecosystems and Man*. Dundee: Department of Biological Science, University of Dundee.

Bellamy, D.J. & Bellamy, S.R. (1967). An ecological approach to the classification of the lowland mires of Ireland. *Proceedings of the Royal Irish Academy* 65, Ser. B, No 6.

Bellamy, D.J. (1971). An ecological approach to the classification of European Mires. *Proceedings, 3rd International Peat Congress, Quebec*, pp. 74–7.

Bibby, C.J. & Lunn, J. (1982). Conservation of reedbeds and their avifauna in England and Wales. *Biological Conservation* 23, 167–86.

Bratton, J. (ed). (1991). *British Red Data Books. 3. Invertebrates other than insects*. Peterborough: Joint Nature Conservation Committee.

Brookes, A. (1981). *Waterways and Wetlands: a practical conservation handbook*. London: British Trust for Conservation Volunteers.

Brown, A., Mathur, S.P. & Kushner, D.J. (1989). An ombrotrophic bog as a methane reservoir. *Global Biochemical Cycles* 3, 205–13.

Burgess, N.D. & Evans, C.E. (1989). *The management of reedbeds for birds*. RSPB Management Case Study. Sandy: RSPB.

Burgess, N.D. & Hirons, G.J.M. (1990). *Techniques of hydrological management at coastal lagoons and lowland wet grasslands on RSPB reserves*. RSPB Management Case Study. Sandy: RSPB.

Burgess, N.D. & Hirons, G.J.M. (1992). Creation and management of artificial nesting sites for wetland birds. *Journal of Environmental Management* 34, 285–95.

Coulson, J.C. (1992). Animal communities of peatland and the impact of man. In *Peatland Ecosystems and Man*. Dundee: Department of Biological Sciences, University of Dundee.

Cowie, N.R., Sutherland, W.J., Ditlhago, M.K.M. & James, R. (1993). The effects of conservation management of reedbeds. II. The flora and litter disappearance. *Journal of Applied Ecology*, **29**, 277–284.

Dierssen, K. (1992). Peatland vegetation and the impact of man. In *Peatland Ecosystems and Man*. Dundee: Department of Biological Science, University of Dundee.

Ditlhago, M.K.M., James, R., Laurence, B.R. & Sutherland, W.J. (1993). The effects of conservation management of reedbeds. I. The invertebrates. *Journal of Applied Ecology*, **29, 265–276**.

Egglesman, R. (1988). Re-wetting for protection and renaturation. *Proceedings, VIIIth International Peat Conference, Leningrad* 3, 251–60.

Ekstam, B., Graneli, W. & Weisner, S. (1992). Establishment of reedbeds. In *Reedbeds for Wildlife*, ed. D. Ward, pp. 3–19. Sandy: University of Bristol/RSPB.

Fojt, W. (1989). *Quick reference to fen vegetation communities*. Nature Conservancy CSD note No. 49. Peterborough: Nature Conservancy Council.

Fojt, W. & Foster, A. (1992). Botanical and invertebrate aspects of reedbeds, their ecological requirements and conservation significance. In *Reedbeds for Wildlife*, ed. D. Ward, pp. 49–56. Sandy: University of Bristol/RSPB.

Gryseels, M. (1989a). Nature management experiments in a derelict reedmarsh. I. Effects of winter cutting. *Biological Conservation* 47, 171–93.

Gryseels, M. (1989b). Nature management experiments in a derelict reedmarsh. II. Effects of summer mowing. *Biological Conservation* 48, 85–99.

Haslam, S.M. (1972a). Biological flora of the British Isles *Phragmites communis* Trin. *Journal of Ecology* **60**, 585–610.

Haslam, S.M. (1972b). *The Reed*, 2nd edition. Norwich: Norfolk Reed Growers Association.

Haslam, S.M. (1973). The management of British wetlands. I. Economic and amenity use. *Journal of Environmental Management* 1, 303–20.

Hudson, P.J. (1992). Herbivore management on ombrogenous mires and dry dwarf-shrub heaths. In: *Peatland Ecosystems and Man*. Dundee: Department of Biological Sciences, University of Dundee.

Joosten, J.H.J. (1992). Bog regeneration in the Netherlands: a review. In *Peatland Ecosystems and Man*. Dundee: Department of Biological Science, University of Dundee.

Jowitt, A.J.D. & Perrow, M.R. (in press). The effects of management in beds of Common reed (*Phragmites australis*), upon small mammal populations; with special reference to Water shrew (*Neomys fodiens*). *Mammal Review*.

Kirby, P. (1992). *Habitat Management for Invertebrates: a Practical Handbook*. Peterborough: JNCC/RSPB/National Power.

Kulczynski, S. (1949). Peat bogs of Polesie. *Memoires Academie Science Cracovie, Series B*, 1–365.

Mackenzie, S. (1992). The impact of catchment liming on blanket bogs. In: *Peatland Ecosystems and Man*. Dundee: Department of Biological Sciences, University of Dundee.

NCC (1989). *Guidelines for selection of biological SSSIs*. Peterborough: Nature Conservancy Council.

Nilsson, L., Nilsson, P. & Sandberg, H. (1988). Effekter au vasskard pa den hockande fagelfaunan i Takern. *Var Fagelvarld* 47, 310–19.

Perrow, M.R. & Jowitt, A.J.D. (in press). The small mammal community in beds of Common

Reed (*Phragmites australis*), with special reference to Harvest Mouse (*Micromys minutus*). *Mammal Review*.

Reiley, J.O. & Page, S.E. (1990). *Ecology of Plant Communities*. Harlow: Longman.

Rodwell, J. (1984). National Vegetation Classification. Swamps and tall-herb fens. University of Lancaster, unpublished report to the Nature Conservancy Council.

Rodwell, J. (ed.) (1991). *British Plant Communities. Mires and Heaths*. Cambridge: Cambridge University Press.

Rogers S.A. & Bellamy, D.J. (1972). Peat exploitaton and conservation, a case history, *Proceedings 4th International Peat Congress, Helsinki*, pp. 219–32.

Rose, F. (1953). Valley bogs. *Proceedings of the Linnean Society*, London **164**, 186–212.

Rowell, T.A. (1986). The history of drainage at Wicken Fen, Cambs, England and its relevance for conservation. *Biological Conservation* **35**, 111–42.

Rowell, T.A. (1988). *The Peatland Management Handbook*. Peterborough: Nature Conservancy Council.

Schimming, C.G. & Blume, H.P. (1987). Anthropogene Bodenversauerung in Schleswig-Holstein. *Mitteilungen der Deutschen Bodenkundlichen Gesellschaft* **55**, 415–20.

Scott, D. (ed.) (1982). *Managing Wetlands and their Birds*. Slimbridge: International Waterfowl Research Bureau.

Shirt, D.B. (1987). *British Red Data Book. 2. Insects*. Peterborough: Nature Conservancy Council.

Smart, P.J. Wheeler, D.B., Willis, A.J. (1986). Plants and peat cuttings. Historical ecology of a much exploited Peatland. *New Phytologist* **104**, 731–48.

Tyler, G. (1992). Requirements of birds in reedbeds. In *Reedbeds for Wildlife*, ed. D. Ward, pp. 57–62. Sandy: University of Bristol/RSPB.

van der Toorn, J. & Mook, J.H. (1982). The influence of environmental factors and management on stands of *Phragmites australis*. I. Effects of burning, frost and insect damage on shoot density and shoot size. *Journal of Applied Ecology* **19**, 477–500.

Vasander, H., Leivo, A. & Tanninen, T. (1992). Rehabilitation of drained peat areas in Southern Finland. In *Peatland Ecosystems and Man*. Dundee: Department of Biological Science, University of Dundee.

Ward, D. (1992a). Management of reedbeds for wildlife. In *Reedbeds for Wildlife*, ed. D. Ward, pp. 65–77. Sandy: University of Bristol/RSPB.

Ward, D. (ed.) (1992b). *Reedbeds for Wildlife*. Sandy: University of Bristol, RSPB.

Ward, R.C. & Robinson, M. (1989). *Principles of Hydrology*, 3rd edition. Maidenhead: McGraw-Hill.

Wheeler, B.D. (1980a). Plant communities of rich-fen systems in England and Wales. I. Introduction. Tall sedge and reed commitments. *Journal of Ecology* **68**, 365–95.

Wheeler, B.D. (1980b). Plant communities of rich-fen systems in England and Wales. III. Fen meadow, fen grassland and fen woodland communities. *Journal of Ecology* **68**, 761–88.

Wheeler, B.D. (1984). *British Fens: a review*. In: *European Mires*, ed. P.D. Moore, pp. 237–81. London: Academic Press.

Wheeler, B.D. (1988). Species richness, species rarity and conservation evaluation of rich fen vegetation in lowland England and Wales. *Journal of Applied Ecology* **25**, 331–53.

Wheeler, B.D. (1992). Integrating wildlife with commercial uses. In *Reedbeds for Wildlife*, ed. D. Ward, pp. 79–89. Sandy: University of Bristol/RSPB.

Wheeler, B.D. & Giller, K.E. (1982). Species richness of herbaceous fen vegetation in

Broadland, Norfolk in relationship to the quality of above ground material. *Journal of Ecology* **70**, 179–200.

Wheeler, B.D. & Shaw, S.C. (1987). *Comparative survey of habitat conditions and management characteristics of herbaceous rich-fen vegetation types.* C.S.D. Research Report No. 764. Peterborough: Nature Conservancy Council.

Grasslands

MALCOLM AUSDEN AND JO TREWEEK

Introduction

The existence of virtually all grasslands in Britain is dependent on some process that prevents their succession to scrub and woodland. Before the influence of people, grasslands would largely have been restricted to areas where woody plants could not grow: on thin or infertile soils (e.g. sugar limestone in Upper Teesdale), where environmental conditions were harsh (as on high ground, cliffs or in very dry conditions such as in Breckland) or in some heavily disturbed areas. From neolithic times (4000–5000 years BP) onwards, clearance of woodland for farming would have allowed plants and animals from the existing grasslands to extend their range. Grazing, cutting and burning of the newly created agricultural grasslands arrested succession and in doing so, maintained suitable conditions for naturally occurring grassland plants and animals and a number of introduced species.

From the nineteenth century onwards, large losses of permanent grassland occurred as a result of the Enclosure Acts and agricultural development which allowed easier establishment of temporary grasslands. Since the 1940s agricultural intensification has accelerated this process. Most old grasslands have now either been ploughed up and converted to arable, or re-seeded with more agriculturally productive species. Of those remaining, the majority have been agriculturally improved using inorganic fertilisers and herbicides with the resultant loss of much of their wildlife interest. To be effective, most inorganic fertilisers have to be applied to free-draining and neutral soils. Drainage and neutralising of acidic soils using lime or marl has therefore usually accompanied their application and these have also usually proved deleterious to wildlife. More vigorous grass growth has allowed earlier mowing which, together with the increased cutting speed of modern mowing machines, has been particularly detrimental to ground-nesting birds.

Grassland communities

Different combinations of environmental conditions and management have created a wide range of grasslands in Britain. These can be categorised on the basis of their soil (acidic, mesotrophic or calcareous) and drainage status ('wet' or 'dry'). Grasslands can also be described as 'improved', 'semi-improved' or 'unimproved' depending on their intensity of agricultural improvement. A further important distinction is between 'meadow' which is cut in the growing season and may be grazed outside it, and 'pasture' that is usually open to grazing all year round.

Acid grasslands

These develop on nutrient-poor soils of low pH and are generally less rich in plant species than mesotrophic and calcareous grasslands. They are widespread over siliceous rocks and peat in the uplands of the north and west of Britain where they form the bulk of rough grazing land. The majority of these upland acidic grasslands are open to grazing all year round, mainly by sheep. Some areas are summer grazed by cattle, although this practice seems much less widespread than formerly. Burning has traditionally also been used in upland areas to burn off dead and unpalatable species (particularly Purple Moor-grass *Molinia caerulea*), to provide a flush of young, palatable grass for grazing.

More species-rich acid grasslands are scattered throughout the warm and dry lowlands of England and Wales. These contain a greater proportion of ephemerals than the upland acid grasslands. They occur on sandy and gravelly soils and are often associated with heathland. The most extensive stands are in Breckland. Also associated with heathland are species-poor grasslands dominated by Wavy Hair-grass *Deschampsia flexuosa*. These are distributed throughout the moderately oceanic parts of the lowlands and upland fringes on free-draining to moderately moist soils. On moister soils in central southern and south western England and usually associated with wet or humid heathland are species-poor grasslands dominated by Bristle Bent *Agrostis curtisii*. These often develop in response to burning and grazing.

Acidic grasslands are poor both in numbers of species and densities of birds. In the uplands they contain breeding Snipe *Gallinago gallinago*, Redshank *Tringa totanus*, Curlew *Numenius arquata*, Lapwing *Vanellus vanellus* and Wheatear *Oenanthe oenanthe*. They are also used as feeding areas by breeding Golden Plover *Pluvialis apricaria* and Merlin *Falco columbarius*. In lowland areas, very short grassland holds relict populations of Woodlarks *Lullula arborea* and Stone Curlews *Burhinus oedicnemus*. Such areas may also provide suitable conditions for a range of scarce, warmth-loving invertebrates at the north-western edge of their European range. The Breckland grass heaths are of particular interest in this respect.

Mesotrophic grasslands

Mesotrophic grasslands are probably best described as grasslands comprising plant species with a preference for soils that are neither too acid nor too basic. They mainly occur on lowland clays and loams of acid to neutral pH and can also develop as a result of a build-up of nutrients following relaxation or cessation of grazing on the more mesotrophic forms of chalk grassland. They have also been extensively created through reseeding acid and calcareous grassland. Mesotrophic grasslands have traditionally been managed as meadow or pasture. Being more or less confined to more fertile soils in the lowlands, a very large proportion have been converted to arable, reseeded or agriculturally improved in other ways.

Hay meadows are left ungrazed from spring (February or March in the lowlands and April or May in the uplands) so the vegetation can be cut for hay between June and August. The regrowth following cutting may be grazed. The precise timing of cutting and details of after-grazing vary between sites and with local customs. In traditional systems loss of nutrients through removal of the hay was compensated for by application of a light dressing of manure in early spring. As a good hay crop was essential for maintaining stock through the winter, hay meadows were usually sited on the best land on the farm, particularly on the rich soils of river flood plains. Hence on farms with little good land, the same fields are likely to have been managed as hay meadows for many years. On farms with a greater proportion of good land, the positioning of hay meadows will not have been so restricted, and hay meadows on these farms are less likely to have had such a long, continuous management history.

Hay meadows also existed at sites where common rights prevented other forms of land use. The most famous of these are on 'lammas lands', where grazing is restricted to the period between Lammas (12 August) and Lady Day (12 February), and the meadow then left to grow for hay. An important feature of such sites is that they are likely to have had consistent management since the rights were introduced in medieval times or earlier.

Unimproved hay meadows usually have a rich and diverse flora comprising a large number of grass species and often containing a large proportion of herbs. Lowland hay meadows provide an important habitat for Fritillaries *Fritillaria meleagris* and several rare species of dandelion *Taraxacum* spp. The absence of grazing during the spring and early summer allows many plants to flower, providing a colourful spectacle that, to many people, symbolises traditional countryside management.

Agriculturally improved or reseeded swards are typically dominated by a small number of highly productive and nutritious species, typically Perennial Rye-grass *Lolium perenne* with other often highly selected strains of grasses and clover *Trifolium* spp. They can be temporary (leys) or permanent (pastures). The grassland may be mown throughout the growing season for silage production, or open to grazing throughout the year. Agricultural management seeks to maintain the dominance of the agriculturally productive species and suppress others, with the result that well-maintained grasslands of this type are inevitably floristically uniform and poor. The invertebrate fauna consequently comprises mainly common and widespread species of little conservation interest. Similarly, the breeding birds of such grasslands do not include the scarcer species found on other grasslands. The grass on leys grows vigorously and the consequently early and repeated cutting destroys bird nests and young.

During winter, improved pastures are used by Greylag *Anser anser*, Eurasian White-fronted, *A. a. albifrons*, Pink-footed *A. brachyrhynchus*, Canada *Branta canadensis*, Brent *B. bernicla* and Barnacle Geese *B. leucopsis*. This habit has become more widespread in recent years as numbers of geese have increased, and has caused conflict with agriculture in localised areas.

Coarse ungrazed grasslands are tall and tussocky. They are typically dominated by False Oat-grass *Arrhenatherum elatius* and contain a variety of tall, bulky perennial herbs. Such grasslands are widespread throughout the lowlands on areas that are regularly but infrequently cut and not grazed afterwards (e.g. many roadside verges). They also occur on steep slopes in the uplands of Northern England. Coarse grassland is able to support an abundance of invertebrates, species richness being enhanced according to the richness of the flora.

Lowland wet grasslands comprise those with a high water table (ill-drained permanent pastures) and those subject to periodic flooding (inundation grasslands). The former consist of grasslands made up of moisture-loving or moisture-tolerant species. Inunda-tion grasslands tend to contain a greater proportion of grasses and ruderal species that can establish themselves on areas disturbed by flooding. Most wet grasslands are grazed; only those dry enough to allow machinery on in summer are cut for hay.

Lowland wet grassland is largely confined to low-lying areas with impeded drainage (alluvial flood plains in river valleys and coastal grazing marshes) and to areas specifically flooded as part of their management (water meadows and washlands).

Improved drainage, especially since the 1940s, and changes in the economics of farming have resulted in the conversion of much former wet grassland to arable. The remaining grassland has in most cases been agriculturally improved with the loss of much of its characteristic wetland-associated flora and fauna. The main interest of much of the remaining improved or semi-improved grassland is its wetland birds. Wetter sites, and particularly the mosaics of wet grassland and swamp created in washlands, hold important and characteristic assemblages of breeding waders: Snipe, Lapwing, Redshank and also Curlew in the north and west. The British breeding populations of Ruff *Philomachus pugnax* and Black-tailed Godwit *Limosa limosa* are more or less confined to a few wet grassland sites. The water-filled ditches running through lowland wet grassland are important for breeding wildfowl and for their aquatic plants and invertebrates (see Chapter 5, Rivers, canals and dykes). In winter wet grasslands, particularly flooded washlands, can attract large numbers of wildfowl.

Wet grasslands, especially those which do not flood, are generally rich in inverte-brates. Seepage and spring-fed grasslands can be particularly important. Inundation grasslands have a more limited invertebrate fauna, though often with rarities, since it tends to be a more specialised fauna that can withstand periods of flooding.

Washlands were built during the seventeenth century to control flooding in the reclaimed Norfolk–Cambridgeshire–Lincolnshire fenland. They consisted of a broad, flat area, either between a natural river and an artificial relief channel or between two artificial relief channels, into which water was directed during periods of high river discharge. High earth banks retained the flood water within the washlands, so preventing inundation of surrounding farmland. The flood waters deposited silt on the washlands which enriched the soil and provided lush summer grazing for stock, mainly cattle. Sheep grazing and cutting for hay was also possible in the drier areas.

Continual improvements in drainage have resulted in reductions in the frequency of flooding on most washlands. This has allowed a major part of 11, and a substantial part of another two of the 24 functional washlands existing in the early nineteenth century to be converted to arable (Thomas *et al.*, 1981). The Ouse Washes in Cambridgeshire–Norfolk, the largest (2276 ha) washland built, remains the only extensively flooded pastured washland.

Water meadows were created under a specific form of management developed during the sixteenth century to provide early grazing for sheep on downland. They consisted of meadows on sloping land or valley bottoms with a system of ditches and channels that allowed calcareous spring or river water to be flowed over them during winter. This water enriched the soil with nutrients, especially calcium. It also warmed it, thereby encouraging early grass growth. Another feature of these meadows was that coarser herbage was often hand-weeded to create a more palatable sward. The resulting grassland was diverse and contained a rich variety of wetland herbs. Water meadows were usually grazed by sheep from March until late April and then left for two months to grow up for hay. After cutting, the meadows were either grazed or re-irrigated for a further hay crop.

Water meadows were most widespread in Wiltshire, Dorset and Hampshire and also occurred in the Midlands. They reached their peak of use between 1700 and 1850, but have since largely become redundant.

Distinctive grassland communities have also developed in response to the effects of concentrated grazing, trampling and dunging. Such conditions were formerly more widespread, particularly on greens and commons, where stock congregated around ponds, stream sides and around settlements. Most greens and commons are no longer heavily grazed, and in the lowlands such conditions are now more or less confined to the New Forest where pony grazing is still widespread. These New Forest 'lawns', and associated trampled, wet hollows support an impressive array of characteristic and rare plants such as Pennyroyal *Mentha pulegium*, Coral-necklace *Illecebrum verticillatum*, Ivy-leaved Bellflower *Wahlenbergia hederacea*, Slender Marsh-bedstraw *Galium constrictum* and Small Fleabane *Pulicaria vulgaris*.

Calcareous grasslands

These have developed on soils rich in calcium, typically overlying chalk or limestone. Their soils are shallow with a pH of 7.0–8.4, and because of the porous nature of the underlying rock, are free draining and well-aerated. Calcareous grasslands are usually particularly rich in plant species and more than 60 species are restricted to them in England (Wells, 1969). In England outcrops of chalk are restricted to the south and east as far north as the Yorkshire Wolds. Large areas of chalk grasslands were traditionally managed as sheepwalks, whereby flocks of sheep roamed freely over the grassland during the day and were folded at night on arable land. Dung deposited during the night transferred nutrients from the already impoverished chalk grassland onto the arable. Cattle grazing also occurred in some areas, particularly Wiltshire and Dorset and some grasslands were also managed as Rabbit *Oryctolagus cuniculus* warrens. Following the reduction in numbers of sheep on downland the effects of Rabbit grazing has assumed greater importance.

Much chalk grassland consists of a short, springy and usually closed turf comprising an often very species-rich, intimate mixture of fine-leaved grasses and low-growing herbs. The warm, dry conditions produced by the short vegetation and light, free-draining soils provide suitable conditions for a large number of nationally rare plants and invertebrates at the northern and western limits of their ranges. Most conspicuous among these are a number of species of orchids and butterflies, particularly blues. More than 300 species of flowering plants and 25 species of butterflies are found on chalk grassland (Wells, 1969).

Many areas of chalk grassland were converted from sheepwalk to arable as a result of technical innovations and the Enclosure Acts. Much of the remaining high quality chalk grassland is now confined to steep slopes where ploughing or agricultural improvement has not been possible, and to ancient monuments, military training areas and areas protected by nature conservation legislation. Even on these sites, the reduction or cessation of grazing, particularly since the First World War and following myxomatosis have resulted in much species-rich chalk grassland becoming dominated by species-poor stands of rank grasses, particularly Tor-grass *Brachypodium pinnatum*, Upright Brome *Bromopsis erecta* and False Oat-grass. This has usually resulted in a decrease in plant diversity and a loss of open conditions favoured by many of the rarer invertebrates and birds associated with chalk grassland. These changes are usually rapid, most low-growing perennials and all annuals often disappearing within four years (Wells, 1969). Litter produced by these dominant grasses (particularly that by Tor-grass and Upright Brome which are particularly slow to break down) also appears to inhibit growth of other chalk grassland species. Many areas of unmanaged chalk grasslands have ultimately succeeded to scrub and woodland.

Chalk grassland is poor both in numbers of species and densities of breeding and wintering birds. Lapwings and Wheatears are typical of short grass areas and remnant populations of Stone Curlews still survive at some heavily grazed sites. Densities and diversity of breeding birds increase with an increase in the proportion of scrub, but with loss of those characteristic of open grassland.

Grasslands on stony and highly nutrient-poor soils in the more continental climates in the lowlands (particularly in Breckland) are the closest representatives in Britain to the steppe grasslands of eastern Europe. These grasslands are typically Rabbit or sheep grazed and often have a history of past disturbance. In Breckland, the generally open sward provides a locus for a large number of continental plants rare or absent from the rest of Britain and is also important for invertebrates, Stone Curlews and formerly for Woodlarks.

Calcareous grasslands also occur on limestone in the uplands. Two distinctive, species-rich communities occur on magnesian limestone in eastern Durham and carboniferous limestone in northern England. The latter contains a large number of northern montane, alpine and arctic alpine plant species, some of which have their major or only British locality in this community. These grasslands locally support important invertebrate communities, including colonies of Northern Brown Argus *Aricia artaxerxesi*. These grasslands are maintained by grazing.

Machair is a Gaelic word used to describe pasture on calcareous wind-blown shell sand and other naturally enriched soils. The most extensive areas of machair in Britain are on the Outer Hebrides. This herb-rich grassland has developed through crofting, a system of part-time small-scale farming. Machair is usually sown as a temporary ley in rotation with arable crops, such as oats. In some areas it is more or less permanent and may be open to sheep or cattle grazing all year round or be managed as hay meadow. Machair has traditionally been fertilised with seaweed.

The intimate patchwork of machair, arable crops and marshy areas provided by crofting supports internationally important numbers of breeding waders at some of their highest densities in Western Europe (Fuller *et al.*, 1986). It is the last stronghold of Corncrake *Crex crex* in Britain and is also important for Twite *Carduelis flavirostris*, and on Islay for feeding Chough *Pyrrhocorax pyrrhocorax*. In winter, machair on Coll, Tiree and Islay is used by internationally important numbers of wintering Greenland White-fronted *Anser albifrons flavirostris* and Barnacle Geese and by the indigenous population of Greylag Geese.

Management of grasslands: an overview

Virtually all grasslands in Britain require some mechanism to prevent succession if they are to remain as grassland. Three such methods, all involving the removal of vegetation, have traditionally been used in agriculture: grazing, cutting and burning. These methods form the basis of grassland management, together with soil disturbance, and hydrological management on wet grasslands. Grasslands usually exist in association with other habitats and the maintenance and creation of these and boundary zones between them and the grassland is likely to be important for the grassland flora and fauna. Other environmental features (e.g. differences in topography) are also likely to be important.

The effects of different management techniques depend on the management history. It is useful to distinguish between grasslands that have been:

1. Left unmanaged for a long period. These have usually been colonised by species-poor swards of coarse grasses or scrub, and have consequently often lost much of their conservation interest
2. Previously managed using a different technique to that being introduced
3. Consistently managed using the same technique.

In the first case, management is likely to be aimed at returning the site to a more desirable state (restorative management). In the other two, it will aim to maintain broadly the same desirable conditions (maintenance management). Successful restorative management should be followed by maintenance management.

Sites which have consistently been managed in a similar way for many years have often developed plant and animal communities of high conservation interest (see Chapter 1). At these sites continuation of the former management regime will usually be the most appropriate means of retaining their interest. If it is decided to introduce new management to an area, then in most cases this should only be introduced to a small proportion of the site, to reduce the risk of inappropriate or unsuccessful management causing extinctions of species or suites of species already present.

With the first two categories of grasslands mowing or heavy grazing should not be introduced throughout the site. In the case of grasslands left unmanaged for a long period, although it may be possible to recreate suitable conditions for some plant species that have survived unfavourable conditions either vegetatively or in the seed bank, this is unlikely to be the case with invertebrates. Most invertebrates have to have the correct conditions year after year and much of the critical period of their life-cycle can be the unnoticed immature stages. Hence although there may be relict populations of invertebrates that require the shorter sward produced by mowing or heavy grazing, the main invertebrate fauna that is likely to have developed on the site will be one suited to longer grass (or scrub-edge communities). Therefore partial treatment is best, concentrating mowing or grazing on parts of the site with residual short grass communities. In the second category of grasslands the main wildlife interest is again likely to be reliant on the former management regime.

The techniques described below can be used in combination, and this has often been done in traditional management (e.g. burning followed by grazing of the regrowth).

Management of grasslands: techniques

Grazing

Grazing is a gradual form of vegetation removal, except at high stocking densities, and therefore less likely to cause large-scale irreparable damage than cutting, burning or large-scale soil disturbance.

All grazing animals are to some extent selective in the plants that they eat. This selective defoliation will affect the species composition and structure of the sward. The overall effect of grazing will be to reduce the quantity of more palatable plant species, particularly tall herbs, and allow grazing resistant plants (mainly low-growing, particularly rosette forming herbs, unpalatable species and grasses) to become more frequent. It will also remove the aerial parts of plants, which can be important to specialist (particularly monophagous) insect feeders. The judicious use of the selective feeding habits of grazers can, in many cases, be used to create/maintain a mosaic of different grassland types and heights and can be fine-tuned and adjusted to the requirements at the site at any particular time (Fig. 8.1). Sward species composition and structure will be important to the invertebrate fauna. Structure will be important to grassland birds and mammals.

Fig. 8.1. Temporary electric fencing is ideal for containing grazing animals during rotational or spasmodic grazing. Although heavy grazing for a short period may be beneficial to the flora, care should be taken to ensure that it does not adversely affect the invertebrate fauna. (Lopham Fen, Suffolk Wildlife Trust Reserve, Norfolk; M. Ausden.)

Grazing also creates physical disturbance to the vegetation and soil. This is important in providing suitable conditions for germinating seedlings, especially annuals, and for invertebrates that require bare, sparsely vegetated or disturbed ground. Stone Curlews and Woodlarks also favour disturbed ground. Too much trampling, most likely to occur during winter or in wet conditions, can cause problems of erosion on slopes, encourage thistle *Cirsium* spp. or Common Ragwort *Senecio jacobaea* infestation and cause increases in trampling-tolerant plants (e.g. rosette-forming species). It will also destroy the nests and young of ground-nesting birds, particularly waders (Green, 1986).

Most of the nutrients removed by grazing are returned to the grassland through the deposition of dung and urine. The nature of and distribution of dung will determine where these nutrients are returned. As well as causing local nutrient enrichment, the avoidance of grazing around dung by animals will result in localised variations in grazing intensity and hence vegetation height.

Many invertebrates are associated with the grazing animals themselves – their discarded fur and wool, carrion, and in particular their dung. The latter is especially important for many species of beetles, especially scarabs, and flies, and provides important feeding sites for some insectivorous birds such as Chough, Yellow Wagtail *Motacilla flava* and Pied Wagtail *M. alba*.

The specific effects of grazing will depend on the *method* of grazing (whether rotational, continuous, seasonal or spasmodic), the *type of grazing animal* (sheep, Rabbit, cattle, horses, ponies or others) and the *intensity* of grazing.

Rotational grazing

In most cases, particularly if grazing is being introduced to or altered at a site, the habitat should be divided into compartments (see Chapter 2, Site management planning) and these grazed in rotation. This is especially important for invertebrates which are very sensitive to inappropriate management. It is also a good insurance mechanism against site extinction for the majority of invertebrates whose habitat requirements are poorly known, if at all. Grazing compartments should be delimited so that the entire population of a species, or especially in the case of insects, that of a particular stage in the life cycle, is not confined to any one compartment. For example, compartments on sloping land should run from the top to bottom of the slope, enclosing a range of different conditions, rather than running horizontally across it. The smaller the compartments, the safer the system will be in this respect, but the greater the resources will be needed to run it. The risk of overgrazing is also greater in smaller compartments.

Many phytophagous invertebrates require not only the presence of specific food plants, but also their correct growth form or stage of development. They may also require suitable surrounding microhabitat. All of these will be influenced by the

Fig. 8.2. Some insects, such as this rare Wart-biter Bush Cricket *Decticus verrucivorus* require a variety of different sward heights in close proximity during different stages of their life cycle. Maintenance of such small-scale variation in sward height is best achieved by judicious use of grazing. (A. Cherrill.)

intensity of grazing. Many insects require a mixture of sward heights in close proximity to provide the whole range of conditions needed during different stages of their life cycle (Fig. 8.2). For example some butterfly caterpillars may require their foodplant to be amongst short grassland. As adult butterflies, though, they may require a taller sward for rest and shelter, and in the absence of flowers in heavily grazed areas, also to provide nectar sources. Some invertebrates even require different sward heights during the same stage of their life cycle. Creation of a patchwork of different swards through rotational grazing is therefore likely to be beneficial for the general invertebrate, especially insect, interest of the site.

Continuous grazing
At low stocking densities this will tend to produce a mosaic of tall and short vegetation, especially on larger sites with greater intrinsic sward variation. High stocking rates are more likely to create a uniformly short turf with few flowers, and will generally be poor for invertebrates, especially at smaller sites with little natural sward variation.

Continuous grazing will be most suitable where creating compartments is impractical or resources are limited, and in heterogeneous habitats where the creation of a uniform, short sward is less likely. It may also be appropriate where a uniform short sward is required for a particular species or range of species.

Seasonal grazing

Winter (October–March) grazing will prevent the build-up of vegetation, thereby arresting succession, but still allow plants to flower and set seed. This will benefit invertebrates that require nectar sources or seedheads. At sites where the sward is already short or vegetation production low, winter grazing alone may be sufficient to maintain it in the desired state. In other cases, annual or irregular spring and early summer grazing may also be necessary. Winter grazing is generally considered least damaging to invertebrates as most of these are then in dormant stages or underground. It is the best time to graze chalk grassland when managing for butterflies (BUTT, 1986). Care should be taken not to graze so heavily or extensively that all dead plant stems and seed-heads are removed, as these can be important overwintering sites for insects, particularly micro-moths, beetles, flies and parasitic wasps Hymenoptera (Fry & Lonsdale, 1991).

Grazing during the growing season (early spring to late summer) favours those plant species able to survive and reproduce under repeated defoliation (grasses and low and rosette-forming species) and distasteful plants. It reduces the abundance of palatable tall herbs and species whose populations rely on plants setting seed. Light summer grazing is unlikely to be deleterious to most invertebrates. Heavy summer grazing (April–September) is likely to be especially damaging to invertebrates, especially butterflies, because it creates a short, uniform sward with few flowers (BUTT, 1986). If summer grazing is necessary, then it should be done at a low intensity for as short a period as possible. As with continuous grazing, there are exceptions to this where maintenance of a short, uniform sward is specifically required.

Grazing to reduce the dominance of particular plant species (usually grasses) is likely to be most successful if carried out during their period of maximum growth, when a smaller proportion of the plants' resources are stored underground.

On wet grasslands, to reduce the proportion of wader nests trampled, stock should be introduced as late in the season as possible, preferably not before early June. Little additional benefit will be gained by delaying stock entry after then, except for any late nesting Snipe. Postponement of the start of grazing is likely to benefit Snipe and Redshank most, since they nest later than other waders. It may, however, be detrimental to Lapwings, as it will allow the vegetation to become too tall for them to forage successfully (Green, 1986). Sward suitability for birds establishing territory at the beginning of the season will be influenced by management during the previous year.

Stock will compete with geese and other grazing wildfowl for food in spring and late

Table 8.1. *Estimated risk of wader nest trampling by stock throughout the whole incubation period*

| Grazing density | Estimated percentage of nests trampled | | | |
	Lapwing	Snipe	Black-tailed Godwit	Redshank
Cows/ha				
0.5	9	18	17	24
1.0	18	34	33	43
2.0	33	57	54	67
4.0	59	82	80	91
8.0	79	>95	95	>95
Sheep/ha				
1.0	13	14	13	19
2.0	17	26	25	33
4.0	25	44	42	56
8.0	52	66	65	78
16.0	67	90	89	>95

Source: From Green (1986).

autumn and winter grazing therefore reduces the amount of grass available for them. Hence, although heavy summer grazing should be used to maintain sward quality, heavy grazing later than early October should be avoided (Owen, 1977).

Spasmodic
At some sites, it may only be possible to graze infrequently and for irregular periods. It may then be tempting to make the most of the opportunity and heavily graze the whole site, particularly if it has been previously neglected. Although this can be beneficial to the flora, spasmodic grazing can be extremely damaging to invertebrates, particularly butterflies, which it can easily eliminate (BUTT, 1986).

Type of grazing animal and intensity of grazing

Different types of grazing animals differ in their grazing behaviour and selectivity. These also vary with age and breed. Grazing behaviour and selectivity will influence both the species composition and structure of the sward. Grazing animals also differ in the nature and distribution of their dung and in their trampling intensity. The latter, together with stocking density influences the availability of bare and disturbed ground and the proportion of nests and young of ground-nesting birds trampled (Table 8.1).

Stocking density also affects the quantity of vegetation removed, which will again

Table 8.2. *Livestock units*

	Livestock units (lsu)		Livestock units (lsu)
Cattle		*Sheep*	
Dairy cow	1.00	Ewes:	
Dairy bull	0.65	Light	0.06
Beef cow	0.75	Medium	0.08
Beef bull	0.65	Heavy	0.11
Other cattle (not intensive beef):		Rams	0.08
0–12 months	0.34	Lambs	0.04–0.08
12–24 months	0.65		
Over 24 months	0.80		
Horses	0.80		

Note: There are a number of different schemes in use for calculating livestock units. Hence care should be taken when comparing units calculated using different schemes. This system is from SAC 1990 and based on data from the Ministry of Agriculture, Fisheries and Food's 'Definitions of Terms used in Agricultural Business Management' (HMSO, London).

influence both the species composition and structure of the sward. The effects of different stocking densities are most marked during the growing season.

Stocking densities are usually expressed in 'livestock units' a system that takes into account the quantity of vegetation removed by different ages and types of stock. An example of one system of calculating livestock units is shown in Table 8.2. Suggested approximate 'medium level' grazing intensities for different grassland types are given in Table 8.3. These should be treated as guidelines only as much will depend on the individual characteristics of the site: aspect, climate, soil depth, season, and the dominant plant species.

Although grazing can often be used with great success in restorative management, it is important to remember that most animals will only consume coarse herbage or browse scrub once more palatable vegetation has been exhausted. Hence it is often necessary to graze such areas very intensively.

Characteristics of grazing by different animals

Sheep
Sheep bite vegetation close to the ground and prefer short, fine swards to coarser herbage. Sheep grazing can therefore result in a combination of under- and over-

Table 8.3. *Suggested 'medium level' stocking rates on different grassland types*

Grassland type	Livestock units/ha/year
Acid	25
Dry mesotrophic	50
Wet mesotrophic	
From mid-May to November to create a mosaic of short swards and tussocks for breeding Snipe, Redshank and wildfowl	100–250
Between mid-July and October on winter flooded areas to create a 5–7 cm high sward for breeding Lapwing, Black-tailed Godwit and wintering Wigeon, Bewick's Swans and other wildfowl	120–370
Aftermath grazing following cutting for hay	50–80
Calcareous	30

Source: From Lane (1992) and Tickner and Evans (1991)

grazing. They are therefore generally unsuitable for restorative management. Sheep cause very little trampling except on loose or sandy soils, or on steep slopes.

At low grazing intensity, sheep grazing produces a varied sward structure which is generally good for invertebrates including many of the typical warmth-loving species on chalk grassland. At higher levels, except on tussocky grassland, it produces a tight, springy sward, generally poor for invertebrates and which can encourage Bracken *Pteridium aquilinum* invasion in some areas (Kirby, 1992). Heavy sheep grazing provides the very short conditions preferred by Stone Curlews, Woodlarks and Wheatears, but is less successful than Rabbit grazing in creating these. If some trampling is required a few cattle can be mixed with the sheep. The actions of Rabbits and Moles *Talpa europaea* in disturbing the soil can also help.

Sheep are less suitable for use on wet grassland than cattle as they are more susceptible to disease (particularly Liver Fluke *Fasciola hepatica* and foot rot) on wet areas. Sheep grazing does not provide the varied sward structure preferred by breeding waders, but does produce an even sward favoured by wintering Eurasian White-fronted geese and Wigeon *Anas penelope*.

During the daytime sheep drop their dung in specific areas. At night, though, they often congregate on areas of short (and often botanically interesting) sward. Dung deposited during the night on these areas will cause nutrient enrichment which may be deleterious to the flora.

Sheep only develop a full set of teeth when three years old and then lose them as they get older. Hence their grazing ability can vary greatly with age. In general, middle-aged sheep are better grazers. Two- to three-year-old store sheep are most suitable for winter grazing on exposed sites and can survive with little supplementary feed. Mountain or hill breeds can cope with more coarse vegetation than lowland breeds and are also well suited for use on more exposed sites. In summer, lowland breeds and older sheep can be used. Short-wooled sheep are preferable in areas with Brambles *Rubus* spp. Sheep are usually easier to handle and manage than cattle.

Cattle

Cattle feed by wrapping their large, rasping tongue around the herbage and cutting it between their lower teeth and upper dental pad as they swing their head. They are able to consume coarser herbage than horses and will feed on taller vegetation than sheep. Their feeding technique and ability to knock down and open up tall vegetation makes them ideal for use on neglected sites. In particular, heavy cattle grazing can be used to reduce the dominance of Upright Brome and False Oat-grass on chalk grassland and break up their accumulated litter. They are less successful in reducing the dominance of the more unpalatable Tor-grass, which once established can be difficult to control.

Cattle are selective in the patches of vegetation that they feed on, and this, together with their heavy trampling, tends to produce an uneven sward consisting of patches of short and tall vegetation and disturbed and bare areas. These conditions are generally better for invertebrates than those produced by sheep (Kirby, 1992). In particular, if summer grazing is necessary, then cattle grazing is better for invertebrates, particularly some butterflies, than sheep grazing. At higher densities, cattle grazing produces a shorter, more uniform sward, but still not usually as tight or short (cattle cannot graze vegetation to less than 1 cm high) as that created by sheep grazing. Too much trampling may break up the vegetation mat and damage the structure of the surface layers of the soil (poaching). This is more likely to occur during winter and in wetter areas or on slopes, and is a greater potential problem with cattle than with sheep.

Cattle drop large, moist pats, wherever they are feeding, producing patches of local nutrient enrichment. They avoid feeding around these pats leaving distinctive areas of ungrazed grassland.

On wet grasslands, moderate to heavy cattle grazing can reduce the dominance of Reed Sweet-grass *Glyceria maxima* and Reed Canary-grass *Phalaris arundinacea*, thereby increasing plant species diversity. Extensive stands of Reed Sweet-grass are of little use to wintering wildfowl and provide poor nesting habitat for most wetland birds (Owen & Thomas, 1979; Fuller, 1982; Thomas *et al.*, 1981; Burgess *et al.*, 1990). On drier areas cattle grazing also reduces the dominance of Reed Canary-grass and Common Couch *Elymus repens*, thereby encouraging species more palatable to grazing wildfowl (Owen & Thomas, 1979).

Table 8.4. *Preferred sward conditions of breeding waders on lowland wet grassland*

	Lapwing	Snipe	Black-tailed Godwit	Redshank
Vegetation height in breeding season (cm)	<15	No clear preference 5–80	<15	5–50
Sward structure	No clear preference	Tussocky	No clear preference	Tussocky
Management best suited to achieving required conditions	Cattle or sheep grazing	Cattle grazing	Cattle grazing or mowing	Cattle grazing

Source: From Tickner and Evans (1991), Green (1986, 1988) and O'Brien (*in litt.*)

Most nesting waders prefer a tussocky sward and thus cattle grazing is generally preferable to that by sheep (see Table 8.4). Also, because of their greater size, fewer cattle are needed to produce the same grazing pressure, thereby reducing the risk of trampling to wader nests and young.

Grazing and some poaching of ditch edges by cattle also benefits the ditch edge flora by increasing the area of marshy and muddy margin available for annuals and other less competitive plant species. Trampling also prevents the development of tall, rank vegetation along the ditch edges, and the encroachment into open water species-poor stands of tall monocotyledons such as Common Reed *Phragmites australis*, Reed Sweet-grass and Reed Canary-grass, whose shading is thought to further inhibit development of a diverse aquatic flora (Thomas *et al.*, 1981). Retention of some patches of tall, emergent vegetation and tussocky species at sites liable to winter flooding is desirable, as these trap seeds and invertebrates carried by flood water producing local food concentrations that can be exploited by feeding waterfowl (Thomas, 1982). Creation of open poached ditch edges also provides feeding areas for breeding waders.

The behaviour of cattle varies to some extent between breeds, and particularly between different ages. Bullocks and other young animals frequently run around in excited groups and are probably a greater trampling threat to ground-nesting birds than more sedate mature milkers. This is a particular problem when young animals have just been released from winter quarters. Keeping them in small groups helps to minimise damage. Store animals, fattening beef animals and sucklers will be the most suitable cattle to use at the majority of sites.

Horses and ponies

Horses and ponies can be very selective grazers and browsers and are capable of completely eliminating individual plant species from a site (Kirby, 1992), so considerable care should be taken when introducing them to areas of high botanical interest. Their degree of trampling is intermediate between that of sheep and cattle. At moderate densities horse and pony grazing tends to produce a patchy mosaic of short and tall swards. This may be a problem on sites of uniformly high botanical interest and may also lead to problems of under and overgrazing. In more complex sites or those of lower botanical interest, though, this is likely to create interesting diversity. The structural mosaic produced by moderate horse and pony grazing will in most cases be beneficial to invertebrates, and is particularly good for butterflies (Oates, 1992). At high stocking rates, particularly in summer, horse and pony grazing can create a very short, botanically uninteresting sward interspersed with tall patches of unpalatable species such as Docks *Rumex* spp. and Common Nettle *Urtica dioica*. Such conditions are very poor for invertebrates.

Fig. 8.3. Moderate levels of horse grazing typically produce a mosaic of short, heavily grazed areas, and coarse rank patches. Note also the large number of ant hills in the background, suggesting that the grassland has not been ploughed for many years. Ant hills provide local variations in microhabitat for invertebrates and plants, further enhancing the diversity of the grassland. (Hoe Rough, Norfolk Naturalists Trust Reserve, Norfolk; M. Ausden.)

Different breeds and individuals vary in their grazing abilities and temperament. In general thoroughbreds and half-breeds are often temperamental and difficult to manage. Native pony breeds such as Dartmoor, Exmoor or New Forest are probably best at creating a mosaic of short and tall swards (Oates, 1992). Ponies are particularly good at opening up rank vegetation and browsing scrub that horses, cattle and sheep have left, and can therefore be useful in restorative management (Fry & Lonsdale, 1991).

Horses usually repeatedly drop their dung in the same area, causing problems of local nutrient enrichment. These regular dunging areas are also avoided by grazing horses and ponies, with the result that they often develop stands of rank, unpalatable vegetation such as Common Nettle, thistles and Common Ragwort.

Goats

Although goats will graze swards, they are by preference browsers and consume a wide range of coarse vegetation. Their selective feeding creates a sward of varied structure. If tethered, their attention can be targeted at particular patches of grassland or scrub. Their ability to browse scrub and consume coarse grasses such as Tor-grass makes them particularly useful in restorative management. They can also be used to create mosaics of tall grassland and scrub.

Age and breed are likely to be important in affecting the grazing/browsing abilities of goats. The main problem with goats is containment.

Rabbits

Rabbits are perhaps the most selective grazers of all. They tend to concentrate on areas of short vegetation, ignoring rank swards. At moderate densities, rabbit grazing produces a patchy mosaic of small, closely grazed areas surrounded by taller grassland. At very high densities it tends to reduce the cover of palatable grasses and herbs and increase the prominence of annuals. This may be a problem on species-rich chalk grassland. Their regular dunging sites can cause considerable local enrichment, allowing the patchy growth of species such as Common Nettle, Burdock *Arctium* spp. and Deadly Nightshade *Atropa belladonna*. Their scrapes and burrows provide bare and disturbed ground important for germinating plants and valuable habitat for many invertebrates (Fig. 8.4). Rabbit burrows, dung and carrion all support considerable assemblages of invertebrates (Kirby, 1992).

Rabbit numbers can vary greatly from year to year and this makes rabbit grazing difficult or impossible to manage, with resultant under- or overgrazing a problem. Because Rabbits seldom feed more than 30 m from cover, though, some control of grazing can be achieved by creating or removing patches of scrub (BUTT, 1986). At many sites rabbit grazing alone is unlikely to maintain the sward in the desired state, although it can often make significant contributions on sites also grazed by other stock.

This lack of control and avoidance of rank vegetation makes rabbit grazing unsuitable for restorative management.

The patchy, short sward produced by moderate grazing provides suitable conditions for warmth-loving invertebrates including those typical of chalk grassland, especially butterflies. On chalk grassland heavy Rabbit grazing results in a limited flora and invertebrate fauna and is generally poor for butterflies.

On grasslands where the main conservation interest are the communities of low-growing plants such as lichens and ephemerals (e.g. some Breckland grasslands), rabbit grazing is the preferred option (Dolman & Sutherland, 1992). It is particularly suitable for lichen-rich grasslands. Rabbits tend to avoid eating lichens, and do not damage them by trampling as stock do. This is discussed in more detail in Chapter 10 (Lowland heathland). Rabbit grazing also provides the short sward and bare and sparsely vegetated conditions required by Stone Curlews, Wheatears and Woodlarks.

Mixed grazing

The effects of different grazing animals can be made to complement each other when trying to produce the required conditions at a site (e.g. if heavily trampled patches are desired in an area predominantly grazed by sheep, then a few cattle can be added).

Fig. 8.4. Bare and sparsely vegetated ground, here created by rabbit burrows and scrapes provides suitable microhabitats for many warmth-loving invertebrates such as this Adonis Blue butterfly *Lysandra bellargus*. (Ballard Down, Dorset; M. Ausden.)

However, as different grazing animals will often eat herbage ignored by others, there is often a danger of mixed grazing creating a relatively uniform sward poor both for plants and invertebrates.

Other considerations when grazing

On some sites, particularly those under heavy grazing regimes, supplementary feeding of stock with hay, straw or minerals may be necessary. Supplementary feeding sites should be located away from areas of high conservation interest and particularly from botanically rich turf, because of the resultant poaching, treading in of feed seed and dropping of dung around them. Siting them on adjoining farmland will mean that nutrients are transferred, via dung, off site (see section on nutrients below). For similar reasons, water troughs should also be placed in areas of lower conservation interest. Siting of feeding points and water troughs can sometimes be used to draw stock into otherwise poorly grazed areas. Even changing the access point for cattle can change the pattern of grazing since cattle often spend more time feeding close to these.

Recent concern has been expressed at the use of Ivermectin (Ivomec as an injection or pour-on formulation), a broad spectrum anti-parasitic drug used to treat cattle, sheep, goats and horses. This drug is excreted in the animal's dung, where its residues can reduce the numbers and variety of the dung associated insect fauna (Wall & Strong, 1987; Madsen *et al.*, 1990). Alternative drugs for the treatment of internal nematode parasites in cattle, which do not appear to adversely effect the dung invertebrate fauna are: benzimidazoles, imadazothiazoles and tetrahydropyrimidines. Alternatives to Ivermectin use for treatment of ecto-parasites in cattle are organophosphates and synthetic pyrethroids, although both of these are toxic to certain other invertebrate groups. Residues of organophosphate dichlorvus in horse dung have been found to have an adverse effect on its beetle fauna.

Cutting

Cutting differs from grazing in being a sudden and largely unselective form of vegetation removal. It produces a sward of relatively uniform height and structure, which in the absence of other strong environmental factors, is often of fairly uniform species composition. Compared with grazing, it encourages tall, bulky palatable herbs and other species usually intolerant of grazing, and those that depend on early (before cutting) set seed, rather than low-growing, grazing resistant species. Unlike grazing, cutting does not create areas of bare or disturbed ground unless this is accidentally caused by machinery. In any case the type of disturbance caused by cutting is likely to cause soil compaction or damage to valuable topographical features such as anthills. Cutting also differs from grazing in its effects on nutrients. Cuttings left *in situ* will cause nutrient enrichment, smother smaller plants and prevent seeds from reaching the soil surface and germinating. They should therefore always be removed. Removal of cuttings will also tend to reduce nutrient levels (see section on nutrient levels below).

The sudden and uniform effects of cutting have the potential to decimate invertebrate populations, particularly those of specialist, monophagous species feeding on the aerial parts of plants. The lack of structural diversity and disturbed ground in grasslands managed by cutting make them generally poorer habitats for invertebrates than those managed by grazing.

Cutting can be used in restorative management to reduce the unwanted dominance of particular plant species, especially unpalatable grasses and rushes that are less likely to be controlled by grazing (e.g. Tor-grass and Tufted Hair-grass *Deschampsia cespitosa*). As with grazing, this is most likely to be successful if carried out during their period of maximum growth. On sites with a dense, impoverished fauna, a flail may be the most suitable method of opening up vegetation for less competitive plants.

If it is decided to introduce cutting at a site, then it should in most circumstances be done rotationally as described when considering rotational grazing. Introducing cutting to previously grazed areas is likely to increase their floristic uniformity and make them less attractive in the long term than non-intervention. Its effects can be varied by cutting areas at different frequencies and times of year and to different heights. This will help to reduce the harmful effects on invertebrate populations at the site, and help maintain a variety of different grassland types. For general invertebrate interest, some areas should be cut 2 or 3 times a year, and others on a 2–3 year rotation (Fry & Lonsdale, 1991). If managing chalk grassland for butterflies, a three year rotation with a single cut between October and March is best, with more frequent cutting only used to maintain specific flora, and even then only very selectively in small areas. Cutting to 8–10 cm will do least damage to butterfly populations on chalk grassland, but this may have to be lower if the vegetation does not respond as required (BUTT, 1986). Sward height preferences for different grassland butterflies are given in Table 8.5. As with grazing, care should be taken not to remove dead plant stems and seed heads unnecessarily.

Mown corridors beside paths can provide the shorter herbage preferred by many invertebrate species, particularly butterflies. Cutting to 5–10 cm in early May, late June, mid August and October is suggested for chalk grassland (BUTT, 1986). Again, this should be done rotationally if possible, each side being cut in alternative years (cf. management of woodland rides for butterflies).

If it is decided to cut an area once a year or less, then it will be least damaging to invertebrate populations if this is done in late autumn or early spring. If the main interest of the grassland is its invertebrates, then only a proportion of the sward should be cut each time and then only to 5–10 cm (BUTT, 1986; Fry & Lonsdale, 1991).

Many meadows that have been repeatedly cut at the same time of year over a long period have developed a distinctive flora of high conservation interest. For the reasons mentioned above, these sites are likely to already have a limited invertebrate fauna and so management should concentrate on maintaining the flora. Plant species will differ in

Table 8.5. *Preferred sward heights of grassland butterflies*

Species	Sward height (cm)
Small Skipper *Thymelicus sylvestris*	15+[1]
Essex Skipper *T. lineola*	20+[1]
Lulworth Skipper *T. acteon*	30+[1]
Silver-spotted Skipper *Hesperia comma*	1–4[2]
Large Skipper *Ochlodes venata*	8–20[1]
Dingy Skipper *Erynnis tages*	2–5[1]
Grizzled Skipper *Pyrgus malvae*	1–7
Green Hairstreak *Callophrys rubi*	4–10
Small Copper *Lycaena phlaeas* Brood 1	10–30
Brood 2/3	1–9
Small Blue *Cupido minimus*	4–15
Silver-studded Blue *Plebejus argus*	2–5[1,2]
Brown Argus *Aricia agestis*	2–5
Common Blue *Polyommatus icarus*	4–10
Chalkhill Blue *Lysandra coridon*	2–6[1]
Adonis Blue *L. bellargus*	0.5–2.5[3]
Duke of Burgundy *Hamearis lucina*	5–20
Dark Green Fritillary *Argynnis aglaja*	8–15
Marsh Fritillary *Eurodryas aurinia*	4–15[1]
Wall *Lasiommata megera*	2–10[2]
Marbled White *Melanargia galathea*	5–20
Grayling *Hipparchia semele*	2–6[2]
Gatekeeper *Pyronia tithonus*	10–20
Meadow Brown *Maniola jurtina*	5–10
Small Heath *Coenonympha pamphilus*	2–5
Ringlet *Aphantopus hyperantus*	15–30

[1]Overwinters as egg or larvae on their foodplant and so heavy winter grazing or cutting is likely to be detrimental.
[2]Prefers very sparse turf.
[3]Prefers short turf with patches of bare ground around its foodplant.
Source: From BUTT (1986) and Fry and Lonsdale (1991).

their ability to withstand particular cutting regimes. Hence the flora of a meadow regularly cut at the same time of year over a long period will only contain those plant species that have been able to withstand that particular cutting regime, and which are less likely to be able to survive under another. The same will apply to the meadow's invertebrates. Hence to maintain the interest of such sites, the former management should be continued. In particular, meadows should not be grazed while growing up for hay. Extension of grazing into late spring should be avoided as this can be damaging to the flora and result in a more grass-dominated community. The effects of changes in cutting regimes may take some time to show, as many of the plants are long-lived and do not have to set seed each year.

As with traditionally managed hay meadows, the most appropriate management of machair is maintenance of traditional management (especially cutting for hay rather than silage), with special considerations when cutting fields which have had calling Corncrakes within 200 m of them earlier in the year. In these cases fields should be cut from the inside of the field outwards, or from side to side so that there is continuous cover with the field edge through which the birds can escape. This will reduce the risk of birds being killed during mowing, but will not of course prevent nests from being destroyed. If possible patches of uncut grass should be left around the perimeter or in corners of the field to provide cover for young birds and preferably only a half or a third of the field cut on any one day. Rough cover should also be left to provide cover for Corncrakes in early spring. Cutting should be carried out as late as possible, while still ensuring that the hay crop does not suffer. Similar considerations apply to fields containing nesting waders and Quail *Coturnix coturnix*. If it is not possible to delay, then cutting should be carried out slowly and with vigilance. Fields should not be harrowed to break up the ground or destroy weeds or rolled in spring as these operations will also destroy nests.

Burning

Like cutting, burning is also a sudden process and tends to produce greater floristic uniformity than grazing. It favours plant species best able to withstand the effects of the burn, notably those with perennating structures protected at or below the surface of the ground (e.g. Bristle Bent, Purple Moor-grass and Mat-grass *Nardus stricta*).

Burning is generally considered very detrimental to grassland invertebrates, the destruction of virtually all above ground vegetation having the potential to rapidly eliminate entire populations at a site. It is particularly damaging to relatively immobile groups such as molluscs, which can be very slow to recolonise even small burnt sites (BUTT, 1986; Fry & Lonsdale, 1991; Kirby, 1992; Oates, 1992). The amount of harm done to invertebrates will depend on the intensity of the burn. Quick, shallow burns cause the least damage, and a variable topography will also reduce the detrimental effects by preventing a deep burn over the entire area.

The only acceptable use of burning is in the removal of unpalatable plants and their litter when restoring neglected grassland. However, even then, extreme caution should be used. For example, one of the few populations of the Red Data Book Wart-biter Bush-cricket *Decticus verrucivorus* was eliminated by burning the patches of Tor-grass that the adults used for cover. Nutrients released during burning can encourage undesirable species including Tor-grass, Rosebay Willowherb *Chamerion angustifolium*, Bramble, Bracken and Birch *Betula* spp. (Wells, 1969; Oates, 1992). After burning, the aftermath should be grazed and the grassland returned to more sympathetic maintenance management as soon as possible.

Burning should never be introduced to an area without very serious consideration. Even previously burnt sites will not necessarily have been regularly managed this way, burning often only having been used to reinstate neglected pasture. If burning is used, then it should only ever be done on rotation and never used over an entire site. It should be carried out in a number of scattered small areas, rather than a single large one, to decrease the likelihood of eliminating entire populations of plants or animals, and to increase the rate of recolonisation from surrounding unburnt areas.

It is illegal to burn grassland between 31 March and 1 November in the lowlands, and between 15 April and 1 October in the uplands without a licence from the Ministry of Agriculture, Fisheries and Food (see MAFF, 1992). Such licences are only issued for exceptional reasons. Burning is also likely to be unpopular with the public.

Soil disturbance

Soil disturbance through scraping, rotovating or ploughing removes or buries the topsoil and its flora, and provides bare subsoil or even bedrock for colonisation by plants and animals typical of early successional stages.

In most established permanent grassland, large-scale soil disturbance should be avoided, since it will inevitably destroy the complex association of plants, and therefore also animals, that have developed on the site. However, disturbance on a relatively large scale may be beneficial to the flora and associated invertebrates where the characteristic flora and fauna is typical of early successional stages; for example, on the Breckland grass heaths or in new grasslands on disturbed ground, such as on old mineral workings. More limited disturbance may be beneficial on sites that otherwise receive very little under their present management (e.g. cutting or sheep grazing).

Soil disturbance is likely to be most beneficial to invertebrates on dry, thin gravelly, sandy, acid or calcareous soils (Kirby, 1992). It can also be used in limited areas to provide suitable bare or sparsely vegetated conditions for nesting Stone Curlews.

Non-intervention

This is likely to change the nature of the grassland by allowing more aggressive plant species, particularly grasses, to become dominant and outcompete (particularly by shading out) other species in the sward. It will also allow the build-up of litter which is likely to inhibit many existing plant species and prevent the establishment of seedlings. An increase in the amount of litter is also likely to make the re-establishment of vegetation dependent on nutrient-poor conditions more difficult.

The tall, rank, grass-dominated swards produced through non-intervention will favour a different range of animals to those of shorter swards; in particular invertebrates dependent on aerial structures of plants and those found in grassland litter. The specialist litter fauna can include a number of rare species, particularly where litter has been present for a long period. Representation of even fairly small areas of longer and rank grassland and scrub edges can therefore be of high conservation value for these invertebrates, given due balance within conservation site management. Tall, rank grassland will also provide suitable conditions for small mammals and these will in turn provide prey for owls, Kestrels *Falco tinnunculus*, harriers *Circus* spp. and mammalian predators. At suitable sites on lowland wet grassland, small areas can be left ungrazed to provide suitable nesting habitat for Gadwall *Anas strepera*, Teal *A. crecca*, Mallard *A. platyrhynchos*, Garganey *A. querquedula*, Shoveler *A. clypeata*, Pochard *Aythya ferina* and Tufted Duck *A. fuligula*.

Over a long period, in most grasslands, non-intervention is likely to allow the development of scrub and woodland, with the resultant loss of the grassland. It should therefore not be practised in grassland of high conservation interest. If the tall, rank

Fig. 8.5. With mesotrophic grasslands, non-intervention usually leads to the development of tall rank swards dominated by False oat grass *Arrhenatherum elatius* and tall herbs such as Hogweed *Heracleum sphondylium*. Such grasslands may eventually become invaded by scrub. Although rank grassland may be floristically poor, it can be an important habitat for invertebrates, particularly species dependent on aerial structures of plants and leaf litter. It may also contain high populations of small mammals. Scrub itself can provide shelter, and also be an important food plant and nectar source for many invertebrates. (Hatch End, Middlesex; M. Ausden.)

Fig. 8.6. Large-scale, regular winter flooding of lowland wet grassland, such as here at the Ouse Washes (Cambridgeshire/Norfolk) can attract large numbers of wintering wildfowl, many of which feed on seeds and invertebrates strained from the water by coarse vegetation. (M. Ausden.)

conditions produced by non-intervention are desired, then it is better to consider long-term rotational management with cutting perhaps every five years instead, with maybe some areas left uncut for up to 10 years.

Water levels

Control of water levels on areas of existing lowland wet grassland can be used to make these more suitable for waders and wildfowl (Fig. 8.6). The success of such techniques is likely to be greater on larger sites, and especially if trying to attract wintering wildfowl, particularly those with a history of use by the species. Wintering wildfowl in particular are very susceptible to disturbance, and regular, large numbers of birds are only likely to be attracted if there are nearby sanctuary areas, for roosting. Freedom from wildfowling is important (Owen; 1977 Owen & Thomas, 1979). Field water levels may be controlled by manipulation of ditch water levels, bunding and pumping. Details of these are given in Burgess & Hirons (1990). Water table height will be important in determining the moisture content in the upper layers of the soil, the extent of surface flooding, the invertebrate fauna of both the soil and vegetation, and the nature of the vegetation.

Although terrestrial grassland invertebrates may survive short duration winter flooding, prolonged winter flooding, and especially flooding during summer, is likely to cause serious impoverishment of the terrestrial and semi-terrestrial fauna. In particular, at sites which have not had a long and continuous history of flooding, it is likely to very seriously reduce the biomass of earthworms, an important food of many waders on wet grassland. Equally, drying out of marshy areas will cause impoverishment of the moist ground invertebrate fauna.

The richest invertebrate faunas on wet grassland on peat tend to occur on areas that remain wet or very moist, even during summer drought. Hence water levels should be kept high enough during summer at least to keep the surface moist. Similarly, lateral seepages from adjacent high ground often provide important microhabitats for invertebrates and care should be taken not to drain these.

On peat sites maintaining high water levels in surrounding ditches will keep the upper soil in a moist enough state to allow Snipe and Black-tailed Godwit to probe for food, particularly earthworms (Green, 1986, 1988). On most clay and silt soils the upper soil cannot be kept in a moist enough state by maintaining high ditch water levels. On these sites shallow flooded areas are probably more important as feeding areas for breeding waders.

If managing specifically for breeding waders on peat soils of little importance to invertebrates, water levels should be kept preferably within 20–30 cm of the soil surface in early March and allowed to fall to 40–50 cm below the soil surface from July to September. Allowing the water levels to drop towards the end of the summer (which at most sites is likely to happen anyway), will allow the grassland to be grazed or cut and so maintain the quality of the sward. If suitable conditions are maintained throughout the spring and summer, then the potential nesting season of Snipe, and to a lesser extent, Black-tailed Godwit, can be prolonged, increasing the chances of successful breeding (Green, 1986, 1988).

Creation of shallow (2–20 cm) flooded areas between March and July will provide suitable feeding conditions for Redshank, Black-tailed Godwit and Ruff (Green & Cadbury, 1987; Burgess & Hirons, 1990). Pool margins are also used by feeding Lapwing and their chicks, particularly towards the end of the breeding season when the vegetation has grown too tall for them to forage elsewhere. The most important feeding areas are usually the shallow margins and damp soil surrounding the floods. It is therefore best to create patchy and irregularly shaped floods that have a large edge to area ratio, rather than flooding large, uniform areas of grassland (Fig. 8.7). This can often be achieved by making use of existing variations in topography, and by allowing flooding to spread out from existing shallow field drains. When creating pools care should be taken not to destroy existing wet grassland plant or invertebrate communities of interest. Possible effects that creation of the pool may have on the drainage of adjacent areas should also be considered. Areas of shallow water and wet ditches also

Fig. 8.7. Shallow summer flooding of lowland wet grassland using existing variations in topography to produce plenty of 'edge' habitat, provides ideal conditions for feeding waders. (Elmley RSPB reserve, Kent; M. Ausden.)

benefit breeding wildfowl, many of which preferentially nest within 20 m of water. Open water is also important for the successful fledging of many species of wildfowl (Thomas, 1980).

During winter, shallow (2–20 cm) temporary or permanent flooding can be created to attract wintering wildfowl and waders by making seeds and invertebrates available for them to feed on (Thomas, 1980). More permanent flooding will benefit inundation-tolerant species such as some sedges, docks *Rumex* spp., buttercups *Ranunculus* spp. and persicarias *Persicaria* spp., whose seeds are important foods for Teal, Mallard, Pintail *Anas acuta*, Moorhen *Gallinula chloropus* and, to a lesser extent, Pochard. It will also benefit Marsh Foxtail *Alopecurus geniculatus*, Creeping Bent *Agrostis stolonifera* and Floating Sweet-grass *Glyceria fluitans* favoured by grazing Bewick's Swan *Cygnus columbianus* and Wigeon (Owen & Cadbury, 1975; Thomas, 1982; Owen & Thomas, 1979).

Nutrient status

Most species-rich swards of high botanical interest have developed in nutrient-poor conditions that have reduced growth rates and allowed a rich variety of slow-growing, stress-tolerant plants to coexist. Increasing nutrient levels by the addition of fertilisers, particularly nitrogenous ones, will favour more vigorous species, especially certain grasses, to the detriment of other plants. They are also likely to result in greater litter production. Litter smothers smaller plants and reduces the availability of gaps in the sward available for germination. As with the effects of non-intervention, the likely increase in nutrient levels and organic matter in the soil caused by the build-up of litter may make the re-establishment of species-rich vegetation dependent on nutrient-poor conditions difficult. Some fertilisers are even thought to have a toxic effect on some plants (Hopkins, 1990). Corresponding changes in the composition and structure of the sward will also adversely affect the existing invertebrate fauna. For these reasons, fertilisers should never be added to botanically rich grassland, the only exception being addition of small quantities of manure to fields where this has been traditional practice. Addition of fertilisers has been the major cause of the increasing botanical uniformity of grasslands as a whole.

Addition of lime or marl to acidic soils will also affect the availability of nutrients by changing the exchange balance of clay–humus complexes. Unless lime or marl have traditionally been applied to a site over a long period they should never be added to areas of botanical interest, or to areas from where they may wash into ditches and streams and affect their fauna and flora.

On lowland wet grasslands application of inorganic fertiliser should be avoided since it increases grass growth making it less suitable for breeding waders and wildfowl. A greater grazing intensity is required to remove the additional growth, increasing the risk of nests and young being trampled by stock. Vigorous grasses encouraged by the application of inorganic fertilisers are likely to out-compete those favoured by grazing Bewick's Swan and Wigeon. They will also make the sward less suitable for Greenland White-fronted and Bean geese which prefer tussocky, unimproved grassland. Such vigorous grasses, particularly highly nutritious species such as rye grasses *Lolium* spp., are favoured by other geese species. Fertilisation of improved (and hence already floristically poor) grasslands has been found to make areas more attractive to feeding Eurasian White-fronted, Pink-footed, Brent and Barnacle Geese. This technique, in combination with management to create suitable sward structure, and scaring of these species of geese from surrounding agricultural land, has been used to create alternative feeding areas for geese.

Herbicides and control of unwanted plant species

Herbicide use should be avoided because of its obvious potential to damage the botanical interest of the site. Many plants considered agricultural weeds (e.g. Creeping Thistle *Cirsium arvense*, Spear Thistle *C. vulgare* and Common Ragwort can support important invertebrate faunas, including a variety of moths and picture-winged flies, especially at sites where the plants have been continually present over a long period.

Many of these plants considered to be agricultural weeds are also important nectar sources for bees, butterflies, hoverflies and other insects. If possible, management should act to control rather than eradicate these plants. Where control is required, this preferably should be done by hand-pulling (e.g. Common Ragwort) or by cutting just prior to flowering and again after a month or so (thistles). In many cases large increases in the numbers of these species are usually symptoms of unsuitable management in other respects.

Retention of other features within the grassland

Topographical features

These will create variations in microclimate which may be important to certain plants and invertebrates. In particular, south-facing slopes provided by, for example, old earthworks or quarries can provide suitable microclimates for warmth-loving invertebrates at the edge of their range in Britain. The value of these features will be enhanced if they provide bare or sparsely vegetated areas that warm up more quickly. Ant hills can be particularly important features and may support plant species such as Common Rock-rose *Helianthemum nummularium* absent from the rest of the sward.

Scrub

Although scrub development may sometimes threaten important grassland, scrub can often be an important element in grassland. Clumps of scrub, particularly in exposed sites, can provide valuable shelter from the wind and also create suntraps, both benefiting invertebrates. They may also be foodplants and provide nectar sources. Hawthorn *Crataegus monogyna* is particularly good in both these respects and Bramble is a good nectar source. Wayfaring tree *Viburnum lantana* and Buckthorn *Rhamnus cathartica* have important specialist invertebrate faunas and Goat Willow *Salix caprea* and Blackthorn *Prunus spinosa* are also good for invertebrates. Scrub-edge will provide habitat for a different range of grassland plants and animals than open grassland. For example Duke of Burgundy butterflies *Hamearis lucina* only oviposit on Cowslips *Primula veris* where they occur on the edge of scrub. Scattered scrub provides less shelter, but may still be important as a food source for invertebrates. It may also provide song posts for birds such as Woodlarks. Hedges surrounding grassland are also important in providing shelter, food plants, nectar sources and hibernation and nest sites for invertebrates. They will also help to reduce possible spray drift from neighbouring farmland.

In some cases it may even be worthwhile to plant scrub on grassland of low conservation interest, in which case consideration should be given to existing topographical features which will enhance its sheltering effect. Suitable tree and shrub species are given in Chapter 9 (Farmland).

References

Burgess, N.D. & Hirons, G.J.M. (1990). *Techniques of Hydrological Management at Coastal Lagoons and Lowland Wet Grasslands on RSPB Reserves*. RSPB Management Case Study. Sandy: RSPB.

Burgess, N.D., Evans, C.E. & Thomas, G.J. (1990). Vegetation change on the Ouse Washes wetland, England, 1972–88 and effects on their conservation importance. *Biological Conservation* **53**, 173–89.

Butterflies Under Threat Team (BUTT) (1986). *The Management of Chalk Grassland for Butterflies*. NCC Focus on Nature Conservation No. 17. Peterborough: Nature Conservancy Council.

Dolman, P.M. & Sutherland, W.J. (1992). The ecological changes of Breckland grass heaths and the consequences of management. *Journal of Applied Ecology* **29**, 402–13.

Fry, R. & Lonsdale, D. (ed) (1991). *Habitat Conservation for Insects – a Neglected Green Issue*. The Amateur Entomologist, vol. 21. Orpington, Kent: The Amateur Entomologist's Society.

Fuller, R.J. (1982). *Bird Habitats in Britain*. Calton: Poyser.

Fuller, R.J., Reed, T.M., Buxton, N.E., Webb, A., Williams, T.D. & Pienkowski, M.W. (1986). Populations of breeding waders *Charadrii* and their habitats on the crofting lands of the Outer Hebrides, Scotland. *Biological Conservation* **37**, 333–61.

Green, R.E. (1986). *The Management of lowland wet grassland for breeding waders*. Unpublished report. Sandy: RSPB.

Green, R.E. (1988). Effects of environmental factors on the timing and success of breeding common snipe *Gallinago gallinago* (Aves: Scolopacidae). *Journal of Applied Ecology* **25**, 79–93.

Green, R.E. & Cadbury, C.J. (1987). Breeding waders of lowland wet grasslands. *RSPB Conservation Review* **1**, 10–13.

Hopkins, J.J. (1990). British meadows and pastures. *British Wildlife* **1**, 202–15.

Kirby, P. (1992). *Habitat Management For Invertebrates: a Practical Handbook*. Sandy: RSPB.

Lane, A. (1992). *Practical Conservation: Grasslands, Heaths and Moors*. London: Hodder and Stoughton.

Madsen, M., Neilsen, B.O., Holter, P., Pederson, O.C., Jespersen, J.B., Vagn Jensen, K-M., Nansen, P. & Gronvold, J. (1990). Treating cattle with Ivermectin and effects on the fauna and decomposition of dung pats. *Journal of Applied Ecology* **27**, 1–15.

Ministry of Agriculture, Fisheries and Food. (1992). *The Heather and Grass Burning Code*. London: HMSO.

Oates, M. (1992). Principles of grassland management with special reference to calcareous grassland. *National Trust Views* **16**, 41–6.

Owen, M. (1977). The role of wildfowl refuges on agricultural land in lessening the conflict between farmers and geese in Britain. *Biological Conservation* **11**, 209–22.

Owen, M. & Cadbury, C.J. (1975). The ecology and mortality of swans at the Ouse Washes, England. *Wildfowl* **26**, 31–42.

Owen, M. & Thomas, G.J. (1979). The feeding ecology and conservation of Wigeon wintering at the Ouse Washes, England. *Journal of Applied Ecology* **16**, 795–809.

Scottish Agricultural College (SAC) (1990). *Farm Management Handbook*. Auchincruive, Ayr: The Scottish Agricultural College.

Thomas, G.J. (1980). The ecology of breeding wildfowl at the Ouse Washes, England. *Wildfowl* **32**, 73–88.

Thomas, G.J. (1982). Autumn and winter feeding ecology of waterfowl at the Ouse Washes, England. *Journal of the Zoological Society of London* **197**, 131–72.

Thomas, G.J., Allen, D.A. & Grose, M.P.B. (1981). The demography and flora of the Ouse Washes, England. *Biological Conservation* **21**, 197–229.

Tickner, M.B. & Evans, C.E. (1991). *The Management of lowland wet grasslands on RSPB Reserves*. RSPB Management Case Study. Sandy: RSPB.

Wall, R. & Strong, L. (1987). Environmental consequences of treating cattle with the anti-parasitic drug Ivermectin. *Nature* **327**, 418–21.

Wells, T.C.E. (1969). Botanical aspects of conservation management of chalk grasslands. *Biological Conservation* **2**, 36–44.

9 Farmland

DAVID HILL, JOHN ANDREWS, NICK SOTHERTON AND
JULIET HAWKINS

Introduction

Farmland, for the purposes of this book, refers entirely to tillage or tillage and livestock (mixed) systems in the lowlands, largely below an altitude of 300 m. Upland pastures and reseeded leys, dominated by livestock rather than tillage are considered in Chapter 11. Farmland which is devoted to cereals, other crops and horticulture accounts for 19.6% of the total land surface of Britain, with cereals alone accounting for 14.9%. This area is sixteen times the combined area given over to all Nature Reserves with statutory protection in the UK (Potts, 1991). For the context of management farmland comprises both cropped land and non-cropped land. The non-cropped habitats are generally remnants of those types discussed in other chapters but their management on the farm is fundamental to improving a farm for wildlife. Furthermore, their management is best considered as an integral part of the farmland landscape in a pragmatic and practical manner if the land manager or farmer is to undertake the work within the farm economy.

The ability to drain and till large areas of land with reduced labour, and the parallel increase in chemical fertilisers and pesticide treatments, followed by crop specialisation and associated monoculture, has led to the greatest detrimental impact on wildlife diversity (Table 9.1). Such 'intensification' is defined as the exclusion of non-crop organisms (pests and competitors) leading to the maximum use of land for food production. Intensification has resulted in fragmentation and direct habitat loss such as hedgerow removal, pond infilling, woodland fragmentation, drainage of wet meadows and saltmarsh. Loss of edges, known to be good for wildlife in most circumstances, has been the result of monoculture farming rather than mixed diverse farming. Intensification practices have involved the abandonment of rotations and undersowing, conversion of permanent pasture to temporary leys, use of inorganic fertilisers, a switch from spring sown to autumn sown crops, earlier hay making and increased cutting of grass for silage early in the summer, and the development of new strains of cereals which grow at higher densities under artificial fertilisation, thus shading out conservationally interesting arable weeds. Finally, wildlife impacts have been most noticeably seen as a result of the directly toxic effects of various pesticides. Whilst the worst of these are now banned, many of the new generation pesticides can have indirect effects on wildlife by operating through the food supply. Table 9.1 gives an analysis of changes to farming practice during this intensification period, the respective ecological impacts and the major wildlife responses. Much of farmland management for conservation therefore involves reversing these impacts in a way which has least effect on the farms' main activity – the production of food, thereby integrating good conservation practice and good farming practice.

Farmland communities

Approximately 300 species of flowering plants are known to occur on lowland farmland in Britain. The total for western and central Europe is over 700. A few species are obviously opportunists and generalists, but many cereal weeds (vestigial remnants of steppe and prairie) are now rare due to herbicides, being restricted solely to parts of farmland which have required more limited chemical inputs and perhaps less mechanical tilling. In context, the cereal ecosystem flora in Britain prior to the use of herbicides and inorganic fertilisers probably contained about 17% of all British flowering species, compared with 10% for woodland (Potts, 1991; Wilson, 1990). Species such as Cornflower *Centaurea cyanus*, Corncockle *Agrostemma githago*, Corn Gromwell *Lithospermum arvense*, Narrow-fruited Corn Salad *Valerianella dentata* and Pheasant's Eye *Adonis annua* are rare arable weed species that are targeted for conservation management.

Arthropod communities of farmland, which are closely associated with plant communities, especially weeds, comprise about 1800 or so species (Potts, 1991), although this is very much an approximation. This represents about 7–8% of the known arthropod fauna in the UK. It has been estimated that the number of species associated with farmland, prior to herbicide use from the 1950s onwards, was more than twice that after herbicide use. Most of the reduction has been caused by removal of the food source of arthropods, except probably for Heteroptera (plant bugs), although the removal of hedges (22% or 175 000 km between 1947 and 1985) destroyed overwintering habitat for predatory species, possibly leading to more pest outbreaks. The effect has been further exacerbated by a decline in undersowing, in which cereal crops are used as a nurse crop to establish grass/clover mixed leys.

Few species of birds (see below) have a distribution restricted to farmland; most farmland inhabiting species are those which have broad habitat requirements having originated from the wildwood or from marshland that was never under forest. Traditional farmland countryside in the UK is dominated by varying mosaics of pasture and arable land with an infrastructure of hedges, copses, woodlands, ditches, ponds, larger water bodies and hedgerows. Forest-derived species such as Blackbird

Table 9.1. *Major farming impacts on wildlife*

Change to farming practice	Ecological impact	Major wildlife response
Changing uncropped land to cropped land	Loss of habitat, fragmentation	Reduced invertebrates, small mammals, raptor feeding areas, birds. Vegetation diversity declines
Fragmentation of woods (large woods to small woods)	Loss of habitat, fragmentation, increased edge	Woodland species decline, large raptors decline
Hedgerow removal	Loss of habitat infrastructure, increased field size	Reduced species dispersal and colonisation (?). Decline in small mammals, owls, passerines, overwintering predatory invertebrates. Removal of berry/nut food source for winter thrushes, etc.
Pond infilling	Loss of specialists' habitat	Decline in newts, frogs, aquatic vegetation, aquatic invertebrates, e.g. dragonflies
Drainage of wet meadows	Loss of habitat	Decline in vegetation and invertebrate diversity, breeding waders
Drainage of saltmarsh	Loss of habitat	Loss of plant species, reduction in geese, waders and wildfowl
Crop type, i.e. mixed to monoculture arable	Loss of edge, ecotones, continuity of land cover	Reduced invertebrate diversity. Decline in Lapwings, Skylarks, Stone Curlews
New crops: linseed, flax, lucerne, evening primrose, oilseed rape	Diversification leading to more ecotones. Greater pesticide applications	Some, e.g. oilseed rape, benefit certain invertebrate groups, more insectivorous birds in summer, more seed-eating birds in winter
Permanent pasture to temporary ley	Loss of ancient unploughed habitat	Reduced vegetation diversity, invertebrates, earthworms, Skylarks, Lapwings, thrushes, gulls, Golden Plovers

Farming practice	Effect	Consequence
Crop rotation system to same crop each year	Reduced temporal diversity. Reduced soil structure	Reduced invertebrate, bird, butterfly diversity
Undersowing with grass/clover to no undersowing	Increased need for N inputs	Reduced summer invertebrates
Crop management techniques: Fertiliser applications	Increased yield/area. Denser crop. N leachate. Few manure stacks left standing in stockyards, etc.	Reduced vegetation diversity. Reduced accessibility for birds, e.g. gamebirds. Decline in water course quality.
Timing of cultivation	Autumn sowing rather than spring. Stubbles not left for long periods. Denser crop	Reduced invertebrates of autumn crops, fewer winter birds because lack of stubbles, spilt grain and seed food supplies
Timing of hay making	Earlier denser grass	Reduced feeding accessibility and breeding for waders. Breeding bird casualties.
Silage making	Early grass cutting	Local extinctions of breeding waders, Corncrakes and gamebirds
New strains of cereals	Denser crop. Reduced growing period	Reduced accessibility for feeding birds. Fewer invertebrates
Pesticide application: herbicides, insecticides, fungicides with insecticidal actions	Direct toxicity	Lethal and sublethal doses on raptors
	Reduced food supplies for invertebrates, small mammals, birds. Increased risk of pest outbreaks	Reduced invertebrate abundance and diversity. Reduced survival of bird nestlings and gamebird chicks
Use of break crop to no break crop	Increased uniformity of farming landscape	Reduced diversity of plants and animals. Low soil invertebrate biomass. Reduced winter birds.

Turdus merula account for 80% of farmland bird numbers, with open field species such as Grey Partridge *Perdix perdix*, Lapwing *Vanellus vanellus*, Skylark *Alauda arvensis*, Corn Bunting *Miliaria calandra* and Yellow Wagtail *Motacilla flava* contributing only about 7–14% of bird numbers on lowland farms in Britain. Approximately 130 species breed on farmland with 62% having small populations (less than 100 000 'pairs' – inverted commas are used because some species do not breed simply in pairs so the generic term is not strictly appropriate). Many birds benefit from mixed farming systems (e.g. Rook *Corvus frugilegus*) which provide continuity of food supply, meet a range of needs for nest sites and roosts, and provide diverse feeding opportunities often on the edge of different crop types or where crops abut pasture (O'Connor & Shrubb, 1986).

The underlying geology and climate is of major importance in determining land productivity and therefore different farming and cropping systems, giving rise to regional variation in wildlife communities. These differences affect the opportunities for conservation management on the farm. The more productive lower lying land in the east of Britain is generally dominated by tillage agriculture; that in the north and west is dominated by permanent pasture (though usually improved) and sheep production, although this is a simplification. The recent past trend towards specialisation to one-product enterprises such as dairy *or* cereals *or* sheep has not favoured mixed farming. More recent emphasis on diversification in which the farming risks are spread, could benefit wildlife. On the whole there are more species associated with farmland in the east than in the west, and more in the south than in the north, as a result of farming type, climate and fertility. Western farms receive generally less extreme weather than eastern farms, which favours better survival of small mammals and small birds, but more precipitation and lower summer temperatures disadvantages many invertebrates. For birds, the winter distribution of species-richness is related to the similarity between tilled land and grassland; more species occur where the balance is 50% tilled:50% grass than where there is a dominance of tilled or grass, because of the improved diversity of land cover. This emphasises the great value of shifting cultivation which was formerly common in European steppe vegetation. So, excluding rare bird species, opportunities for enhancement of farmland for birds are greater in the midland–southern counties of Wiltshire, Hampshire, Sussex, Oxfordshire, Northamptonshire, Berkshire and Shropshire (O'Connor & Shrubb, 1986).

Management principles

The major principle for conservation management of farmland is related to maximising biodiversity, that is the diversity of species and groups in the landscape, and conserving specific habitat features used by a few scarce species. Under current farming regimes both intensification and specialisation work against this. However, much past conservation advice has ended up benefiting only the usual and the common. This need not be the case in the future since we have increasing knowledge of the needs of many species, although we are still ignorant about the bulk of invertebrates, some birds and other vertebrates and some plants. When designing farm conservation management, always design for particular species if possible, but always also apply general principles to benefit the unknown species. For example, retain existing habitats in traditional management, increase areas of existing semi-natural habitats, and create structural variation and plant composition variation within new habitats.

As described above only a small number of species have a large proportion of their population restricted to farmland; these include certain rare arable weeds, and birds such as Corn Bunting, Corncrake *Crex crex*, Lapwing, Stone Curlew *Burhinus oedicnemus*, Grey Partridge and Quail *Coturnix coturnix* – species which originated from the vast grassland steppes of former times. For these species specific management prescriptions can be adopted rather than maximising biodiversity on the farm. Other groups which are the focus of habitat management on the farm and for which advice is usually sought are invertebrates (as food for other groups, e.g. birds although they should also be viewed in their own right), small mammals and their predators, seed-eating small birds, gamebirds, and species such as Barn Owl *Tyto alba* which can be encouraged to nest in retained old trees and maintained but isolated barns and other outbuildings.

Maximising biodiversity focuses on the inter-relationships between land used for production (cropped) and marginal areas (non-cropped). The way that the marginal areas are interspersed amongst the productive areas, and the juxtaposition of different crop types, each with different structures (e.g. tall cereals, short vegetable crops) gives rise to the 'landscape mosaic'. Tapper & Barnes (1986) show the importance of landscape pattern and crop juxtaposition for Brown Hares *Lepus europaeus*. The landscape mosaic can be improved for wildlife by increasing crop diversity, making the area of land under tillage and under grass of similar proportions, by mixing livestock within the arable and vegetable regime, together with maintenance and enhancement of all semi-natural and marginal habitats on the farm. Increasing the number of habitat patches in this way also leads to an increase in the amount of 'edge' habitat, and generally edges can support a greater number of species than core areas, although such species tend to be commoner ones and small fragments will not benefit larger species of predators which require large home ranges. The importance of land use and land cover diversity on species richness at the landscape scale has been shown strikingly for birds. Generally the number of bird species on farmland in the south of England increases from about 12 species per 2×2 km square in areas of monoculture cereals with low land cover diversity, to about 40 or so species in areas with the greatest diversity of crop types and semi-natural habitats interspersed on the land.

Much wildlife on farmland depends on the existence of non-cropped habitat such as small woods, hedgerows, ponds, damp meadows and small marshes. Intensification of agriculture has generally led to a break up of existing mosaics by removing these habitats. These habitats have become fragmented. To an extent, some fragmentation can lead to an increase in mosaics and landscape structure. Extreme fragmentation, however, leads to the habitat being isolated. For example, a small wood surrounded by a huge area of cereals will generally be of less value to wildlife than a landscape of cereals broken up by small copses each inter-linked by hedgerows. Populations in fragmented and isolated habitats are not only reduced or subdivided, they are exposed to ecological changes associated with induced edges. Maximising biodiversity described above is also about reducing the fragmentation of existing habitats and indeed reversing this recent history.

Monoculture cereals should be divided up by planting other crops which will also increase the edge to surface area ratio, which incorporates both size and shape variation, thereby benefiting wildlife. Many studies have found that habitat area and number of bird species, for example, are positively correlated. Therefore, given the job of managing a farm for conservation, effort should be put into reducing habitat fragmentation and isolation by planting and creating woods, copses, ponds, hedges and different crop types in a mosaic which will, however, create more edge to encourage a greater species diversity. Consideration should be given to improving landscape structure by the positioning of these, taking account of prevailing nutrient conditions and exposure.

Habitat corridors: are these a myth?

Much emphasis has been placed recently on the importance of habitat corridors such as hedgerows, ditches and dykes, river banks, etc., particularly by groups giving farm conservation advice. Yet there is little information on which to substantiate this claim except for certain small mammals and some invertebrates. The larger species of small mammals are generally less tolerant of habitat fragmentation than are smaller species, but habitat corridors have been shown to allow greater movements of small mammals between patches (Bennet, 1990). Habitat corridors only benefit the most opportunistic species however, although connections between good habitat blocks are seen as an important management objective and one which is often taken up and used in the farmland situation. Connections should be 'made of' the same habitat, e.g. tree planting between woods, grassland between pastures, heath between heath, ditch between ponds. Even then many species are sedentary by habit or physically incapable of moving except over very long timescales.

Hedgerows and ditches are the main example of habitat corridors in farmland, providing connecting infrastructures across the farm. Although not evidence of hedgerows as corridors, they are important for certain groups at certain times of year. They provide overwinter shelter for predatory arthropods. They provide food, shelter and nesting places for birds such as Whitethroat *Sylvia communis*, Lesser Whitethroat *Sylvia curruca*, Linnet *Carduelis cannabina*, Yellowhammer *Emberiza citrinella*, Redwing *Turdus iliacus* and Fieldfare *Turdus pilaris* at various times of the year, and for butterflies. Old hedges with a good physical structure, many shrub species, few or no gaps, relatively wide at the base and with scrubby residual vegetation at the base of the hedge which stays through the winter, support more wildlife than gappy short hedges of a few species, narrow at the base with little or no residual cover. The ways of managing them to this state are discussed in a later section.

Pesticides: herbicides, insecticides, fungicides

Apart from the effects on birds of prey and owls, some years ago, of substances such as DDT and certain rodenticides, the main effect of pesticides on farmland has been to reduce broadleaved plants and the invertebrates which either live on them (so that their food supply is removed) or which are directly killed. This effect works its way through the farmland food chain. Many birds living on farmland eat and depend on invertebrates (either soil, epigeal, or foliage-dwelling), seeds or berries, so their food supply is affected by pesticides. The Game Conservancy Trust's work on Grey Partridge and Pheasant *Phasianus colchicus* has demonstrated how herbicides have reduced the preferred invertebrate food of young chicks by removing weeds which are host plants of their favoured insect prey, leading to lower chick survival. Tree Sparrow *Passer montanus* has also been shown to be affected by reduced food supplies as a result of pesticide use. There is no easy solution. Partial answers in a conventionally farmed system are to leave areas of the crop unsprayed or selectively sprayed, to put part of the land under set-aside, to establish unsprayed grass–wildflower strips, or undersowing crops which builds up the invertebrate community. Some of these prescriptions will be dealt with in the next section.

In addition to these effects there are other, more subtle ones such as chemical treatment of the internal parasites of stock which can cause dung fauna, an important source of food for some other species, to be impoverished.

Management prescriptions

There are seven main areas for which farmers and land managers regularly seek advice about habitat management on the farm. These are, for typical lowland farms: tree and woodland planting, creation and restoration of ponds, hedges, establishment of grass tracks and rides, creating herb-rich meadows, set-aside and management of the cropped area (e.g. use of Conservation Headlands which are areas of the crop nearest the field boundary which receive selective pesticide sprays so as to enable insects eaten by gamebird chicks to survive), and how to apply for government schemes. The last area is outside the scope of this book but it is important to find out what schemes are available before planning habitat management on the farm (see sources of advice below). This section deals with both modifying the cropping regime to benefit wild-life, and the creation or management of non-cropped habitat which includes field margins, beetle banks (non-cropped habitat designed to encourage overwintering by beetles), arable ditches and dykes, hedgerows, small farm woods, and game crops (crops planted for game species). The management of herb-rich meadows is dealt with in Chapter 8.

The first important task is to prepare a farm conservation plan. Both new and old aerial photographs and maps should be pulled together as well as finding out if any rare species occur on the farm. At the earliest stage possible it is vital, yet not often undertaken, to seek archaeological advice from the County Archaeologist about such features as ancient lanes, hedges and boundaries, barrows, cropmarks, deserted villages and moats, and old farm buildings. Management prescriptions may be affected by the presence of important archaeological features and it is important that their location is known.

If there are rare species, the priority, given resources available, should be to concentrate effort on them rather than trying to go for diversity across the farm. The relevant country agency (e.g. English Nature) should be consulted if there are SSSIs on the farm and all areas of conservation interest should be marked on a preliminary site map. Next, the current farming system should be described, together with the distribution of soil types and the location of any rights of way. It is important to know the level of resources available such as the farmer's skills, the machinery he has available and the commitment he has to conservation (does he shoot or is he interested in birds or butterflies for example?). Conservation management may be fitted into slack times on the farm so it is important to know when these are likely to occur.

At the start, preferably on a walk around the farm, it is useful to ask some questions about existing habitats for example: *Farm woodland* – how old is it and has it been coppiced or is it neglected? Does the farm have the necessary machinery to manage it back to something useful for wildlife? Where are all the old trees on the farm? In terms of planting is there a particular problem with any pest species which could influence decisions on tree protection? *Wetland* – where is it and is it isolated such that management could connect it to another non-cropped habitat? Valuable wetland should never be destroyed in order to create a farm pond and one common mistake is to dig out a hole in a meadow which floods – it is more useful as a wet meadow than a pond ever could be. Nor should trees be planted on old wet meadows. *Hedges* – where are the best in terms of age, thickness and scrub species? One large, thick, well-managed hedge may be better for wildlife than two thin, gappy, badly managed hedges. *Grassland* – what is the potential for converting some of the arable to hay and wildflower meadow? Is there existing unimproved pasture or meadow (see Chapter 8).

Fig. 9.1. Some new crops, such as sunflowers shown here, can attract large flocks of finches (in this case, greenfinches) in the late summer prior to harvest. Some areas of the farm such as difficult corners or along farm woodland boundaries should be planted up with these types of crops and left unharvested to provide winter food for birds. Sunflowers are also eaten by badgers, which knock the plants down at the time when the seeds just become ripe. (D.A. Hill.)

Management prescriptions should be considered at three levels. First, the prescribed management should aim to end up with communities which one would expect at that geographical location. It would not be wise to attempt to introduce and manage a type of community alien to the particular location of the farm. Second, at the spatial scale of the landscape, the management prescriptions should fit in to the surrounding landscape. Third, the layout of habitat patches in relation to each other should be drawn up by using aerial photographs and maps, although in practice the farm cropping patterns will determine the layout of habitat patches at the planning stage rather than vice versa. The individual field treatments should be carefully mapped out and each field treated as an individual management block. Table 9.2 shows a basic habitat checklist, together with the habitat's function, natural history and wildlife management required.

Modifying the cropping regime

There are a number of ways to modify the cropping regime including a change to organic farming, growing new and different crops (Fig. 9.1), and changing to mixed

Table 9.2. *A basic habitat checklist for the farm showing its farm function, the farm management applied to it, the groups of organisms which live in it and the conservation management required to enhance and encourage its wildlife potential*

Habitat	Farm function	Farm management	Natural history	Conservation management required
Pasture	Grazing, hay, silage	Weed control, fertilisers and lime, stocking density, farmyard manure, reseeding	Plants, butterflies, other insects, winter thrushes, breeding waders, winter waders	Retain diversity of species by using as little fertiliser and herbicide as possible. Retain damp areas and corners. Prevent colonisation of scrub. Refrain from planting trees on areas with rich flora. Avoid reseeding
Woods, copses	Timber, firewood, etc. Game conservation	Clear felling and replanting. Coppice thinning and standard rotation. Ride/glade maintenance	Plants, butterflies, birds, mammals. Historical and archaeological interest	Plant and/or retain native species. Manage part at a time to allow recolonisation. Use coppice with standard management where suitable. Maintain wide rides. Cut alternate sides each year. See Chapter 12
Hedges	Stock-proof fences, shelter, game conservation	Laying, cutting, coppicing, annual trimming	Plants, butterflies, birds, mammals, reptiles. Old hedges. Historical value	Maintain thick bottoms and reasonable height. Retain hedgerow trees and allow for replacements. Laying/coppicing at regular intervals where appropriate

Feature	Use	Management	Wildlife	Recommendations
Water courses	Drainage, irrigation, stock barriers	Mechanical and chemical weed control. Regular cleaning	Plants, dragonflies, fish, birds, mammals	Manage part at at time to allow recolonisation. Ensure they contain water throughout year. Ensure gentle profiles
Ponds	Drinking stock, irrigation, angling, emergency water, shooting	Occasional dredging or chemical control	Plants, dragonflies, fish, amphibia, birds, mammals	Manage part at a time to allow recolonisation. Ensure they contain unpolluted water throughout year but with periodic draindown. Prevent use as rubbish dumps. Avoid over shading by surrounding trees. Cut back trees on south and west sides
Lanes, roadside verges	Access	Mowing, possible use of herbicides	Plants, butterflies	Cut as late as possible after flowering and remove cuttings if possible. Avoid using herbicides
Farmstead, farm buildings	Housing	Treatment of timber, etc. New buildings	Birds (Swallow, House Martin, Barn Owl, bats)	Leave places for nesting owls, bats, etc. Put up nest boxes where possible. Use timber treatment chemicals which are harmless to bats. Careful use of rodenticides

Source: Adapted from FWAG advisers' notes.

farming. These involve major changes in farming policy by the manager and have important financial implications. In essence the intensity of farming remains the same, but mixed systems are subjectively considered more beneficial to wildlife largely because landscape structure is improved and fragmentation is reduced. The benefits of organic farming remain to be fully validated. The conversion to organic farming usually involves taking up crop rotations, mixed farming, incorporation of crop residues into the soil, and the use of animal and green manures. Short fallows are encouraged for weed control, insoluble mineral fertilisers are used, and crop protection is developed through careful use of rotations, mechanical cultivation, mulching, biological control of pests, use of resistant crop varieties, and restricted use of plant-derived insecticides (although some of the latter can be extremely toxic). In contrast conventional farming relies heavily on high input of artificially produced chemical fertilisers and pesticides. A thorough review of organic farming practices is given in Lampkin (1991).

Many conventional farming practices lead to a reduction, over time, in the amount of organic matter in soils, which therefore has a strong indirect effect on populations of soil organisms. Groups such as earthworms and leatherjackets are important food sources for birds, particularly during winter. Various species of wintering thrushes, waders, gulls and corvids are known to prefer fields with high densities of soil invertebrates, such as old pasture, particularly where farmyard manure is used to fertilise the grass. Cereal stubble and ley fields support moderate densities of birds, with bare till, winter cereals and oil seed rape fields being little used (Tucker, 1992). A large amount of organic matter in soil has effects on soil structural stability, porosity, infiltration of water, consistency, and workability. Organic farming systems promote the use of natural manures rather than artificial fertilisers. Both cattle slurry and pig slurry, once weathered, have been shown to significantly increase the biomass and numbers of earthworms in arable soils when compared with inorganic fertilisers. Grass systems show less of an effect because of the associated high production of root litter when compared with arable systems, such that food is not limiting to earthworms and the application of fertiliser does not lead to increases in numbers or biomass. Herbicides will also reduce the amount of organic matter available to soil invertebrates.

Management of field margins

For the purposes of management we shall consider the field margin as the area between the hedge and the crop; hedgerow management will be considered later. Good, conservationally interesting field margins are comprised of grass strips (for agricultural access and recreation), wildflowers with fine-leaved grasses on light soils or coarse, tufted vegetation on heavier soils, and 'Conservation Headlands'. They are used extensively by a diversity of wildlife groups; they hold many insects beneficial to the farmer as well as pest species.

Grass margins

It is important to first map the field margins on the farm to look at how they form a network over the land. New field margins can be established or existing ones improved. If there is a good seed bank in the soil natural regeneration can be used for establishment. If natural regeneration is not appropriate (poor seed bank), a perennial grass strip can be established by sowing a new field margin. Less vigorous fine grasses such as Red Fescue *Festuca rubra*, Meadow grasses such as *Poa pratensis* (some are agricultural weeds), Crested Dogstail *Cynosurus cristatus* and Timothy *Phleum pratense* can form a single base on light soils. Cocksfoot is a useful species which quickly forms a tussocky sward giving cover to insects and small mammals. Use by insects is much influenced by sward structure and by richness of nectar sources and food plants. If a wildflower margin is developed, use a good local stock of locally occurring species (speak to the local Farming and Wildlife Adviser for details of source). The final objective should be a grass mixture with broadleaved species with a recommended ratio by weight of 85:15 grasses to broadleaves, sown at 25–40 kg/ha. For sowing coarse tufted grass margins a guide might be to use 20% False Oat Grass *Arrhenatherum elatius*, 35% Cocksfoot *Dactylis glomerata*, 35% Tall Fescue *Festuca arundinacea* and 10% Yorkshire Fog *Holcus lanatus* on dry soils. On wetter soils replace 5% of False Oat Grass with Tufted Hair-grass *Deschampsia caespitosa*. Sowing of coarse grass margins should be at 5 kg/ha. Autumn sowing is more successful on heavy soils. On less fertile soils a nurse crop of Westerwold's Rye-grass *Lolium perenne* creates a closed sward which helps the margin in the first season. A fine grass mixture may require 5 cuts per season initially whereas the coarse grasses will require 2 or 3 cuts, although more than 2 or so cuts can have detrimental effects on butterflies.

The general spreading of fertilisers and pesticides should be avoided on all field margins. Weed control where necessary should be by spot treatment with herbicide to affected areas only, following label recommendations. If hedge bottoms contain rampant Sterile Brome *Anisantha sterilis* or Cleavers *Galium aparine*, herbicides could be used in late spring to early summer to remove them prior to establishing a conservation field margin if all other species have been removed by sprays.

On arable/grass rotations there is an opportunity to leave a permanent grass strip to a minimum width of 1 m. This, together with the use of a sterile strip (regularly rotovated or chemically treated strip up to 1 m wide between the field boundary and the crop), will prevent the problem of invasive weeds and provide a clean, easily managed crop edge. On arable field margins the edge is often the source of many pernicious weeds which invade the crop. This is true of sprayed out or disturbed field edges but not where a margin of perennial grasses and broadleaved plants exists. It is vital therefore that pesticide drift does not remove the 'stable' plant community of an established conservation field margin in order that a long-term weed problem is not created. Stable field margins are best created when fields are ploughed, by turning in a furrow towards the field, making a clear and lasting edge to the cropped area. Future ploughing should maintain this edge thus preventing soil being thrown on to the established vegetation. It is recommended that a width of 1 m at least is left undisturbed by cultivations, sprays or fertilisers beside all arable crops, even if conservation headlands are used. The grassy part of the field margin immediately next to the hedge (Figure 9.2) is used as a nest-site by gamebirds and for overwintering by beneficial insects and spiders. It should be at least 1 m wide and one should allow the build-up of dead grass material to provide cover, topping the vegetation every 2–3 years to avoid scrub encroachment. Too much dead grass, however, can suppress broadleaved plants and prevent regeneration resulting in a grass-dominated sward poor for invertebrates.

Maintaining grass margins relies on cutting or grazing. On most soils cuttings should be removed (by grazing or making hay) once the plants have set seed otherwise they will swamp the broadleaved plants, although some insects which overwinter in seed-heads and stems may be lost. Fine grass margins on light soils may be cut in July/August, although two cuts (June and October) may be necessary on very fertile soils, preferably removing the cuttings. Cutting in June, however, runs the risk of damaging ground-nesting birds such as Grey Partridge or Yellow Wagtail and even cutting in July can be damaging to second broods of ground-nesting species. Coarse grass margins require less maintenance, i.e. once every two years they should be flail-mown to prevent scrub encroachment by cutting in July–September to a height of 12 cm which will encourage tussocking.

Sterile or barrier strips

A sterile or barrier strip of about a metre width can be used between the grass strip and crop. Its purpose, if the problem exists, is to prevent invasion of the crop by Cleavers and Sterile Brome. It is maintained by rotovation or residual herbicides in February/early March. Spray drift should be avoided by shielding the spray nozzle down to ground level using a skirt which extends from the ground to at least 28 cm above the boom.

Conservation Headlands

A technique recently developed by The Game Conservancy which has found much support as a result of significant increases in gamebird chick survival, butterflies and other invertebrates, is the 'conservation headland'. The area between the crop-edge and the first tramline (usually 6 m wide according to boom width) is treated with selective pesticides to control grass weeds, Cleavers and diseases whilst allowing most broadleaved weeds and beneficial insects to survive. Ploughing of headlands is recommended especially on heavy soils or where grass weeds are a problem. It is important to avoid turning furrows onto the grassy strip as this area can create ideal conditions for annual weeds. The siting of the Conservation Headland is important:

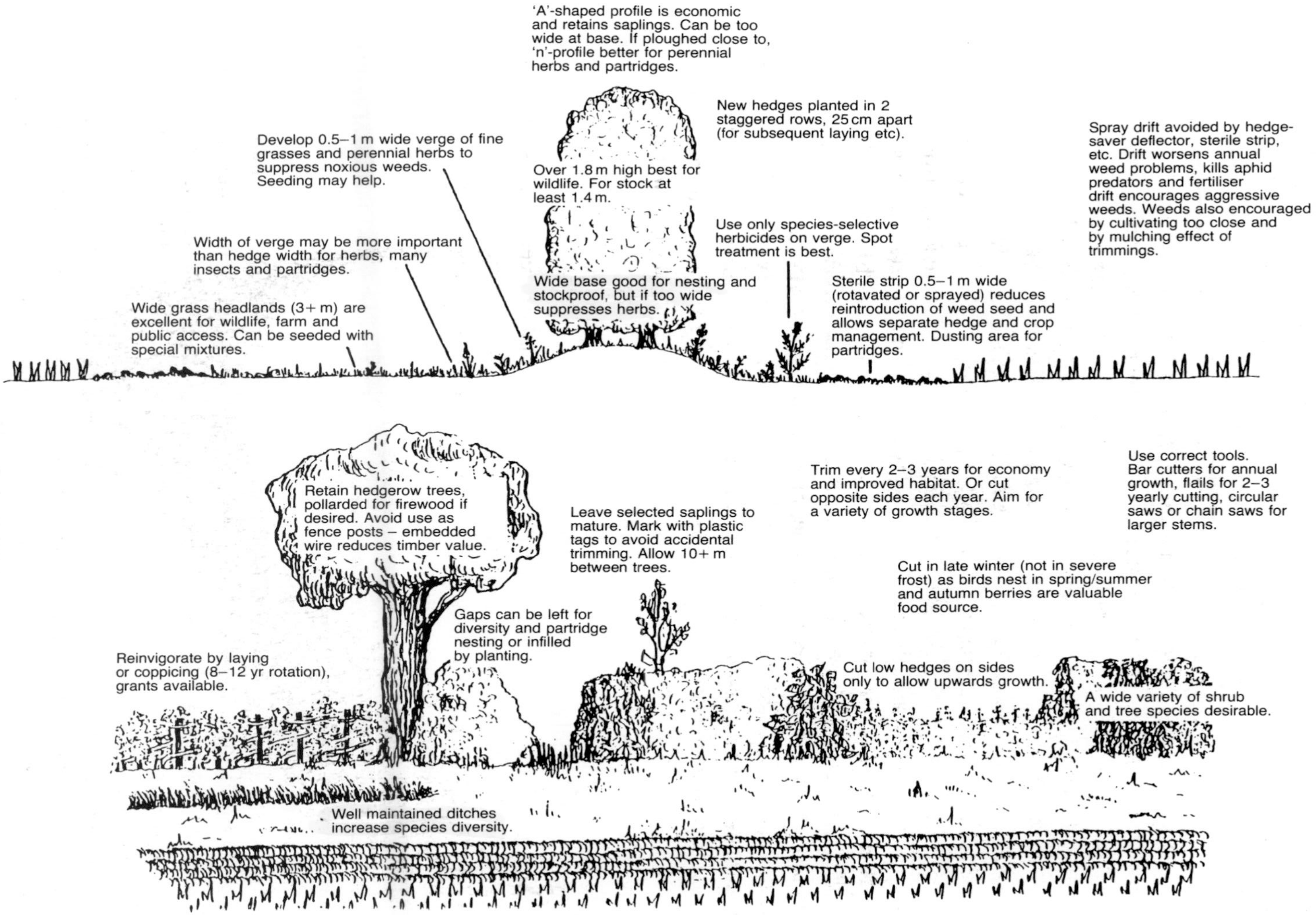

Fig. 9.2. Hedge and field margin management.

'A'-shaped profile is economic and retains saplings. Can be too wide at base. If ploughed close to, 'n'-profile better for perennial herbs and partridges.

Spray drift avoided by hedge-saver deflector, sterile strip, etc. Drift worsens annual weed problems, kills aphid predators and fertiliser drift encourages aggressive weeds. Weeds also encouraged by cultivating too close and by mulching effect of trimmings.

New hedges planted in 2 staggered rows, 25 cm apart (for subsequent laying etc).

Over 1.8 m high best for wildlife. For stock at least 1.4 m.

Use only species-selective herbicides on verge. Spot treatment is best.

Wide base good for nesting and stockproof, but if too wide suppresses herbs.

Sterile strip 0.5–1 m wide (rotovated or sprayed) reduces reintroduction of weed seed and allows separate hedge and crop management. Dusting area for partridges.

Use correct tools. Bar cutters for annual growth, flails for 2–3 yearly cutting, circular saws or chain saws for larger stems.

Trim every 2–3 years for economy and improved habitat. Or cut opposite sides each year. Aim for a variety of growth stages.

Cut in late winter (not in severe frost) as birds nest in spring/summer and autumn berries are valuable food source.

Cut low hedges on sides only to allow upwards growth.

A wide variety of shrub and tree species desirable.

Develop 0.5–1 m wide verge of fine grasses and perennial herbs to suppress noxious weeds. Seeding may help.

Width of verge may be more important than hedge width for herbs, many insects and partridges.

Leave selected saplings to mature. Mark with plastic tags to avoid accidental trimming. Allow 10+ m between trees.

Wide grass headlands (3+ m) are excellent for wildlife, farm and public access. Can be seeded with special mixtures.

Retain hedgerow trees, pollarded for firewood if desired. Avoid use as fence posts – embedded wire reduces timber value.

Gaps can be left for diversity and partridge nesting or infilled by planting.

Reinvigorate by laying or coppicing (8–12 yr rotation), grants available.

Well maintained ditches increase species diversity.

Table 9.3. *Summary of recommendations for pesticide use on cereal headlands*

	Autumn-sown cereals	Spring-sown cereals
Insecticides	Only until 15 March	No
Fungicides	Yes (not pyrazophos after 15 March)	Yes (not pyrazophos)
Herbicides (grass weeds only)	Avadex BW, Avadex BW granular, Avenge 2, Cheetah R[1], Commando[1], Hoegrass, Muster, Roundup	Avadex BW, Avadex BW granular, Avenge 2, Cheetah R[2], Commando[1], Hoegrass, Muster, Roundup
Herbicides (broadleaved weeds)	No (some exceptions, e.g. Cleavers). If broadleaved weeds are a problem seek advice	No
Growth regulators	Yes	Yes

Note: [1]Commando is a Shell trademark.
[2]Wheat only.
Source: From Game Conservancy (1990).

choose headlands next to good nesting cover for gamebirds and avoid headlands infested with those difficult weeds already mentioned. Figure 9.3 shows the layout of a conservation headland and Table 9.3 gives a summary of recommendations for pesticide use on cereal headlands. Plate 10 shows a conservation headland in full bloom, with an abundance of arable weeds providing habitat for beneficial cereal arthropods and gamebird chicks and other birds which feed on them.

Rare arable weeds are a specific group which should benefit from good conservation management practice, and many have been significantly reduced in range and abundance since the 1950s with the advent of herbicides. Fields with a long history of arable cultivation will have larger and more diverse seed banks of annual broadleaved plants than those recently ploughed, and lighter soils may offer more potential for a diversity of arable weeds than heavy soils. Fallow margins to encourage the rare arable weeds described in the introduction should be on relatively infertile light soils and made by breaking the old stubble and disturbing the top 15 cm of soil with a cultivator, rotovator or power harrow. To control grass weeds from infesting the fallow strip specific gramnicides should be used; mowing is less favoured because it encourages the tillering of annual grasses and perennial plants at the expense of the broadleaved annuals.

Another specific example of field margin management is for Barn Owl, which has suffered a significant population decline and fragmentation in the past 60 years. Useful feeding habitat can be developed by laying down a network of perennial grass strips and

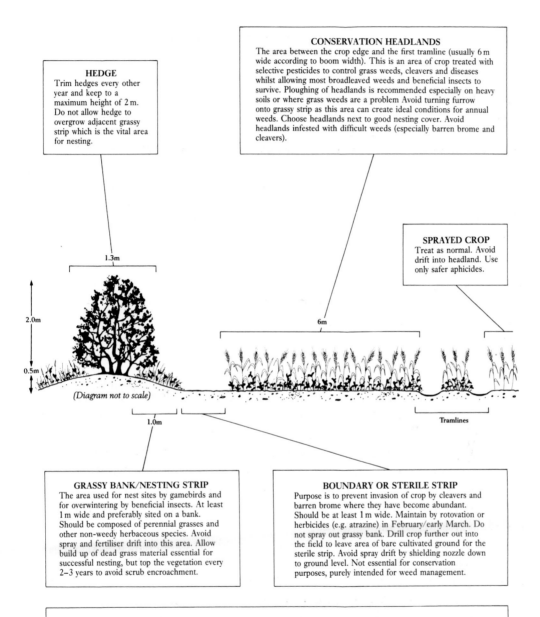

HEDGE
Trim hedges every other year and keep to a maximum height of 2 m. Do not allow hedge to overgrow adjacent grassy strip which is the vital area for nesting.

CONSERVATION HEADLANDS
The area between the crop edge and the first tramline (usually 6 m wide according to boom width). This is an area of crop treated with selective pesticides to control grass weeds, cleavers and diseases whilst allowing most broadleaved weeds and beneficial insects to survive. Ploughing of headlands is recommended especially on heavy soils or where grass weeds are a problem Avoid turning furrow onto grassy strip as this area can create ideal conditions for annual weeds. Choose headlands next to good nesting cover. Avoid headlands infested with difficult weeds (especially barren brome and cleavers).

SPRAYED CROP
Treat as normal. Avoid drift into headland. Use only safer aphicides.

1.3m

2.0m

0.5m

(Diagram not to scale)

6m

Tramlines

1.0m

GRASSY BANK/NESTING STRIP
The area used for nest sites by gamebirds and for overwintering by beneficial insects. At least 1 m wide and preferably sited on a bank. Should be composed of perennial grasses and other non-weedy herbaceous species. Avoid spray and fertiliser drift into this area. Allow build up of dead grass material essential for successful nesting, but top the vegetation every 2–3 years to avoid scrub encroachment.

BOUNDARY OR STERILE STRIP
Purpose is to prevent invasion of crop by cleavers and barren brome where they have become abundant. Should be at least 1 m wide. Maintain by rotovation or herbicides (e.g. atrazine) in February/early March. Do not spray out grassy bank. Drill crop further out into the field to leave area of bare cultivated ground for the sterile strip. Avoid spray drift by shielding nozzle down to ground level. Not essential for conservation purposes, purely intended for weed management.

MACHINERY
A specially designed sprayer is now available which can selectively spray a 6 m strip along the headland while treating the main crop with standard chemicals. Each part of the machinery is independent of the other, thus saving the need for a separate run along the Conservation Headland. While spraying sterile strips it is vital to prevent drift into crop and hedge bottom. A very useful device, which applies the chemical safely and accurately from the tractor, has been designed for this purpose.

Fig. 9.3. Layout of a Conservation Headland.

field margins which encourages high densities of Short-tailed Vole *Microtus agrestis*. After establishment of a coarse grass sward it should be cut once every three years, cutting one-third of the total area on the farm each year. Leave old trees in hedgerows as they are their preferred nesting sites, but also consider siting or retaining old makeshift buildings adjacent to field margin grass strips. Nest boxes can be erected on trees lacking sizeable holes or on existing buildings. Sites chosen should be well away from roads – a major source of death of Barn Owls, and facing away from the prevailing wind. Boxes should have 18 cm holes, and there should be a clear flight-path from open ground straight into the box.

Rare arable weed conservation

Some of Britain's rarest plants are those annual weed species (ruderals) that grow alongside crops in our arable fields. Over the last 40 years many once common species have become very rare. Research by The Game Conservancy described the factors which made the particular species vulnerable and those agricultural practices responsible for these drastic declines. This research has produced some preliminary guidelines for rare weed conservation in cereal crops.

The initial findings show that, to encourage the return of wildflowers from existing seed stocks on field margins and the wildlife that depends on them, farmers should:

> select areas of fields which are not heavily infested with highly competitive weeds such as Sterile Brome or Cleavers

> leave cereal stubbles as long as possible after harvest because some rarer species produce their seed in the early autumn

> sow spring varieties of cereals in March or April; sow winter wheat in October or November rather than September

> apply no nitrogen fertiliser to the selected area – it only encourages the aggressive, unwanted weeds

> apply herbicides to the selected area only in very exceptional circumstances and then only products which are highly specific to the aggressive weed species

> do not create a sterile strip with weedkillers or by cultivation, because the outermost strip of the field is frequently the most botanically diverse.

Beetle banks

Work carried out by The Game Conservancy in the early 1980s showed how those polyphagous beetles and spiders most important as predators of aphids, overwintered in non-cropped cover such as hedgebanks and fence lines rather than out in the open fields. Good field boundary habitats could harbour over 1000 predators per square metre of perennial tussocky grass. This work identified which species of perennial mat-forming grasses were most used and why predator survival has increased in these habitats compared to other grass cover types or the open field situation itself.

Joint work by The Game Conservancy and the Department of Biology at the University of Southampton developed these findings to allow farmers to recreate this habitat in the centres of their larger fields by creating 'beetle banks' which hold predatory insects over winter enabling a quicker colonisation of the surrounding crop in spring and hence more rapid predation of pest insects such as aphids, thereby helping the farmer (Fig. 9.4). Beetle banks thus recreate *in situ* all the benefits of hedgebank cover without planting and establishing a hedgerow.

Construction of beetle banks is very simple. During normal autumn cultivation, a ridge or earth bank is created approximately 0.4 m high and 1.5–2 m wide, by careful two-directional ploughing. A gap should be left of about 25 m at each end of the bank to allow farm machinery to move around the edge of the field. Larger fields might be divided into two smaller ones by extending the ridge right across the field. Ridges should be drilled or hand sown with Cocksfoot at 3 g/m^2, Yorkshire Fog at 4 g/m^2, or a 50:50 mixture of the two with sowing rates adjusted accordingly. Sow either in the autumn of construction or in the following spring, first removing any weeds with a broad-spectrum non-residual herbicide such as glyphosate, at the recommended field rate to remove any opportunistic weeds which may have appeared. For example, a square 20 ha field (450 × 450 m), which has established boundaries with raised underbanks and an abundance of tussocky grasses such as Cocksfoot and Yorkshire Fog, should have one ridge in the centre of the field. Fields of 30 to 50 ha should have three or more ridges. This is believed to provide the optimum amount of beetle bank without taking up too much cropped land. A sterile strip can minimise the risk of a grass herbicide, which is being applied to the crop, from affecting the grasses on the ridge. To prevent the slight risk of the grasses spreading from the bank into the crop, remove the flower heads by 'topping' in summer, taking care not to damage the vegetation structure. The ridges will soon be considered by overwintering insects and spiders. If farming practices change the ridges are easy to remove or recreate later.

Ditches and dykes in the arable landscape

Arable ditches and dykes should be regarded as special types of field margins (Fig. 9.5). If neglected they may develop into dry land, filling with silt and opportunistic tall herbs, reducing their value to wildlife. Damage to wildlife when ditch cleaning is minimised by doing the work in autumn, not during spring and summer which affects breeding birds, aquatic insects and seed setting. Use tractor-mounted buckets or flails to remove silt and vegetation and avoid cutting into the bottom of banks which damages roots and seeds and leads to erosion. Ditch vegetation should be cut rotationally every 3–4 years (if they have a tendency to dry out), and up to every 8 years if they hold water

(a)

(b)

Fig. 9.4. (*a*) A field showing the position of a 'beetle bank' designed to encourage overwintering predatory insects, which can rapidly move into the field in spring to attack crop pests such as aphids. (*b*) The predatory carabid *Pterostichus melanarius* which take advantage of these beetle banks.

Fig. 9.4 cont. (c) A larval sawfly *Tenthredinidae* spp. which exploits plant species in Conservation Headlands and beetle banks. (M. Woodburn, N.W. Sotherton.)

permanently, with some stretches left uncut on both sides of the bank to aid recolonisation by animals and plants. South-facing slopes which receive most sun should shelve as gently as possible to encourage plant growth. If space is a limiting factor have one side of the bank gently shelving and the other steep, rather than both shelving moderately steeply.

Hedges along ditches and dykes are also important for wildlife and these should be maintained. Also, encourage those species which need a stable water supply by excavating areas of deeper water, making small ponds in the ditch bottom which will retain water if the ditch itself dries out. At all cost, chemical control of weeds should be avoided because this can encourage dominant species such as reed to takeover (see Chapter 1 for a discussion on dominance). Similarly, avoid silage effluent, slurry and yard washings, pesticide and fertiliser drift getting into the ditch. Remove cuttings which would otherwise impede drainage, deoxygenate water and encourage the development of algae. When removing spoil the verge vegetation should be protected from being damaged and covered and do not use it to fill in damp hollows which are valuable wildlife habitats. Finally, when farming the adjacent field do not cultivate to the very edge of the bank because soil will spill into the ditch and cause clogging, leading to greater ditch maintenance in the future.

Hedgerows

The primary functions of hedgerows are as field and farm boundaries, to control livestock and to regulate grazing. They provide shelter for stock, crops, buildings and game and reduce soil erosion. They enhance the landscape aesthetically and act as corridors linking new habitats to existing ones, thereby benefiting wildlife (but see section on corridors). This section concerns the planting of new hedges and the management of these as well as existing ones for wildlife.

New hedges are normally established on the flat rather than on sloping ground, in a cultivated strip 60 cm wide and 25 cm deep. For overwintering predatory arthropods and for game, hedges are best planted on a bank (see Figure 9.2). On poor soils a dressing of farmyard manure should be incorporated. Plants should be well-rooted and grown having been transplanted at least once, and 45–60 cm high. Planting is best done in good weather between October and March but be aware of the increased exposure to rabbit and hare damage if planted in autumn. Glyphosate can be used before planting in order to create a weed-free strip. Plants should be set into the ground at the same depth as they were in the nursery, in a double row, staggered to produce a dense solid hedge with stems suitable for laying. At a 25 cm spacing, 9 plants will be used per metre of

Fig. 9.5. Arable ditch and dyke management for conservation showing the retention of ditch vegetation to benefit invertebrates and birds such as reed and sedge warblers. (D.A. Hill.)

Fig. 9.6. Seeding of a wildflower mixture alongside an existing tall hedgerow provides warm growth conditions and a rich food source for invertebrates. (D.A. Hill.)

hedge. Chlorthiamid will keep the plants free from competing broadleaved perennials if used for 3–4 years in February or March. It is a good idea to strim weed growth at the side of the new plants (beyond the cultivated strip) to reduce competition. Alternatively black polythene sheeting mulch can be used around the base of each plant in order to control weeds and retain moisture, although it can look unsightly. Further, individual protection against livestock, rabbits and hares, is important.

Species to use include Hawthorn *Crataegus monogyna*, Blackthorn *Prunus spinosa*, Holly *Ilex aquifolium*, Hornbeam *Carpinus betulus* and Hazel *Corylus avellana*, although be aware of Blackthorn spreading too vigorously out from the line of the hedge. Good stockproofing is achieved by using thorny species such as these for 45–70% of the hedge. Hawthorn is the most versatile but does not like very wet conditions. Blackthorn prefers heavier soils and tolerates more exposure than Hawthorn but has a tendency to sucker which will require additional management. Hazel is suitable for drier and more fertile soils and is fast growing although it does not produce strong side growth. It responds better to coppicing than to laying or trimming. For wildlife benefits (although this has not been scientifically proven) it is advisable to plant one or two other species in association with the main hedge such as Buckthorn *Rhamnus catharticus* (food plant of Brimstone butterfly), Holly (food plant of Holly Blue butterfly), Dogwood *Cornus sanguinea*, Field Maple *Acer campestre*, Crab *Malus sylvestris*, Guelder Rose *Viburnum opulus* and certain willows, and should be planted in 1.2 m tree shelters, although some willows will establish from wands. Early management of the hedge by heavy pruning in the first year will encourage bushy side growth. Laying will also encourage bushy growth low down and help stockproofing. Lightly trim the side of the hedge annually until the required size and shape is achieved, followed by cutting every third year, ideally at the end of the winter, avoiding periods of hard frost, and after berries have been eaten by birds and small mammals. Maintenance costs are reduced if trimming is done on a 2–3 year rotation.

The shape of the desired hedge may be less important to wildlife than is often claimed. An 'A' shaped hedge which has a wide base on which the ground vegetation becomes sparse is poor for most wildlife. The width of the grass bank and strip up to the cultivated part of the field is often most important in combination with the hedge (Fig. 9.6). The idea of a 2–3 year rotational trimming programme is to give a variety of heights and side growth, with more flowers, berries and fruit and consequently greater feeding, nesting and sheltering opportunities for wildlife. Trimming alternate sides of hedgerows every other year produces a similar effect. Any of the three profiles as shown in Figure 9.7 can be considered. The best tool for trimming hedges in good condition

Rectangular or box shaped

Chamfered

A-shaped

Fig. 9.7. Three hedge profiles.

with little thick stem material, is the cutter-bar. The flail is designed for heavier growth and should be used on small and medium-sized woody growth. It is overused on overgrown hedges and gives rise to a ragged and shattered appearance (Fig. 9.8). When using a mechanical flail, start at the top of the hedge and work downwards. In this way, cut material will be further chopped and mulched. Saws should be used for dealing with overgrown hedges rather than using the flail.

In existing hedges and new ones once they have grown up, laying can be carried out to improve stockproofing of the hedge. Laying every 10–30 years is recommended and is best carried out in the winter months between mid-November and early March. Hedgerows reach an optimum state for laying when stem growth reaches between 5–10 cm diameter at the base. Beyond this stage, coppicing (see below) may be more suitable together with the filling-in of any gaps with new plants. If hedges have been allowed to grow beyond the trimming stage, laying is a useful tool but should be delayed by up to 5 years in order to allow shrubs sufficient height for laying. Laying a hedge requires either experience or some training and much practice. The laid hedge should not be trimmed for a few years in order to enable the new base shoots to grow.

Coppicing is useful to rejuvenate a hedge which has grown too old, thick and gappy to lay. Growth is cut down to about 7 cm above the ground and left to regrow on a cutting rotation of 7 to 20 years. If cut down to above this level the hedge will grow with a bushy head and a thin base. Cutting is done by hand during winter months in spells of mild weather. Gaps in hedges are best planted up with suckering species such as Dogwood and Blackthorn which spread into the gaps, although the latter will require careful management to prevent encroachment into the field.

Hedgerow trees should also be planted in the hedgerow design, or to fill large gaps in existing hedges. Standing trees within a hedgerow are one of the most important features maximising the number of territories of Yellowhammer and mature standing trees with dead limbs are important for hole-nesting species such as Barn, Tawny *Strix aluco* and Little Owl *Athene noctua*, Great Spotted Woodpecker *Dendrocopos major* and various passerines, including Tree Sparrow (which especially prefer pollard willows). Transplants or whips of native species such as Oak, Ash, Wild Cherry, Field Maple, Lime or Beech can be used, provided they are protected by tree shelters or similar if rabbit damage is expected. Young saplings can alternatively be allowed to develop rather than trimming them back with a hedge cutter. Walk the length of the hedge prior to trimming and mark the saplings with a coloured tag so that it can be avoided by the hedge cutter. The distance between trees in hedgerows should be more than 10 m and irregularly spaced to ensure a more 'natural' appearance for added landscape value. It is claimed that too many trees in a hedge can be detrimental to ground-nesting species because their main avian predators sit and watch for evidence of partridge nests in the hedge base before going down to take eggs. It is believed partridges actively avoid nesting too close to trees.

Fig. 9.8. Hedgerow management. The first example (*a*) shows a well-trimmed, properly cut hedge with a good structure down to the base. The second example (*b*) shows a poorly managed, gappy hedge where the emphasis has been on 'tidiness' rather than on wildlife conservation. Hedgerow trees have been shattered in the process. The example in (*b*) offers little value to wildlife and is unable to retain stock. (D.A. Hill.)

Small farm woods

Most farms have some existing woodland or areas of ground suitable for planting with trees, often less than 5 hectares in size. Trees should not be planted in areas of existing wildlife interest where they might degrade the local ecology. Here we consider only the establishment of new farm woods rather than the management of existing ones (see Chapter 12). Small farm woods can be used to derive income from timber production, to provide shelter for stock or crops, to increase the sporting value and appearance of the landscape, and to provide habitats for plants, birds and other animals. We shall consider largely the latter. Problems in trying to establish a wood can be caused by pests (rabbits, hares, deer), incursion by livestock, competition from roots, weeds, weather (drought, wind, frost), waterlogged soils, poor plant stock and bad planting design. The larger the wood the better for wildlife although there are parts of the UK, for example parts of Devon, where a pattern of small woods is vital to the region's landscape character.

Before planting the wood, drains and pipelines should be located and checked for blockages. Hawthorn, Blackthorn, Elder, Hazel and Privet are the only common hardwood plants that can be planted directly across water and gas pipelines. Broad-leaved trees should not be planted within 10 m of them. Authority for planting close to civil pipelines is also required. Rides, rights of way and boundaries should then be marked out and weeds on the site controlled using a recommended herbicide in 1 m diameter circles in the planting situation. Where there are only grasses and herbs, application of glyphosate in the autumn or propyzamide in the winter will give reasonable control of weeds well into the next growing season. It should be used 3–7 days before planting. However, it is also effective when applied in the growing season (March to August) to the leaves (it is then translocated to the roots). If there is woody growth a clearing saw or tractor-mounted swipe should be used. It is important to calculate the amount of labour involved and the farm machinery available – where brashing and pruning is planned after establishment a specially designed curved saw is worth using. Trees should be ordered well in advance and thought should be given to applying for grant aid (see suggested reading). If a fence is to be established, some training will be required or the job can be contracted out to someone else.

At the design stage it is important to decide on the area to be planted, the spacing widths and hence the number of trees required. Often the most important element of a new wood is its glades, rides and sunlit waters. Choosing species will depend largely on the soil type. A guide to choice is given in Table 9.4. For conservation, broadleaves are best, but planting of them is not generally recommended above elevations of 250 m. Avoid siting the new wood in a frost hollow unless you intend to use hardy species such as Alder, Birch, or Hornbeam. When the plants arrive check them and return any shrivelled ones, then soak the rest. If you do not intend to plant them straight away, keep them in bags in cool, moist, unfreezable locations or cover with straw or sacking, or

Table 9.4. *Trees used for planting on the farm depending on soil type*

	Chalk	Clay	Sand	Damp	Coastal	Upland
Large trees						
Beech *Fagus sylvatica*	*		*			*
Ash *Fraxinus excelsior*	*	*		*	*	
Hornbeam *Carpinus betulus*		*	*			
Small-leaved Lime *Tilia cordata*	*	*	*			
English Oak *Quercus robur*		*	*	*	*	
Sessile Oak *Quercus petraea*		*	*	*	*	
Red Oak *Quercus rubra*[1]			*			
Grey Poplar *Populus canescens*		*		*		
Black Poplar *Populus nigra*		*		*	*	
Aspen *Populus tremula*		*	*	*	*	
Horse Chestnut *Aesculus hippocastanum*[1]		*	*			
Sweet Chestnut *Castanea sativa*[1]			*			
Sycamore *Acer pseudoplatanus*[1]	*	*			*	*
Norway Maple *Acer platanoides*[1]	*					
White Willow *Salix alba*		*		*	*	
Crack Willow *Salix fragilis*		*		*	*	
Southern Beech *Nothofagus spp*[1]		*	*			
Grey Alder *Alnus incana*[2]		*	*	*		
Italian Alder *Alnus cordata*[2]	*					
Larch *Larix spp*		*	*			*
Corsican Pine *Pinus nigra*	*		*		*	
Scots Pine *Pinus sylvestris*			*			*
Douglas Fir *Pseudotsuga menziesii*[1]		*	*			*
Sitka Spruce *Picea sitchensis*[1]						*
Norway Spruce *Picea abies*[1]	*	*				*
Small to medium trees						
Field Maple *Acer campestre*	*	*			*	
Whitebeam *Sorbus aria*	*				*	
Rowan *Sorbus aucuparia*	*		*		*	*
Wild Cherry *Prunus avium*	*	*				
Yew *Taxus baccata*	*	*	*			
Holly *Ilex aquifolium*	*	*	*			
Birch *Betula pendula*			*			*
Hazel *Corylus avellana*		*	*			
Alder *Alnus glutinosa*		*		*		*
Walnut *Luglans regia*		*				

Table 9.4. (*cont.*)

	Chalk	Clay	Sand	Damp	Coastal	Upland
Wild Service Tree *Sorbus torminalis*		*				
Shrubs						
Hawthorn *Crataegus monogyna*	*	*	*	*	*	*
Guelder Rose *Viburnum opulus*	*	*		*		
Wayfaring Tree *Viburnum lantana*	*					
Dogwood *Cornus sanguinea*	*	*			*	
Privet *Ligustrum vulgare*	*				*	
Gorse *Ulex europaeus*			*		*	*
Broom *Cytisus scoperius*			*			
Blackthorn *Prunus spinosa*	*	*	*	*	*	
Buckthorn *Rhamnus catharticus*	*					
Goat Willow/Sallow *Salix caprea*	*	*	*	*	*	*
Grey Willow *Salix cinerea*	*	*	*	*	*	
Elder *Sambucus nigra*	*	*				
Alder Buckthorn *Frangula alnus*		*	*	*		
Spindle *Euonymus europaeus*	*					

Note: [1]Non-native species which should not be used in preference to native species.
[2]Orchards only.
Source: ADAS/FC (1986).

alternatively set them on a slant in a trench 30 cm deep, covering them with soil.

Wider spaced planting than for commercial forests is recommended, at a distance of 3–4 m between trees (this may depend on the grant aid being sought). Tree shelters can be used to help growth in some species. Deciduous species should be planted in late autumn up to a few weeks before the new leaves emerge in spring, and when the ground is not frozen. For small transplants notch planting is the best method whereas for larger and container-grown plants pit planting is necessary (Figure 9.9). In the first summer after planting, if there are very dry spells of more than 2 or 3 weeks it is helpful to water the plants using 20 litres per tree once per week. Too little water can encourage surface rooting and too much can wash away nutrients or cause waterlogging around roots. Weeds will be suppressed and soil moisture can be retained by using a mulch of straw, bracken or bark or polythene matting. When planting in the lowlands, fertilisers are not normally needed to help tree growth. Plant in compatible groups (i.e. the same growth rate) to ease management of trees and shrubs. Planting a range of species with different (economic) maturation dates will eventually produce an uneven aged wood, e.g. Cherry at 60 years, Ash at 90, Oak at 120 and 150.

To give protection to the new trees from browsing animals it is often necessary to fence the whole area or to guard trees individually. Rabbit fencing comprises 31 mm hexagonal mesh 0.9 m high with 0.15 m turned out and covered with turf. To deter hares, an extra single trip wire is added 0.15 m higher. Stock fencing can be single wires but should be placed at least 1.5 m away from planted trees. Deer fencing needs to be 2+ m high and is expensive. The more the area of wood to be protected resembles a circle, the cheaper will be the cost of fencing. For long thin sites and widely scattered clumps, individual protection will prove more cost-effective. If deer do become a problem, leaving badly frayed trees rather than removing them should be considered, as

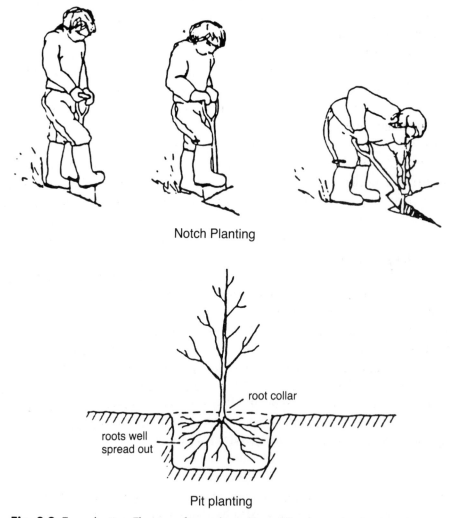

Fig. 9.9. Tree planting. The size of tree planted has different cost implications. Larger specimens require more costly pit planting, but have greater survival than notch-planted whips, which are relatively cheap to plant.

deer may re-fray only these ones, allowing others to grow on healthily. Alternatively a chemical repellant can be used. For individual trees spiral guards (0.6–0.75 m high) can be used for trees with a stem less than 3 cm, pushing the bottom of the guard into the ground. For all guards the specified height should be measured on the uphill side. Alternatively PVC tubing or tree shelters (1.2 m high) of corrugated or reinforced plastic can be used. The latter are best with transplants 15–40 cm high and promote growth by acting as a mini greenhouse, but it is important to check for chaffing on the top edge of the shelter. Taller tree shelters can force growth of the tree however, leading to frailty. Plastic mesh, wire mesh and timber guards can also be used, the latter particularly in a parkland setting where protection from livestock is most important. Further, stake or ties may be used where larger trees need support. The comparative cost-effectiveness between fencing and individual tree protection (where stock are present), is such that deer-, stock- and rabbit-proof fencing is progressively cheaper after a break-even point of about 2.5 ha. In areas where there is no farm stock the break-even point is about 3.5 ha. The cheapest method, at under half the cost of the others, is to plant an already stock-proofed area without using tree shelters. With broadleaves, however, mammal damage could be too severe.

Regular mowing of grass around newly planted trees retards the tree's growth, as cutting causes the grass to grow more vigorously thereby competing with the trees for soil moisture and nutrients. Therefore it is best to chemically weed against species such as Bramble or dense grasses, treating a radius of 1 m around the base of each tree being careful to keep glyphosate spray off broadleaved trees. One treatment per year in late winter or early spring is all that is needed. For small-scale plantings bark mulches can be used to suppress grass and weed growth.

It is especially valuable to leave a fringe of uncultivated ground along the new woodland bank and to encourage wildflowers in order to create a woodland edge of high value to wildlife, particularly butterflies.

Special variants on the small wood design are prescribed for game, largely pheasants. Indeed, the sporting value attributed to pheasant shooting in the UK has been given as one of the most important reasons why landowners plant new woods. The important factors are the provision of winter 'holding' habitat, flushing points from which pheasants are pushed to gain height for sporting shooting, provision of nesting cover on the edge of the wood and on the edge of rides within the wood. The Game Conservancy offers specific advice to landowners on woods for game.

Game crops

Game crops are crops planted explicitly to encourage game, rather than to be harvested (Fig. 9.10). They have three purposes: to increase nesting and holding cover for birds (pheasants and partridges) and provide an extra 'drive' (i.e. where birds are driven from cover to fly over the guns) on a shooting day; to draw birds to an area where they offer

Fig. 9.10. Game crops in a landscape (*a*) used to hold principally pheasants, but smaller seed crops such as millet (*b*) can provide food for flocks of finches in winter, which would at one time, have fed extensively on wild-seed threshings spilt on the autumn stubbles and left through the winter. Modern farming methods have removed this source of food for birds, or reduced the time spilt grain remains available before being ploughed in. (Game Conservancy Trust.)

better sport; to provide supplementary feeding for game. The Game Conservancy has provided a booklet on their design and layout. Crop types planted for game include Artichokes, Beans, Buckwheat, Canary Grass, Cocksfoot, Kale, Millet, Maize, Sugar Beet and many others. Those which produce an abundance of small seeds such as Millet are also used extensively by small seed-eating birds during winter. If the crop is left to seed itself in subsequent years, an herbaceous ground flora can develop, which becomes particularly valuable to invertebrates in summer and hence those birds and small mammals which eat them. Self-seeding game crops may need re-establishing after 3 or so years, but they represent a relatively cost-effective conservation enhancement for seed-eating birds and small mammals during winter.

Set-aside

On the whole, land set aside from cereal farming is richer in wildlife than when it is cropped. A variety of opportunistic flowering plants at first colonise, with butterflies, grasshoppers, small mammals and some birds, e.g. Kestrel *Falco tinninculus* and Barn Owl *Tyto alba*, being in a position to benefit. Two types of management are commonly practised by farmers. First, land can be set aside permanently for perhaps 10 years or more. This may be land which is naturally poor for producing crops, and hence the best for encouraging wildlife habitat. Wet areas or those susceptible to flooding, or dry sandy areas, offer opportunities to restore a scarce habitat such as heath, downland or flood meadow using techniques described under appropriate chapters of this book.

Second, the land can be put into rotational set-aside in which if 15% of the land were set aside each year, each field would be fallowed on average for one year in six. Whole fields can be set aside, or strips around the field boundary (Fig. 9.11) or in the centre. Some farmers may use rotational set-aside to reduce the abundance of persistent weeds such as Blackgrass. Except for fast-maturing common weeds, seed production under rotational set-aside would be minimal and the timing of cultivation, i.e. if in the spring, is deleterious to wildlife. However, annual rotation could be excellent for some wildlife if the land is allowed to 'tumble-down'. Unploughed and unherbicided stubbles used to be an important winter foraging area for many birds including finches, buntings, partridges and Pheasants. In some areas stubbles are much used by geese in autumn. Allowing stubbles to develop into set-aside rather than ploughing them in may produce a rich variety of plants but the species which flourish at first, exploiting open ground and freedom from competition, will be later overwhelmed by ranker vegetation which will deprive them of light and bury them in litter. Plant cover in tumble-down set-aside differs significantly depending on soil type and is also influenced by the previous cropping regime. Even within individual fields, there can be considerable variation in species composition and vegetation structure, providing conditions which support a much richer variety of invertebrates and birds than a sown grass cover.

Grazing (preferably by cattle rather than horses or ponies) or mowing is essential for

(a)

(b)

Fig. 9.11. Set-aside. This shows a comparison of the installation of (a) field boundary set-aside with (b) whole field set-aside. The larger the blocks of set-aside, the greater the potential for encouraging top predators such as raptors and owls. (J. Clarke.)

the development and retention of a rich plant community, with grazing being much better for invertebrates because the transition between short turf and longer tussocks is 'softer' whereas mowing creates an abrupt change and uniform habitat over the whole field. There are, however, restrictions on the types of grazing animals which are allowed – only those that are not marketed for food being acceptable.

References

ADAS. (1986). Practical work in farm woods series. London: MAFF Publications, HMSO.

ADAS. (1986). Hedgerows. London: MAFF Publications, HMSO.

ADAS/FC. (1986). New planting. Practical work in farm woods 7. Alnwick, Northumberland: MAFF.

Andrews, J.H. & Rebane, M. (1994). *Farming and Wildlife: a Practical Management Handbook*. Sandy: RSPB.

Beckett, G. & Beckett, K. (1979). *Planting Native Trees and Shrubs*. Norwich: Jarrold.

Bennett, A.F. (1990). Habitat corridors and the conservation of small mammals in a fragmented forest environment. *Landscape Ecology* 4, 109–22.

Diamond, J.M. (1976). Island biogeography and conservation: strategy and limitations. *Science* 193, 1027–29.

Evans, J. (1984). *Silviculture of Broadleaved Woodland*. Forestry Commission Bulletin 62. London: HMSO.

Farming and Wildlife Advisory Group. *Farming and Field Margins*. Stoneleigh, Warwickshire: National Agricultural Centre.

Forestry Commission. (1990). *The Establishment of Trees in Hedgerows*. Research information note 195. Farnham, Surrey: Alice Holt.

Galbraith, H. (1988). Effects of agriculture on the breeding ecology of lapwings *Vanellus vanellus*. *Journal of Applied Ecology* 25, 487–503.

Game Conservancy Trust (1990). *Cereals and Gamebirds Research Project. Guidelines for the management of field margins (Conservation Headlands and field boundaries)*. Fordingbridge, Hampshire: Game Conservancy.

Game Conservancy Trust. (1990). *Helping Nature to Control Pests. Advice leaflet in association with the University of Southampton*. Fordingbridge, Hampshire: Game Conservancy.

Hill, D.A. & Robertson, P.A. (1988). *The Pheasant: Ecology, Management and Conservation*. Oxford: Blackwell Scientific Publications.

Lampkin, N. (1990). *Organic Farming*. Ipswich: Farming Press.

O'Connor, R.J. & Shrubb, M. (1986). *Farming and Birds*. Cambridge: Cambridge University Press.

Potts, G.R. (1986). *The Partridge: Pesticides, Predation and Conservation*. London: Collins.

Potts, G.R. (1991). The environmental and ecological importance of cereal fields. In *The Ecology of Temperate Cereal Fields*. Proceedings of 32nd symposium of the British Ecological Society with the Association of Applied Biologists, pp. 3–21. Oxford: Blackwell Scientific Publications.

Simberloff, D. (1982). Refuge design and island biogeographic theory: effects of fragmentation. *The American Naturalist* 120, 41–50.

Simberloff, D. & Cox, J. (1987). Consequences and costs of conservation corridors. *Conservation Biology* **1**, 63–71.

Sotherton, N.W. (1991). In Conservation Headlands: a practical combination of intensive cereal farming and conservation. *The Ecology of Temperate Cereal Fields*. Proceedings of 32nd symposium of the British Ecological Society with the Association of Applied Biologists. Oxford: Blackwell Scientific Publications.

Tapper, S.C. & Barnes, R.F.W. (1986). Influence of farming practice on the ecology of the Brown Hare (*Lepus europaeus*). *Journal of Applied Ecology* **23**, 39–52.

Tucker, G.M. (1992). Effects of agricultural practices on field use by invertebrate-feeding birds in winter. *Journal of Applied Ecology* **29**, 779–90.

Wegner, J.F. & Merriam, G. (1979). Movements by birds and small mammals between a wood and adjoining farmland habitats. *Journal of Applied Ecology* **16**, 349–57.

Wilson, P.J. (1990). *The Wildflower Project: A Summary*. Fordingbridge, Hampshire: The Game Conservancy Trust.

Lowland heathland   10

PAUL M. DOLMAN AND REG LAND

Introduction

Western European lowland heathland is a distinctive habitat found on nutrient-poor soils, particularly acidic podsols. Catastrophic losses of heathland have occurred throughout western Europe, through conversion to farmland, afforestation, urban development and succession. Despite large-scale losses of heathland in Britain, estimated at 75% between 1800 and 1983, *c.* 57 000 ha remain representing about 20% of the European resource (Farrell, 1989). Within Britain the survival of a number of species depends on lowland heathland, for example the Smooth Snake *Coronella austriaca*, Dartford Warbler *Sylvia undata* and numerous specialist invertebrates. Other rare species, such as Woodlark *Lullula arborea*, Nightjar *Caprimulgus europaeus* and Sand Lizard *Lacerta agilis* depend heavily on heathland but also breed in other habitats.

Lowland heathland communities

Lowland heathland communities found below 300 m are distinct from upland heather moorland. They are characterised by sandy mineral soils of generally lower nutrient status than moorland peat soils. Although we concentrate on 'dwarf-shrub' heathland dominated by heathers, particularly Heather or Ling *Calluna vulgaris* but also *Erica* species, and gorse (*Ulex* species), we also consider heathland habitat dominated by grasses and lichens. The ecological processes and land use practices which created and maintain these grass heath and lichen heath communities are the same as those which created and maintain dwarf-shrub heathland.

Many heaths in southern England consist of a complex of communities, with dry heath on higher free-draining sandy areas, characterised by *Calluna* and Bell Heather *Erica cinerea*, grading into a valley system in which wet heath and valley mire communities occur. In these *Calluna vulgaris* is generally replaced by Cross-leaved Heath *Erica tetralix*, *Sphagnum* mosses, and Purple Moor-grass *Molinia caerulea*. These systems should be considered as a single hydrological unit. However, the vegetation, ecological processes and land use history of dry heathland and wet heath or mire communities are distinct. Consequently, this chapter considers the management of lowland dry heathland, while wet heath and mires are considered with fens and bogs in Chapter 7.

In addition to variations within sites due to grazing, soil or hydrology there are important regional differences in the vegetation and characteristic species of lowland heathland, partly due to regional differences in soil and land use history, but largely related to a climatic gradient (Webb, 1986; Rodwell, 1991). Annual rainfall is lowest in eastern England, particularly Breckland, and this is reflected in the presence of many continental and Mediterranean species absent from other areas. The climate of the London Basin, Weald, Hampshire Basin and eastern Dorset is less extreme but is still sunny and warm. These areas support a diverse and important heathland invertebrate fauna characterised by southern species reliant on hot microclimates (Stubbs, 1983) as well as other species such as Sand Lizard and Smooth Snake. Moving further into south-western Britain the climate is increasingly oceanic, with higher rainfall and mild winters, reflected in the presence of species such as Cornish Heath *Erica vagans* and Dorset Heath *E. ciliaris*. In contrast, lowland heaths in the Midlands and northern Britain are transitional to upland communities. A classification of lowland heathland communities is presented in the National Vegetation Classification (Rodwell, 1991).

The origins of heathland

Some maritime heathland may be entirely natural, maintained by salt spray, wind and summer drought. Elements of heathland communities may also have occurred naturally in woodlands on sandy soils where grazing by large herbivores created open conditions. However, most lowland heathland is a semi-natural habitat, created and maintained by certain human land use practices. Although localised clearance occurred in the Mesolithic (10 000–5000 BP), large-scale creation of lowland heathland was initiated later by farmers clearing woodland on sandy soils. In some areas, such as Breckland, this process began in the Neolithic (5000–4000 BP), while heaths in central southern England were largely created in the late Bronze Age (from 3600 to 3000 BP).

Following clearance, woodland regeneration was prevented by stock grazing. This facilitated the leaching of soluble minerals through the soil and the development of acidic podsols. A pattern of land use emerged with heaths used as permanent pasture within a mixed farming economy. Historically grazing has been the major human influence in maintaining heathland. Grazing of wood-pasture in the Middle Ages led to the development of further heathland so that its area was probably maximal in the late seventeenth century. Other practices which created disturbance and removed organic material also helped maintain heathland. For example, people cut gorse for firewood and fodder, turf for fuel, Bracken *Pteridium aquilinum* for livestock-bedding, and heather for thatching, fodder and road building, while tracks and quarries created areas of disturbed sand. In Breckland shifting agriculture periodically disturbed and rejuvenated the heaths, with areas cultivated and then abandoned due to exhaustion of soil fertility and trends in the agricultural economy. Many of these activities are now being reintroduced, adapted or mimicked by managers in an attempt to halt the adverse changes now occurring on most heaths.

Problems in heathland management

Succession

With the notable exception of the New Forest, where during 1990–1992 commoners still pastured 3500 ponies and 2000 cattle, grazing of heaths by livestock declined during the nineteenth century and virtually ceased in most areas by the mid twentieth century. This, combined with the abandonment of other traditional land use practices such as cutting gorse and Bracken has allowed succession to occur. Closely grazed vegetation and bare sandy areas have been lost as dwarf shrubs or grass formed dense, closed, vegetation. On many sites the invasion of Bracken and scrub and the development of secondary woodland has led to the loss of characteristic heathland commu-

Fig. 10.1. Wet and dry heath in need of some grazing in order to diversify sward structure and provide more habitats for invertebrates. Encroachment of pine *Pinus sylvestris* and birch *Betula pubescens* are a common problem on these sites and can be expensive (£2000/ha) to remove in order to restore heath back to its original status. (J. Andrews.)

nities. As documented for plant species in the Breckland heaths (Dolman & Sutherland, 1992) and for Sand Lizard and Smooth Snake on southern heaths, vegetation change has led to the loss of many local populations. Despite the efforts of conservationists losses to succession continue and halting and reverting this process is likely to be the manager's prime concern.

Increased nutrient levels

Heathland vegetation is dependent on the low availability of soil nutrients, particularly phosphorus and nitrogen (Chapman *et al.*, 1989; Dolman & Sutherland, 1992). Traditional grazing management, the harvesting of plant material and in some regions rabbits, continuously removed nutrients from the ecosystem. These practices have now ceased while natural inputs of nutrients continue, leading to their gradual accumulation in soil and vegetation. In addition, air pollution has greatly increased the concentration of nitrogen in rain. Although this deposition is low in the Southwest Peninsula and Dorset, it is high in the Weald, while the concentration of nitrogen in rainfall in East Anglia is higher than in any other lowland area (Dolman & Sutherland, 1992). On Dutch heaths the widespread replacement of dwarf shrub communities by grassland has been attributed to nitrogen eutrophication. Similar changes have been noted in Britain, for example the spread of Wavy Hair-grass *Deschampsia flexuosa* on heaths in the Weald. Although it is hard to assess the relative importance of increased nutrient availability, reduced grazing and other changes in management in bringing about such vegetation change, increased nutrient availability is certainly detrimental to conservation objectives. Management techniques which deplete nutrients will at least favour, and in cases may be essential to, the long-term conservation of heathland communities.

Fragmentation

In addition to severely reducing its area, agriculture, afforestation and urban development have fragmented the heathland that remains. For example, in the mid eighteenth century the Poole Basin heaths in Dorset consisted of eight major blocks totalling *c.*40 000 ha, but by 1978 these had been fragmented into 768 blocks totalling 5832 ha, of which only 14 were larger than 100 ha while 476 were less than 1 ha (Webb & Haskins, 1980). Fragmentation poses serious problems such as increased extinction rates of species on small fragments, reduced chances of re-colonisation and the greater influence of edge effects (Webb, 1989; Webb & Vermaat, 1990). However this does not justify neglecting the management of small fragments or allowing their loss through development. Instead it emphasises the crucial importance of linking remaining fragments within larger areas of re-created heathland.

Heathland management

Overview

On previously unmanaged sites a priority is to restore degenerate areas to heath vegetation, by scrub control, Bracken control and nutrient stripping. Subsequent management is aimed at the long-term maintenance of heath communities.

When restoring unmanaged sites, initially prevent further encroachment of scrub and Bracken. Clear areas which retain heathland vegetation rather than areas of mature scrub or dense Bracken. Control should be seen as an ongoing commitment. One-off or short-term solutions are unlikely to be successful and may ultimately waste resources as scrub and Bracken are able to re-invade if control ceases (Lowday & Marrs, 1992; Gimingham, 1992). At Arne, Dorset, Bracken control has focused on pushing back the advancing Bracken front rather than attempting to clear dense stands with deep litter, and no new areas are sprayed until regrowth in all areas treated in previous years has been spot-sprayed (Pickess *et al.*, 1989).

Although clearing scrub and Bracken is often a high priority, the retention of limited areas is important. Scrub and trees, particularly birch, support an important inverte-brate fauna. Different species are associated with scrub of different ages while others specialise on isolated trees (Stubbs, 1983; Kirby, 1992). Rich invertebrate commu-nities, including many species absent from extensive open heath, may occur on the margin of heathland sites in mosaics of heath vegetation, herb-rich grassland, Bramble *Rubus fruticosus* agg., Hawthorn *Crataegus monogyna* and oak (Stubbs, 1983). Scrub is also critical for other key species. For example, Nightjar prefer the interface between open heath and scrub and have benefited from creation of glades and scalloped heath/ wood margins (see Box 10.1). Dartford Warblers use pines whose lower branches touch the *Calluna* canopy and in some localities are dependent on dense bushy Common Gorse. Consequently this is managed on a rotation at sites such as Arne, Dorset (Pickess *et al.*, 1989).

Within a site consideration should be given to the arrangement of those areas of scrub to be retained or cleared. Isolated blocks of heathland should be linked by the removal or thinning of intervening scrub and woodland. Some isolated trees, scattered scrub and narrow shelter belts with sheltered scalloped margins should be left for invertebrates and other key species (Kirby, 1992).

Following a restoration phase, management moves on to long-term maintenance. Ideally this should be achieved by grazing, complemented by soil disturbance and, if necessary, cutting. On dwarf shrub heaths burning may also be used. In this chapter we outline the alternative management options, while costings for equipment and manage-ment operations are given by Gimingham (1992). This chapter considers these techniques individually; however, they were combined in traditional pastoral econo-mies which are the model for heathland management.

Box 10. 1. Management of lowland heathland for Nightjars

Nightjar populations are declining throughout much of Europe and have experienced a widespread and severe decline in Britain. Nightjars breed in sheltered woodland edge habitats. Lowland heaths are of major importance to the British population, in addition to restocked forestry clearfells. Nests are made on bare ground, often amongst *Calluna* or Bracken. Management of heathland for Nightjars highlights the importance of trees and scrub to a variety of wildlife.

Management at the RSPB Minsmere reserve aimed to increase the number of breeding Nightjar following a decline (Burgess, Evans and Sorensen, 1990). A wide edge zone of scattered trees on open heathland was created by selectively felling woodland, while retaining scattered birch coppiced on a 10–15 year rotation. Belts of trees were retained and woodland margins were managed to form a wavy edge, increasing shelter, habitat diversity and the length of wood–heath interface. Glades up to 1.5 ha were created within the birch woodland. Open areas are managed to maintain heath vegetation and prevent succession. At Minsmere it was considered that Nightjar selected nest sites on bare ground at the base of small birch trees. Consequently patches of soil, 2 m diameter, were created in heather next to 1–3 m tall birch saplings, however the effectiveness of this was difficult to determine. After an initial lag, Nightjar numbers at Minsmere increased following management, from eight churring males in 1978 to 40 in 1989. Similar management on heathland in Surrey has also led to an increase in breeding Nightjar numbers (R. McGibbon, pers. comm.).

Calluna has a life-cycle consisting of 4 phases: pioneer (3–10 years), building (7–13 years), mature (12–30 years) and degenerate (over 30 years) (Webb, 1986). Different heathland species require patches of vegetation at different stages of this succession. In Britain, many rare heathland invertebrates are at the northern limit of their distribution and consequently require the microclimate and habitat structure of open early successional vegetation and pioneer heather. However, many web-spinning spiders and insects such as the Emperor Moth, *Pavonia pavonia*, rely on well-developed vegetation. Similarly, although Sand Lizard need areas of bare sand for egg laying, they also require mature to degenerate *Calluna* for basking and shelter from predators. To conserve the full spectrum of heathland species it is necessary to provide a full range of habitat structure, from bare sand and pioneer *Calluna* through to mature and degenerate phases. In the past, traditional land use practices provided this diverse mosaic of different microhabitats.

Ideally *Calluna* should go through its life cycle without intervention, with young seedlings developing in gaps created by dead and dying plants or with regeneration occurring vegetatively. Selective grazing of grasses and tree seedlings may reduce competition with dwarf shrubs, thereby favouring heather regeneration. However, if the site is not grazed, non-intervention does not guarantee continued *Calluna* cover, instead degenerate stands may be invaded by scrub, grass or Bracken.

In the absence of grazing, cutting or burning may be used to rejuvenate *Calluna* and maintain its dominance. Vegetative regeneration may be poor in older *Calluna* so that regeneration from seed is increasingly important. However, this is slower, leaving bare ground available for colonists. Consequently, if an ungrazed, over-mature stand is managed by burning or cutting then birch, Bracken or grasses may invade. Because of this, short burning rotations have been advocated to keep most *Calluna* in the pioneer to early mature phases, ensuring rapid vegetative regrowth and continued dominance. However, this does not provide the range of habitat structure which should be an objective in heathland conservation management.

If burning or cutting are to be used, areas to be managed, block size, number, shape and rotation should be identified within a long-term plan. If a rotation of 15–20 years is used then some *Calluna* should be excluded and left to over-mature, particularly in areas used by reptiles, Dartford Warblers and specialist invertebrates. These areas should be separated to minimise possible losses through accidental fire. Similarly grazing should be excluded from some areas to allow the development of tall ungrazed and untrampled *Calluna*. As a general rule it is better to provide a mosaic of vegetation structure and age at a small rather than a large scale. This favours species which require different microhabitats within their territory or range and those which can only survive within a site if individuals are able to re-colonise newly managed areas from nearby patches of suitable habitat.

Creation or maintenance of pioneer vegetation, bare ground (see Fig. 10.2), sandy

Fig. 10.2. Firm, warm, bare sand patches on lowland heath are very important for solitary bees and wasps as nesting sites. Shown here is the sand wasp *Ammophila sabulosa* dragging a caterpillar into its nest burrow in a bare sandy bank. (R. Key.)

banks and vertical sand faces on south facing slopes is extremely important. These provide the microclimate and habitat structure vital to many invertebrates, particularly burrowing wasps, predatory beetles and thermophilous ants, as well as reptiles and early successional plants. For example, the Silver-studded Blue butterfly, *Plebejus argus*, depends on disturbance from burning, cutting, heavy grazing, quarrying or tracks. This is related to its dependence on the ant *Lasius niger* which is abundant in warm open conditions, and the restriction of feeding larvae to an exceptionally warm microclimate. Open areas may be maintained by rotovation, by a moderate level of footpath use or by handpulling vegetation. Timing of management should take account of any breeding birds or reptiles. In Dorset rotovation is restricted to the 7–31 May to minimise risk to hibernating or egg-laying Sand Lizards.

Bracken control

Bracken control can be expensive, so attempts should be made to offset costs wherever possible. Control is by two main methods, cutting or spraying, which may be used singly or in combination. Practical details of these methods are provided in Box 10.2. Bracken has substantial reserves of carbohydrate stored in rhizomes. Consequently it is very hard to eradicate, as shown by work carried out in Breckland by the Institute of Terrestrial Ecology. In this study, various Bracken control and heathland restoration treatments were assessed on *Calluna* heathland at Cavenham Heath and on grass heath vegetation at Weeting Heath (Lowday and Marrs 1992, Marrs and Lowday 1992 a,b). Although cutting and spraying all reduced the level of Bracken biomass none of these treatments eradicated bracken, even after 10 years.

Where heather remains below Bracken, spraying is generally preferable to cutting, although it may be possible to cut above the heather. Spraying may also be preferable on rough terrain. The herbicide Asulam (trade name Asulox) is effective against Bracken. It has low toxicity and an insignificant effect on soil micro-organisms. It does not damage established *Calluna*, but gorse (both Common and dwarf species) may be scorched and young plants killed (Pickess *et al.*, 1989), while ferns and some grasses are also susceptible. Stock should not be grazed on treated areas for approximately 2 weeks after spraying. Single spraying has a large (reducing biomass by >95%), but ephemeral effect on Bracken with recovery to 50% within 6 years (Lowday & Marrs, 1992).

Cutting may be preferable for dense long-established Bracken. Extensive areas may be treated, for example at Brettenham Heath *c.*180 hectares are treated annually by a single tractor operator (P. Wright pers. comm.). The re-establishment of heath vegetation is often more rapid on cut than sprayed plots. This is attributed to disturbance of the litter layer. However, this disturbance may also allow species such as birch to establish from seed. Annual cutting reduces biomass relatively slowly, to 20%–30% in 7 years and to <10%–20% after 10 years. Cutting twice a year is more effective, reducing biomass to below 10% within 3 years and maintaining it at this level

Box 10. 2. Practical details of Bracken control

Small areas may be cut by hand with a scythe, strimmer or Allen scythe. Larger areas should be cut with a tractor mounted swipe or forage harvester. On grass heath on difficult terrain with many bushes 4–5 ha may be swiped per 7–8 hour day, increasing to *c.*8 ha day[-1] on flat areas, while forage harvesting rates are approximately a third of this (P. Wright, pers. comm.). On dwarf shrub heathland in Dorset rates for tractor swiping Bracken are *c.*2 ha per 7 hour day, at £66 per ha (Auld, Davies & Pickess, 1992). Cut in late July if once a year, in mid June and late July if twice a year and in May, July and August if cutting three times. In sparse stands, rolling using agricultural machinery, bruising by hitting fronds with a stick and hand pulling are alternatives to cutting. When cutting Bracken using high speed cutters masks to the appropriate standard should be worn to avoid inhalation of carcinogenic tissue.

At Thursley Common, in Surrey, trials are being conducted into the use of soil disturbance following cutting. This is carried out on a hot day in late summer exposing rhizomes to desiccation and then to winter frost which can kill rhizomes. Disturbance treatments combined with cutting have a greater effect on Bracken density than cutting alone while rotovation appears superior to chisel ploughing (S. Nobes, pers. comm.). The disturbance has also created suitable breeding habitat for Woodlark which have increased in density.

Asulam should be sprayed when fronds are about fully extended, but have not had time to recharge carbohydrate stores in the rhizomes, generally mid July to early August. Spray when the weather is dry, wind speed 1–3, and with no rain forecast for 36 hours following application. In addition hot, still days should be avoided as evaporation can be excessive. The use of an adjuvant to increase absorption and rain resistance markedly improves effectiveness and is recommended (Andrews, 1990). If cut in the first year, regrowth consists of a higher density of shorter fronds which may be sprayed more easily and effectively in the second year.

For continuous stands an ultra low volume applicator (ULVA) system may be used, such as the Micron ULVA, as these use small volumes of liquid dilutant, are light and easily portable; however, they cannot be used for spot spraying (Pickess *et al.*, 1989). Gimingham (1992) does not recommend their use because of spray drift problems. One person applying Asulam and adjuvant using an ULVA may cover 1 ha in 2.5 hours at a cost of £118 per hectare (Andrews, 1990). At Arne spot-spraying is carried out with a 1.25 l Hozelock ASL Polyspray 2, taking care to avoid non-target species, particularly gorse. Extensive areas may be tractor sprayed with a boom of up to 6 m wide (Gimingham, 1992) at a rate of 4–5 ha per 7 hour day (Auld *et al.*, 1992a).

(Lowday & Marrs, 1992). At both Brettenham Heath and Thursley Common dense Bracken has been effectively and efficiently controlled by initially cutting at a rate of 3 times a year, decreasing to two cuts a year and then to a holding treatment of one cut a year when only sparse fronds remain (M. Wright pers. comm.; S. Nobes pers. comm.).

When treating areas of long established Bracken, deep litter should be removed wherever possible. For large-scale restoration a forage harvester or JCB should be used, see Figure 10.3. At a number of sites Bracken litter has been mechanically removed and sold for horticultural purposes. For example at North Warren, Suffolk, litter was removed by forage harvesting following rotovation and sold for £1.15 m^{-3} (R. Powell pers. comm.). However, in the Netherlands levels of cadmium and lead precluded the sale of Bracken litter for horticulture (C. Tubbs, pers. comm.). In Surrey, Bracken litter has been removed by burning with no noted adverse effects (S. Nobes pers. comm.).

Heath vegetation may regenerate naturally following stripping of Bracken litter. For example, at Arne, Dorset, Bracken litter (15–20 cm deep) was stripped exposing mineral soil without seed addition. Within 4 years *Calluna*, Bell Heather, Bristle Agrostis *Agrostis curtisii*, Purple Moor-grass and Pill Sedge *Carex pilulifera* had established naturally (Pickess *et al.*, 1989). However, active measures to develop heath vegetation may be important if natural regeneration fails. Methods for collecting and applying heather seed are outlined below for heathland re-creation. At Cavenham Heath the rapid establishment of *Calluna* cover through litter disturbance and seeding

Fig. 10.3. When controlling dense stands of Bracken the removal of litter by a forage harvester can favour the development of heathland vegetation characteristic of nutrient-poor soil rather than rank grassland. (Purbright Common; R. McGibbon.)

did not inhibit Bracken recovery once control measures ceased, despite the widely held assumption that this will occur, similarly at Arne further spot spraying of Bracken was necessary (Pickess *et al.*, 1989).

Scrub control

A combination of mechanical control and herbicide treatment is essential if scrub is to be effectively controlled, particularly as most species resprout after cutting. Disposal of cut material may cause a problem. Burning on site is the least desirable method and other approaches should be considered, such as removal from the site for use as firewood, timber or pulpwood or stacking in existing dense scrub.

Scrub up to 2 m tall may be controlled by foliar spraying with selective herbicides such as triclopyr (trade name Garlon) applied by ultra low volume applicator (ULVA) or fosamine ammonium (trade name Krenite) applied by Knapsack sprayer or mist blower. The Control of Pesticides Regulations 1986 and Health and Safety Executive guidance should be consulted before herbicide use. Trees or large scrub should be cut and the stumps painted with triclopyr, glyphosate (trade name Roundup) or ammonium sulphamate (trade name Amcide). However some may still regenerate, for example up to 40% of large stumps of Rhododendron, *Rhododendron ponticum*, painted with ammonium sulphamate may resprout. A higher success rate can be achieved by drilling holes of 1 cm diameter 8–10 cm deep into the stump surface and filling these with supersaturated ammonium sulphamate solution. Care should be taken in using glyphosate and ammonium sulphamate as both damage heath vegetation including *Calluna*. Alternatively stumps may be hand-pulled, dug up or winched out. Subsequent regrowth should be spot sprayed with glyphosate, applied at about the late leaf bud stage, using a knapsack sprayer, drench gun or rope-wick applicator, seedlings which regenerate in the disturbed ground should be hand-pulled or weed-wiped. When weed-wiping Rhododendron with glyphosate, Mixture 'B' should be added to assist penetration of the waxy leaves (Becker, 1988). MAFF recommend that cattle and ponies be excluded from Rhododendron control areas as the cut material is considered toxic.

Nutrient depletion

Depletion of nutrients from areas of heathland vegetation will help to maintain typical communities, particularly during periods when grazing pressure is low. Both Bracken and scrub, particularly Common Gorse *Ulex europaeus*, but also birch *Betula* spp. and oak *Quercus* spp., can increase soil fertility. Because of this, merely removing dense stands may not result in the development of typical heathland communities characteristic of nutrient-poor soils. Instead undesirable nitrophilous species such as Yorkshire Fog *Holcus lanatus*, Wavy Hair-grass, False Oat Grass *Arrhenatherum elatius* and Rosebay Willowherb *Chamerion angustifolium* often become established. Consequently, nutrient depletion is also advisable following control of scrub or dense Bracken. It is also important in the restoration of heathland on former farmland. Although a number of possible techniques for nutrient depletion exist there has only been limited critical assessment of their relative effects on soil nutrients, vegetation composition and Bracken re-invasion.

Work on grasslands suggests that depletion of nutrient capital by cutting and removal of vegetation may take tens of years or longer to achieve; however, the effect of cutting and removal of dwarf shrub vegetation on heathland nutrient budgets has not been assessed. It has been suggested that continuous fallowing of soil, for example by repeated rotovation, may allow the loss of soluble nutrients by leaching. In calcareous Breckland grass heath repeated annual rotovation (10–14 times) reduced soil organic matter suggesting that leaching losses had occurred, and had beneficial long-term effects on vegetation (Dolman & Sutherland, 1992). However, trials of repeated rotovation in nutrient-poor calcareous grass heath did not find any effect on nutrients in the first three years of treatments suggesting that rotovation may need to be repeated many times (Dolman & Sutherland, 1994).

At Blackheath, Surrey, experimental management was carried out in an area of degenerate heathland dominated by Wavy Hair-grass and with no regeneration of the over-mature *Calluna*. Effects of rotovation and turf removal were compared. Turfs were removed to about the bottom of the root mass; this corresponded to the junction between organic topsoil and mineral sand. Both rotovation and turf stripping resulted in good regeneration of *Calluna* from seed. Turf stripping removed nutrients in the humus layers and exposed nutrient-poor subsoil, while rotovation mixed turf and organic rich litter into the topsoil. After three years the turf-stripped area retained much bare sand and had very little grass, while Wavy Hair-grass was spreading vigorously throughout the rotovated area (R. McGibbon, pers. comm.). In addition in the Netherlands, where replacement of heather vegetation by grassland is attributed to high nitrogen deposition, cutting and removal of turfs consisting of live biomass, litter and the top 2 cm of organic soil has also been successful in regenerating *Calluna* (Helsper, Glenn-Lewin & Werger, 1983). Removal of turfs and topsoil provides a mosaic of pioneer vegetation and bare sand adjoining untreated areas. This is particularly beneficial to specialist invertebrates and early successional plant species including mosses and lichens. If turfs are stripped too deeply the *Calluna* seed bank in the organic layer may be entirely removed; however, in such cases regeneration may be achieved by seeding.

Turf stripping may be carried out using a multiscraper attached to a tractor, while large areas may be treated using a long arm excavator, see Figure 10.4. At Arne, Dorset, litter and the upper 2.5 cm organic horizon were removed from 8 ha of former heathland following clearance of mature naturally regenerated pine. Litter and organic soil were first loosened with a polypropylene/steel brush driven by the PTO of a tractor, then manually swept into rows and collected by trailer using a custom-made vacuum

Fig. 10.4 Turf-stripping using a long-arm excavator in order to remove rank vegetation and hence nutrients prior to re-seeding the heath. (Albury Heath; R. McGibbon.)

unit operated from the tractor PTO (B.P. Pickess, pers. comm.). Large-scale treatment presents problems of disposal, local requirements for soil could be exploited such as capping of land-fills or road construction. Mass regeneration of birch from seed is often a problem after soil disturbance. This is best controlled by grazing, otherwise hand-pulling or chemical control is necessary.

Firebreaks

Accidental fires and arson pose a serious threat to heather heathland. Intense summer fires can severely degrade a heath, killing *Calluna* rootstocks, delaying regeneration and leading to topsoil erosion and invasion of grasses and Bracken. The impoverishment of the lichen and moss flora of old *Calluna* has been attributed to repeated accidental fires. Unchecked accidental fires may burn a large part of a site. If this occurs crucial habitats such as mature and degenerate *Calluna* may be lost, leading to local extinction of Sand Lizard, Smooth Snake and many specialist invertebrates. Consequently, a fire plan will be needed, with permanent firebreaks at least 10 m wide maintained by rotovation or close cutting. These may benefit species which require early successional habits, for example, Woodlark breed on firebreaks in Surrey.

Grazing

Grazing was the most important aspect of historic land use and management. On grass heath, grazing is the only viable long-term management option. On large heather-dominated heathlands grazing is the most cost-effective management method for maintaining a diverse habitat structure and suppressing succession. Selective grazing may suppress grass, favouring *Calluna* and creating open microhabitats. In contrast to management exclusively by cutting or burning, grazing favours invertebrates associated with dung.

Grazing can be achieved with a commercial grazer although this may not allow the flexibility or intensity of grazing required for optimum site management. Alternatively a management body may purchase stock. At the outset, economic, technical and welfare issues must be thoroughly considered. Transport costs may be high if many small sites are grazed by the same stock. The cost, type and scale of fencing needs to be considered as well as implications for public access and legal aspects of grazing common land. Water requirements must be planned in advance, while impacts on ponds, streams and fringing vegetation should be considered and controlled. Stock may require supplementary feeding. If this is undertaken on site localised poaching and nutrient enrichment can occur, which is particularly serious on small sites. Box 10.3 gives details of a sheep grazing project run in the Suffolk Sandlings.

There is a wide choice of grazing animal. Historically, sheep grazing and commercial Rabbit warrening were important on East Anglian heaths, while in southern England

Box 10. 3. Grazing management by the Suffolk Sandlings Project

Following a reclamation phase the Sandlings Project used sheep grazing as part of its maintenance programme (FitzGerald 1992). A flock was purchased and managed by project staff and volunteers: 150 Beulah Speckled-face ewes, which are relatively hardy but also productive, were purchased (at a cost of £5760, 50% grant-aided) and crossed with Blue-faced Leicester rams (cost for 4 rams: £1081, 50% grant-aided) to produce saleable lambs. Heathland is grazed at 1 ewe + followers ha^{-1} during May–September, the flock is transferred to meadow and marshland reserves for September–December and is in-wintered from Christmas for lambing in February. During 1990 105 ha of heathland was enclosed, using flexible three strand electric fence (cost £3000, 50% grant-aided) which may also be used on other reserves. Water is supplied by a tractor with mobile bowser. Sheep are sponsored on an annual basis, in 1990 at £30 per ewe and £100 per ram raising £5600. Other income in 1990 was: £6720 from lamb sales, £900 ewe premium and £312 from wool sales. Additional running costs were £4985 for feed, transport and veterinary services, giving an overall economic surplus towards capital costs and salaries.

Although 1990 was a drought year with poor grazing, all lambs were raised (1.38 per ewe) and were well grown and healthy, while the condition of ewes after weaning leads to the conclusion that in most years the flock will retain reasonable body condition while raising lambs from summer heathland grazing.

cattle, sheep and ponies were used. Stock used for management should be hardy and easy to overwinter. Cattle, particularly Galloway, native pony breeds and hill sheep such as Beulah Speckled Face, Swaledale, Scottish Blackface, or rare sheep breeds such as Soay, Hebridean or St Kilda, are all suitable. There is a need for further monitoring of the long-term effects of different grazing regimes in a variety of lowland heathlands.

A period of intensive cattle grazing may be suitable for restoring old stands of *Calluna* (Gimingham, 1992). However, care should be taken with such one-off intensive grazing as it is important to maintain a full range of vegetation structure within a site. Cattle graze pine, eating stems as well as needles. Winter grazing of cattle on humid heath is effective in controlling Purple Moor-grass as dead litter and rank grass are eaten (R. McGibbon, pers. comm.). Traditional pony grazing continues in the Lizard Peninsula and in the New Forest combined pony and cattle grazing maintains the single largest area of lowland heathland in Britain (Tubbs 1992). Ponies selectively graze Purple Moor-grass and may be useful in reducing its dominance, favouring regenerating *Calluna*. If grazed on heathland largely dominated by dwarf shrub vegetation, native pony breeds may require supplementary feeding or periods of grazing on grassland, particularly in winter. Experience in Surrey shows that ponies do not eat pine, are reluctant to browse birch, and only lightly graze oak or sallow *Salix* spp., consequently unless grazed so hard that they lose condition they are not considered useful for scrub control (R. McGibbon, pers. com.). In contrast, goats have been used to reduce scrub on degenerate heathland. In Surrey they browsed birch in leaf, ate pine in winter and later ring-barked pine and birch killing trees, as well as grazing Purple Moor-grass (R. McGibbon, pers. com.). However, they also heavily grazed *Calluna* to the extent that old leggy bushes could be killed.

Sheep are often favoured for heathland management as they are considered to be more manageable than other stock (Gimingham, 1992). Grazing by sheep and a moderate level of trampling of degenerate stands favour *Calluna* regeneration, by regrowth from stem bases, layering and seedling regeneration in bare areas. Sheep may graze pine in addition to birch (C. FitzGerald, pers. comm.). Regeneration from cut birch stumps may be selectively grazed by sheep in early spring and summer, successfully killing trees without the need for chemical stump treatment or follow up spraying of regrowth with herbicide. This has been demonstrated both in Surrey and Dorset (Pickess, pers. comm.). Trials in Surrey found that Hebridean sheep will eat Purple Moor-grass (R. McGibbon, pers. comm.), although it is considered that most breeds will not (C. Tubbs, pers. comm.).

Precise stocking rates cannot be given as these depend on *Calluna* productivity and phase, the effects of nutrient availability and climate on the competitive balance between grass and *Calluna* and the relative proportions of grass heath and *Calluna* within a site. However, for sheep a rate of 1–2 ewes ha^{-1} is generally used (Gimingham, 1992; FitzGerald 1992). If more than 30–40% of the annual growth increment of *Calluna* is grazed it may be replaced by grassland or Bracken. Some authorities consider ponies and cattle to be preferable as, in mixed heathland, sheep are more likely to convert extensive areas of *Calluna* into grass heath (Tubbs 1992). This may occur at stocking densities necessary to control scrub; for example, if present all year 2–3 sheep ha^{-1} may convert *Calluna* into grass swards (Tubbs 1992). Conversely undergrazing does not maintain *Calluna* in the building phase, but allows it to over-mature and degenerate and therefore may not control succession. *Calluna* is particularly susceptible to grazing damage during September–October and is least vulnerable in winter. Where grass is available this is preferentially grazed in the summer with *Calluna* being grazed mainly in the winter (Gimingham, 1992).

In addition to management by stock, sites may be affected by Rabbit grazing. Rabbits create a mosaic of microhabitats, with closely grazed vegetation, clumps of unpalatable ruderal species, and areas of disturbed soil (see Figure 10.5). They can create lichen-rich heath with many annuals and winter annuals (Dolman & Sutherland, 1992) which favours invertebrates requiring the hot microclimate of this open, disturbed, early seral vegetation. Rabbit grazing is particularly effective in suppressing regeneration of Common Gorse. This may be advantageous in many situations, however it may cause problems in areas such as the New Forest where Gorse is valued for Dartford Warblers, invertebrates and shelter and winter feed for ponies. In most situations rabbit populations are now unlikely to exert sufficient pressure to lead to the loss of *Calluna* cover. In degenerate stands rabbits may encourage regeneration from seed.

The presence of rabbits within some part of a heathland site is therefore likely to be beneficial in most situations; however, if high density populations spread throughout a *Calluna* heath then the whole area may be closely grazed with a loss of structural diversity. The major tasks for a site manager who wishes to encourage rabbits are to prevent their persecution; to reduce predation by controlling predators and if necessary, control their spatial distribution by culling. If rabbit populations are to be encouraged in areas adjacent to agricultural or forestry land then fencing or localised control will be necessary.

Burning

Periodic rotational burning of *Calluna* is commonly used to manage extensive heathland, but is less appropriate on sites less than *c*.160 ha due to the need to retain a mosaic of age classes (Andrews, 1990). Although Webb & Haskins (1980) and C. Tubbs (pers. comm.) consider that periodic burning was part of traditional grazing management in Dorset and the New Forest, Gimingham (1992) considers that there is little evidence that burning was a regular practice on heathland in southern England. Unlike grazing, a well-controlled burn may remove accumulated litter, favouring specialist invertebrates of bare sand (Stubbs, 1983). Controlled burning appears compatible with maintaining lichen-rich *Calluna* heathland (C. Tubbs, pers. comm.).

Burning produces an even-aged stand of heather, consequently sites managed exclusively by rotational burning consist of relatively large-scale mosaics rather than the small-scale patchwork which is an ideal objective. Following an intense burn, most heathland invertebrates will need to recolonise the area by dispersal from nearby unburnt areas, consequently only small blocks should be burnt. At Arne blocks range from 0.1–1 ha (Pickess *et al.*, 1989), but others recommend 0.5–2 ha (Gimingham, 1992). Ideally patch margins should be scalloped or meandering. This creates diverse and sheltered microhabitats important for invertebrates and facilitates recolonisation and seed dispersal into the block due to the increased edge. Wet heath and grass heath should generally be excluded from the burning rotation.

Burning results in the loss of some nutrients through smoke and leaching of minerals from ash, thereby favouring heathland vegetation. Detailed nutrient budgets have been calculated for Dorset heaths but it is now known that the ability of soil from other heathland areas to retain phosphorus can be much higher (Chapman *et al.*, 1989), so that losses through leaching may be lower. Consequently cutting and removal of heather and litter may be more effective than burning in removing nutrients in areas other than Dorset (Chapman, 1992).

Regeneration is most effective when building-phase *Calluna* is burnt and declines with the age of the stand. The length of rotation will depend on various factors, particularly the rate of growth of the heather and grazing levels; English Nature recommend a rotation of 6–12 years for 'Eastern Heaths', Gimingham (1992) recommends 10–15 years, while longer rotations have been used, for example over 20 years in the New Forest (C. Tubbs, pers. comm.) and 30 years on very nutrient poor soils at Arne, Dorset (Pickess *et al.*, 1989). Back-burning, burning into the wind, gives a hot burn. When managing old *Calluna* which is not expected to regenerate vegetatively, back-burning may favour regeneration from seed through the removal of accumulated litter and the exposure of bare sand. Burning with the wind produces a cooler burn which may not consume litter. Compared with back burning this is less effective at depleting nutrients and creating bare sand for invertebrates, but favours vegetative regeneration particularly in younger *Calluna*. Where *Calluna* regeneration is absent or slow, prolonged exposure of bare ground may lead to erosion and invasion by Bracken and in the absence of grazing, grasses or birch. In such cases disturbing litter and topsoil by rotovation often increases seedling germination, alternatively *Calluna* may be reseeded as described below.

Areas to be burnt may be separated from adjacent heath by meandering firebreaks of at least 4 m, produced by rotovation or cutting. A temporary firebreak may be produced by spraying sodium alginate onto vegetation. This requires special machinery, particularly agitators, but could be cost-effective if labour is short (Pickess *et al.*, 1989).

(a)

(b)

Fig. 10.5. Heavy grazing, in this case by rabbits, can have a huge impact on plant composition and vegetation structure on nutrient-poor heaths. These pictures show the effect of excluding rabbits on (*a*) grass- and (*b*) heather-dominated heaths at Ropers Heath in the Brecklands of Norfolk and Arne in Dorset respectively.
(J. Andrews.)

Burning in February is preferable, and is prohibited after March. Burning is regulated by the Heather and Grass Burning Regulations 1986 (and amendments) explained in a code of practice (MAFF, 1992) which should be consulted for practical guidance.

Cutting

Cutting has similar effects on *Calluna* as burning and is a useful alternative where the latter is unacceptable due to small site size or public perceptions. In *Calluna* there is generally good regeneration from buds below the cut as more are left undamaged compared with burning. On older *Calluna* cutting may be more effective than burning which frequently kills plants. There is a difference of opinion as to whether to cut in spring or autumn. The former may cause more physical damage but it allows rapid regeneration and could reduce surface erosion, however it is advisable not to mow when there is frost. Instead of cutting uniform blocks, a narrow sinuous strip may be cut, giving a greater edge and facilitating recolonisation by dispersal from adjacent areas. Overlapping strips mown in different years produces a diverse structural mosaic.

It is important to remove cut material, this will remove some nutrients from the site. In contrast, leaving material in place after flailing or swiping effectively adds an organic mulch to the site. This may directly increase levels of soil nutrients and retention of soluble phosphorus, which will in turn favour the development of rank grassland and scrub (Chapman *et al.*, 1989; Gimingham, 1992). In addition the litter and organic matter may retain moisture compared with drier sandy soils, favouring Wavy Hair-grass or Purple Moor-grass. On some heaths in the Weald swiping without removal is thought to have accelerated the loss of heather and development of rank grassland (K. Hearn, pers. comm.).

A suite of machinery is available for cutting heathland, but careful evaluation of cost-effectiveness, financial resources, labour skills and the area and terrain to be cut is necessary. Reciprocating mowers may be useful in small areas and are manoeuvrable. A double-chop forage harvester can be used with a trailer to collect litter for removal, but this requires smooth terrain. A forage harvester used without a trailer spreads fine cuttings over the site. Tractor-mounted swipes and flails may be used on more uneven ground but leave cut vegetation and litter. Flails disturb soil litter and expose areas of bare sand, favouring seed germination and pioneer invertebrates. If litter and cuttings cannot be removed they may be raked into wind rows resulting in more localised concentrations of deep litter (Gimingham, 1992).

Heathland re-creation

There is increasing interest in re-creating heathland on arable farmland and forestry plantations. Its value in linking small heathland fragments and in creating buffer zones around existing sites cannot be over-emphasised. Re-creation is likely to become increasingly important, due to incentives to landowners through Environmentally Sensitive Areas and Countryside Stewardship. The importance of targeting habitat re-creation near or adjacent to existing heathland fragments is increasingly recognised. Most re-creation focuses on the re-establishment of *Calluna* heathland. However in the Breckland ESA attempts are also being made to restore grass heath vegetation. Techniques for heathland re-creation are also applicable to other situations such as habitat restoration after mineral extraction. No definitive techniques for restoration have yet been developed. Those prescriptions which have been developed await critical testing in comparison to alternative treatments while much *ad hoc* restoration is being undertaken, often without sufficient monitoring. Below we outline a few current approaches.

Attempts to re-create heathland on arable farmland may be hindered by residual soil fertility, particularly high levels of extractable phosphorus due to fertiliser inputs. This may result in communities dominated by undesirable species. Before attempting re-creation soil fertility should be assessed by chemical tests. Arable reversion prescriptions generally include a period of phosphate stripping by the cultivation and harvesting of cereal crops. Initial assessment of this approach, at Ropers Heath in Breckland, showed that harvesting spring barley in the first year followed by two years of rye with no fertiliser inputs reduced soil fertility to some extent (Marrs, 1986). However, it was not clear to what extent extractable phosphorus had been reduced by this treatment or how far it would need to be reduced before nutrient-poor heathland communities could develop. Subsequent sheep and Rabbit grazing has now created short, mesotrophic grass heath, while *Calluna* is spreading from adjacent heath. Prescriptions for reversion by cereal cropping now include limited additions of nitrogen, to increase uptake of phosphorus by crops and its subsequent removal.

Areas of heathland adjacent to Stoborough Heath, Dorset, were converted to arable in the late 1960s and abandoned in the late 1970s. In one of these old fields *Calluna*, *Erica* and Dwarf Gorse *Ulex minor* established, leading Smith *et al.* (1991) to conclude that where nutrient levels are very low and there is a seed source then heath vegetation may be established without the need for nutrient stripping by cereal cropping. Experiments on grassy fields which lacked heath vegetation showed that cutting had no significant effect on the development of heathland vegetation, but turf stripping and rotovation with the addition of heather cuttings both aided heathland restoration (Smith *et al.*, 1991).

In re-creating heathland the opportunity should be taken to create microhabitats such as depressions, gullies and sunny banks to diversify vegetation and benefit reptiles and invertebrates. In many cases active measures may be necessary to re-establish heathland vegetation. For grass heaths techniques developed for re-creating other types of grassland may be applicable. A number of techniques have been developed for

Fig. 10.6. A species of conservation importance Wild Gladiolus *Gladiolus ilyricus* growing on Beaulieu Heath, Hampshire. (J. Geeson.)

re-creating *Calluna* heathland, generally by collecting seed from existing heathland, whether in the form of litter, soil or heather cuttings and applying this to the recipient site. These techniques are described in detail elsewhere (Putwain & Rae, 1988; Pywell 1992), however, a brief summary is presented here.

Heather tops may be collected between September and December, before seed falls, using a double-chop forage harvester and trailer when creating firebreaks or cutting stands as part of a management programme. Further seed collection at other times of the year increases the range of heathland species applied in the restoration. Spreading on the recipient site can be done by hand or by muck spreader, either immediately after collection or with material stored in dry conditions over-winter. The ratio of cut to seeded land is approximately 1:2–3. Alternatively, litter may be collected from existing heathland, by hand or by suction devices used for clearing leaves, and used to seed recipient sites. Collection in early spring may be preferable as the previous year's seed will have fallen. One disadvantage of this method is the possible incorporation of excessive amounts of gorse or birch seed. Dried litter can be stored for 3–4 years.

Translocation of topsoil may be an option, especially where the donor site faces destruction. Before stripping it is advisable to assess the seed bank profile with depth. Topsoil may be stored in shallow heaps for 2–3 months in winter but for only a few weeks in summer before viability of seeds of many species may be lost. Topsoil should be spread to a depth of 2–3 cm, matching drainage and topographical characteristics between the donor and the recipient areas. Advantages are that establishment is often rapid, a greater assemblage of heath species are obtained and the cost is low. Disadvantages are the short storage time, especially in summer, and the scale of soil removal may not be acceptable for the donor site.

Protection of the new site from erosion, desiccation and loss to windblow is important. Options include creating a lightly compacted ridge and furrow by a Cambridge roller, spreading forestry brash or cut *Calluna*, or using grass companion crops. It may be necessary to protect the recipient site from grazing for up to 5 years as well as controlling successional species and grasses. Low inputs of lime may be required to achieve the optimal pH for *Calluna* of 3.8–4.

Although it may be possible to translocate heathland vegetation, the re-created heath will not contain a complete faunal assemblage and should not be seen as a substitute to the effective protection and management of the original site.

References

Andrews, J. (1990). Wildlife habitat management: The management of lowland heathlands for wildlife. *British Wildlife* 1, 336–46.

Auld, M.H.D., Davies, S. & Pickess, B.P. (1992a). Restoration of lowland heaths in Dorset. *RSPB Conservation Review* 6, 68–73.

Auld, M.H.D., Pickess, B.P. & Burgess, N.D. (eds.) (1992b). *Proceedings of Heathlands Conference. II. History and Management of Southern Lowland Heaths*. Sandy: RSPB.

Becker, D. (1988). *Control and Removal of* Rhododendron ponticum *on RSPB Reserves in England and Wales*. Sandy: RSPB.

Burgess, N., Evans, C. & Sorensen, J. (1990). Heathland management for Nightjars. *RSPB Conservation Review* **4**, 32–5.

Chapman, S.B. (1992). Plant nutrient relationships in the ecology and management of heathlands in lowland Britain. In *Proceedings of Heathlands Conference. II. History and Management of Southern Lowland Heaths*, eds. M.H.D. Auld, B.P. Pickess & N.D. Burgess. pp. 13–17. Sandy: RSPB.

Chapman, S.B., Rose, R.J. & Basanta, M. (1989). Phosphorus adsorption by soils from heathlands in southern England in relation to successional change. *Journal of Applied Ecology* **26**, 673–80.

Dolman, P.M. & Sutherland, W.J. (1992). The ecological changes of Breckland grass heaths and the consequences of management. *Journal of Applied Ecology* **29**, 402–13.

Dolman, P.M. & Sutherland, W. J. (1994). The use of soil disturbance in the management of Breckland grass heaths for conservation. *Journal of Environmental Management*, **41**, 123–140.

Farrell, L. (1989). The different types and importance of British heaths. *Botanical Journal of the Linnean Society* **101**, 291–9.

FitzGerald, C. (1992). The reintroduction of grazing on the Suffolk Sandlings. In *Proceedings of Heathlands Conference. II. History and Management of Southern Lowland Heaths*, eds. M.H.D. Auld, B.P. Pickess & N.D. Burgess. pp. 27–33. Sandy: RSPB.

FitzGerald, C., Martin, D. & Auld, M. (1985). *Report of the Sandlings Project 1983–1985*. Saxmundham: Sandlings Group, Suffolk Wildlife Trust.

Gimingham, C.H. (1992). *The Lowland Heathland Management Handbook*. Peterborough: English Nature.

Helsper, H.P.G., Glenn-Lewin, D. & Werger, M.J.A. (1983). Early regeneration of *Calluna* heathland under various fertiliser treatments. *Oecologia* **58**, 208–14.

Kirby, P. (1992). *Habitat Management for Invertebrates: A Practical Handbook*. Sandy: RSPB.

Lowday, J.E. & Marrs, R.H. (1992). Control of bracken and the restoration of heathland. I. Control of bracken. *Journal of Applied Ecology* **29**, 195–203.

MAFF (1992). *The Heather and Grass Burning Code*. London: HMSO.

Marrs, R.H. (1986). Techniques for reducing soil fertility for nature conservation purposes: a review in relation to research at Ropers Heath, Suffolk, England. *Biological Conservation* **34**, 307–42.

Marrs, R.H. & Lowday, J.E. (1992a,b). Control of bracken and the restoration of heathland. II. Regeneration of the heathland community. III. Bracken litter disturbance and heathland restoration. *Journal of Applied Ecology* **29**, 204–11, 212–17.

Pickess, B.P., Burgess, N.D. & Evans, C.E. (1989). *Management Case Study: Heathland Management at Arne, Dorset*. Sandy: RSPB.

Putwain, P.D. & Rae, P.A.S. (1988). *Heathland Restoration: a Handbook of Techniques*. Liverpool: British Gas plc (Southern) and University of Liverpool Environmental Advisory Unit.

Pywell, R.F. (1992). Heathland Translocation and Restoration. In *Proceedings of Heathlands Conference. II. History and Management of Southern Lowland Heaths*, eds. M.H.D. Auld, B.P. Pickess & N.D. Burgess. pp. 18–26. Sandy: RSPB.

Rodwell, J.S. (ed.) (1991). *British Plant Communities. Volume 2: Mires and Heaths*. Cambridge: Cambridge University Press.

Smith, R.E.N., Webb, N.R. & Clarke, R.T. (1991). The establishment of heathland on old fields in Dorset, England. *Biological Conservation* **57**, 221–34.

Stubbs, A.E. (1983). The management of heathland for invertebrates. In *Focus on Nature Conservation. Number 2: Heathland Management*, ed. L. Farrell, pp. 21–35. Shrewsbury: Nature Conservancy Council.

Tubbs, C.R. (1992). The management of heathland in the New Forest, Hampshire. In *Proceedings of Heathlands Conference II. History and Management of Southern Lowland Heaths*, eds. M.H.D. Auld, B.P. Pickess & N.D. Burgess. pp. 13–17. Sandy: RSPB.

Webb, N.R. (1986). *Heathlands*. London: Collins.

Webb, N.R. (1989). Studies on the invertebrate fauna of fragmented heathland in Dorset, UK, and the implications for conservation. *Biological Conservation* **53**, 253–64.

Webb, N.R. & Haskins, L.E. (1980). An ecological survey of heathlands in the Poole Basin, Dorset, England, in 1978. *Biological Conservation* **17**, 281–296.

Webb, N.R. & Vermaat, A.H. (1990). Changes in vegetational diversity on remnant heathland fragments. *Biological Conservation* **53**, 253–64.

11 Upland moors and heaths

DESMOND B.A. THOMPSON, ANGUS J. MACDONALD
AND PETER J. HUDSON

Introduction

This is a chapter about myths and magic. The myth is that the uplands of Britain are well managed and self-perpetuating with a beauty and wildlife fashioned by the kinder elements of nature and humans! The magic is that in just a few regions – amongst some of the hills and on just a few estates – there are heaths and bogs, woodlands and grasslands, abounding in plants and animals that bring the meaning of 'sustainable biodiversity' alive. For the good keeper of an estate, manager of a National Nature Reserve or the inspired visitor the distinction is strikingly obvious. How then do we spread this magic across the uplands?

The uplands are defined here as the third of Britain's land surface lying above the upper reaches of enclosed farmland (Ratcliffe & Thompson, 1988). They may appear to be the product of millennia of deforestation and wilful burning which has produced a distinctively open, apparently stunted range of habitats managed principally for forestry, sheep, Red Grouse *Lagopus lagopus* and Red Deer *Cervus elaphus*. But, they still retain much of natural interest and conservation value, and a sense of wildness not found elsewhere.

As a land manager one has to ask four questions:

1. What am I to manage the land for?
2. How do I achieve these goals?
3. Where do I begin?
4. Over what timescale should I measure my success?

This chapter provides a synoptic answer to these questions, and indicates where further information can be obtained.

Upland communities

The uplands are composed mainly of dwarf shrub heaths, grasslands and peat bogs, with more rugged mountain areas to the north and west, but more gently contoured rolling hills to the east and along the Pennines. When forest cover was at its greatest, about 5000 years ago, most of the ground below 700 m in the south and east, and below 450 m in the north and west, was wooded, apart from parts of the Outer Hebrides, Orkney, Shetland and Caithness (Birks, 1988). However, a change to a wetter, cooler climate and clearance by Mesolithic hunter-gatherers was also occurring at this time, increasing the extent of open heaths, rough grasslands and bogs now so characteristic of the British uplands (Birks, 1988; Simmons, 1990).

International importance

The assemblages and particular variation of habitats and wildlife in the British uplands as seen today are unique, principally for four reasons. First, the climate is highly oceanic with small seasonal variation in cloudiness, rainfall and temperature, high humidity and precipitation, little sunshine and high winds. Second, there is a spectacular mixture of plants drawn from different regions of the world varying from the high Arctic, through boreal and temperate areas, to alpine and continental parts. Third, there is the long history of grazing-range management for sheep and Red Deer, which represents a continuity of slowly evolving management activity stretching back many thousands of years. One development of this, management for the Red Grouse, is virtually unique. Fourth, the variation in geology, soils and topography gives rise to marked life zone gradients within small regions which are paralleled in the rest of Eurasia only across great latitudinal gradations.

A quarter of the 81 upland plant communities described by the National Vegetation Classification (Rodwell, 1991, 1992) are globally rare or particularly well represented in the British Isles in a world context (Thompson & Sydes, in press). The bird assemblages on blanket bogs and in high mountain areas are unique in species mixture, and important as outliers of largely arctic Eurasian distributions. At least six birds breed at remarkably high densities, or in large concentrations: Golden Eagle *Aquila chrysaetos*, Peregrine Falcon *Falco peregrinus*, Dotterel *Charadrius morinellus*, Red Grouse, Ptarmigan *Lagopus mutus* and Raven *Corvus corax*. Well-burned grouse moor has a remarkably high density of Red Grouse and Golden Plover *Pluvialis apricaria*, while sheep walk supports high density populations of scavenging and predatory birds (Ratcliffe & Thompson, 1988, Ratcliffe, 1990).

Moors and heaths

The main moorland communities are dominated by Heather *Calluna vulgaris*[1] with Bilberry *Vaccinium myrtillus* and Heather with Hare's-tail Cottongrass *Eriophorum vaginatum* (Fig. 11.1). These represent the classic Heather moor and Heather-dominated blanket bog, respectively. The former is most extensive in the eastern Scottish Southern Uplands, the southern Pennines and the North York Moors. In the eastern Highlands, Bearberry *Arctostaphylos uva-ursi* is often abundant in heather communities on the better soils (Fig. 11.2). In western Scotland the classic dry heather moor gives way to wet heaths with Purple Moor-grass *Molinia caerulea*, Deer Sedge *Trichophorum cespitosum* and mosses. Cross-leaved Heath *Erica tetralix* is found most extensively in the wetter communities throughout the Welsh Uplands and in the western Highlands of Scotland (Fig. 11.3). Where grazing pressures are higher, especially towards the south, Bilberry, Cowberry *Vaccinium vitis-idaea*, and Crowberry *Empetrum nigrum* (Fig. 11.4) often become more prominent. This is particularly noticeable in parts of the Yorkshire Dales, Peak District and Welsh uplands.

[1] In this chapter 'Heather' refers specifically to the plant *Calluna vulgaris*, while 'heather' refers to ericaceous dwarf shrub vegetation of which *Calluna* is a dominant or prominent component.

Fig. 11.1. Hare's tail Cottongrass in flower. It can become completely dominant where there has been prolonged heavy grazing and burning, and atmospheric pollution, as in the southern Pennines. Sheep and grouse relish the nutritious young flowering shoots which appear in early spring. (A.J. MacDonald.)

Fig. 11.2. Bearberry is a common associate of Heather on the moorlands of the eastern Highlands. It is a creeping subshrub and temporarily colonises gaps among the taller Heather or where the Heather canopy has been removed by burning. (A.J. MacDonald.)

These broad vegetation types can be divided into a much greater number of identifiable plant communities. In total, the National Vegetation Classification (Rodwell, 1991, 1992) describes 10 woodland, 15 heath, 29 bog and 27 grass- or sedge-dominated communities which occur in the uplands. These often occur in complex mosaics and, in the open range conditions which normally apply, management can rarely be directed at single plant communities (see Thompson *et al.*, 1995a).

Objectives for conservation and management

Why do we need to manage upland moors and heaths? There are at least two answers.

First, what you obtain from your upland estate depends upon what you have to start with, your overall goal and how you pursue this. Are you managing for large bags of Red Grouse and, if so, have you the right mixture of habitat mosaics, sufficient bog flushes for broods, and a means of reducing the impact of the nematode parasite *Trichostrongylus tenuis*? Or are you managing the hill principally for Red Deer, for sheep, or as an area to attract visitors? Perhaps you are concerned with conserving the broad

Fig. 11.3. Cross-leaved Heath is a common dwarf-shrub in wet heaths and bogs.
(A.J. MacDonald.)

Fig. 11.4. Crowberry is a sprawling dwarf-shrub that occurs widely in the uplands. It can be the sole remaining dwarf-shrub in the Cottongrass dominated moors of the southern Pennines. (A.J. MacDonald.)

range of wildlife, in which case you may wish to enhance the more natural and structurally diverse features, such as relics of native woodland, diverse dwarf shrub heaths, subalpine scrub and tall herb communities.

Second, there is mounting concern about the way in which the uplands of Britain have dwindled in character, with the more natural habitats altered or becoming wholly unnatural. Since 1945 just under a quarter of the British upland land surface has been transformed. Much of this has occurred through substantial afforestation and agricultural reclamation (Fig. 11.5), but also through the steady attrition of heather moorland into grassier and less diverse moors (Fig. 11.6), so obvious along the thousands of fence-lines where grazing pressures from sheep have been higher on one side than on the other (Figs. 11.7, 11.8).

In England and Wales, 70% of the existing heather moorland is estimated to be at risk of loss, with over 50% in 'poor' or 'suppressed' conditions liable to further loss and damage if high numbers of grazing sheep persist (that is, at over 2 ewes per hectare (Bardgett & Marsden, 1992)). Comparable figures for Scotland are not yet available, although the Scottish Natural Heritage National Countryside Monitoring Scheme estimates, from aerial photograph interpretation, that some Scottish regions have suffered substantial heather losses between the 1940s and the 1970s, e.g. losses of dry

Fig. 11.5. There has been widespread loss of moorland habitat around the periphery of the uplands as a result of heavy grazing and agricultural improvement, as can be seen in this view from Skiddaw. (D.B.A. Thompson.)

Fig. 11.6. Fragmentation of heather-dominated vegetation by heavy grazing over a prolonged period in the Moorfoots of the Southern Uplands, Scotland. (A.J. MacDonald.)

Fig. 11.7. The effects on the vegetation of differential grazing pressures can clearly be seen in this fenceline example from Shetland. (A.J. MacDonald.)

heather moorland of 63% in Dumfries & Galloway (Scottish Natural Heritage, 1992) and 26% in Grampian (Nature Conservancy Council and Countryside Commission for Scotland, 1988). The most important cause of loss was afforestation, but conversion to unimproved grassland by chronic heavy grazing was the next most important factor.

The percentage of plant communities which have been modified, or are susceptible to modification by various agents of change, are as follows: afforestation 55%, agricultural reclamation 24%, heavy grazing pressure 23%, recreation-related disturbance 28% and acidic deposition 28%. However, when we quantify the proportion of communities that has actually suffered at least 5% loss or modification since 1945 then some of these figures change quite markedly to: afforestation 52%, heavy grazing pressure 23%, agricultural reclamation 11%, recreation-related impacts 4% and acidic deposition 4% (Thompson and Sydes, in press). This emphasises the major impacts of afforestation and heavy grazing pressures from sheep on community diversity across the Uplands.

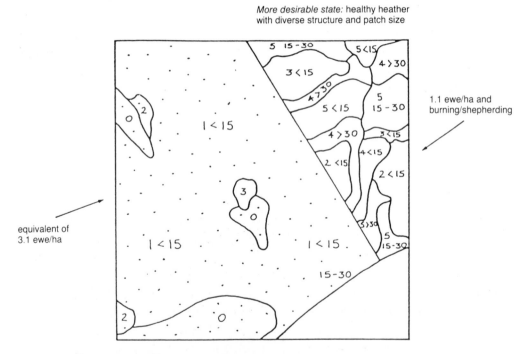

Fig. 11.8. An actual 1 km × 1 km square of moorland showing Heather condition in each of two different management units. The first figure gives percentage cover of Heather: 5 (<75%), 4 (75–51%), 3 (50–26%), 2 (25–10%), 1 (10–1%) and 0 (absent). The second figure gives the average height of Heather: >30 cm, 15–30 cm, <15 cm. (From Thompson *et al.*, 1995b).

Of the birds associated with heather moorland (at least 35 species), 35% have declined since the 1970s mainly due to afforestation, persecution and habitat deterioration under heavy grazing. Nine of these species have declined because of expanding afforestation, 6 because of persecution, 5 because of heavy grazing pressures, 4 because of land drainage, and 3 because of predation pressure (Thompson *et al.*, 1988; Bibby, 1988).

The nub of this is that important constituents of our natural heritage – the natural biodiversity of our land – are being changed into communities less complex and less diverse than their natural counterparts. There are whole areas and suites of communities now turned over to conifer forestry or to rather bland grasslands that may never return to their former state. Some regard these changes as good for society, with huge tracts of 'waste-land' being made more productive in agricultural and even socio-economic terms. But for those in society moved by the wild beauty of nature and the more pristine landscapes this is an untenable situation, particularly in view of the economic costs involved!

Management practices, and therefore problems encountered, depend centrally upon management objectives. This chapter gives guidance based on the assumption that the manager's overall aim is to maintain or enhance the distinctive natural characteristics of his or her 'patch'. The plant communities making up heather moorland are pivotal in the management of the uplands if these are to maintain their distinctive features and the full range of benefits which they can provide for the whole population.

Management: problems and prescriptions

Burning

Good burning practice maintains the open and diverse character of our moorlands. The more recently burned areas provide grazing for sheep and deer (Fig. 11.9), particularly throughout the winter, and for Red Grouse throughout the year. The older, taller stands provide cover for nesting grouse and other birds. The mixture of different heights of vegetation is important for many moorland species (Table 11.1 and Figure 11.10). The mosaic of heather stands of different ages on a carefully fire-managed moor (Figs. 11.11, 11.12) also provides a range of habitats for a remarkable diversity of invertebrates – though the flora can be rather depauperate (Ratcliffe & Thompson, 1988; Usher & Thompson, 1993). Some species seem to benefit from a very intimate mixture of heather of different ages whilst others may benefit more from a very coarse scale mosaic (Fig. 11.10). On any large block of moorland it is likely that a range of fire sizes will produce the optimum results for conservation. Purely practical concerns may produce this by default if effort is concentrated where the opportunities for burning and its benefits are greatest (see Phillips *et al.*, 1993; Thompson *et al.*, 1995a).

Table 11.1. *Preferred heather heights for foraging, sheltering or nesting of various species*

	Preferred height of foraging	Preferred height for shelter or nesting
Mountain hare	<15 cm	<30 cm + 50% canopy
Sheep	<20 cm	
Red Deer	>25 cm	>25 cm
Red Grouse	10–30 cm	20–30 cm
Black Grouse	20–30 cm?	>40 cm + dense cover. Cover does not have to be of heather
Merlin		Tall (>30 cm?)
Hen Harrier		>60 cm
Twite		>15 cm
Golden Plover	Short	<15 cm
	<10 cm?	Will use vegetation up to 25 cm

Source: Moss, Miller & Allen 1972, Orford 1986, Piccozzi 1986, Ratcliffe 1976, Roberts & Green 1983, Rowan 1921, Savory 1986, Thirgood & Hewson 1987, Thompson *et al.* 1995a,b, Watson 1977, Williams 1981.

Fig. 11.9. Vegetation regenerating after a recent managed fire often attracts high densities of grazing animals. If the area burnt is too small in relation to stocking levels, grazing can be high enough to prevent successful regeneration of dwarf-shrubs. (A.J. MacDonald.)

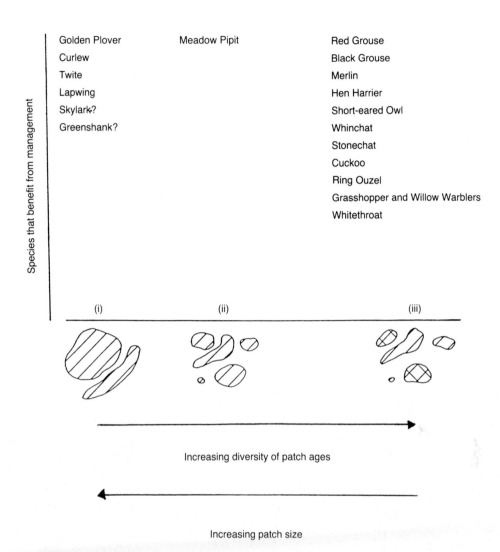

Fig. 11.10. A qualitative description of how large patches of moorland (i) burned in one year, (ii) burned in smaller patches with a more variable mosaic, and (iii) burned over two cycles rather than just one cycle, might affect bird species richness. (Adapted from Usher & Thompson 1993.)

The rate of heather growth, and the maximum heather height required, determines the duration of the burning rotation. For Red Grouse a simple guide involves measuring the number of years heather takes to reach 30 cm. If it takes 10 years then burn on a 10 year cycle as in the southern and eastern uplands. If it takes 25 or more years, as in the western Highlands and Islands, then burn over a quarter-century cycle.

Burning takes place in winter. The legal heather burning ('muirburn') season in Scotland below 450 m is 1 October–15 April, extendable to 30 April, and above 450 m it is October–30 April, extendable to 15 May. In England and Wales the season for heather and grass burning is 1 November–31 March, or 1 October–15 April on land falling within the Severely Disadvantaged Less Favoured Areas, or at other exceptional times under licence from the Ministry of Agriculture, Fisheries and Food or the Welsh Office Agriculture Department. In the west, there are on average only 5–15 days when weather conditions are suitable for heather burning during a burning season, but in the east there may be 10–25 days (Phillips *et al.*, 1993). You should plan to be able to take full advantage of these opportunities for burning.

Initiating a burning plan on a large block of moorland where burning management has been neglected can present problems when grazing animals are also present. If only small areas are burned they may attract such high densities of grazing stock that regeneration of the heather may be prevented. During the first years of the first burning cycle larger areas than would normally be recommended must be burned to prevent such damaging grazing concentration. Figure 11.12 provides an illustration of this for an area where the entire planned burning cycle will be 18 years.

Some areas on moorlands are best excluded from burning management. These include the following:

Fig. 11.11. Well planned and controlled use of fire can produce an intricate mixture of vegetation in patches of varying sizes and heights. This favours many groups of heathland animals and helps maintain the maximum diversity of heathland plants. (A. J. MacDonald.)

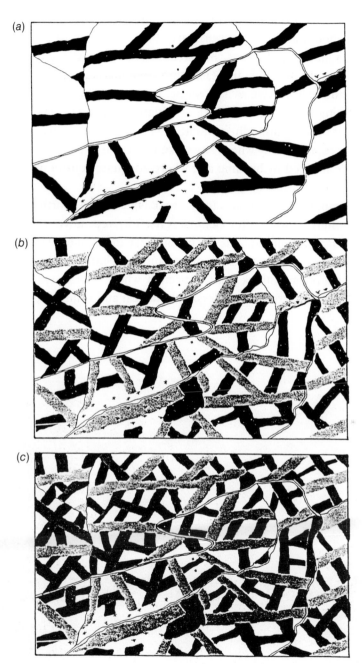

Fig. 11.12. The creation of an intricate mosaic of heather patches of different ages on a block of moorland where management by burning has been long neglected, with the aim of establishing an 18-year burning rotation. (*a*) First year's burning programme, 20% covered in 50 m fires; (*b*) years 4–12, a further 40% is burned making *c*. 60% total by year 12; (*c*) year 12, two-thirds of the way through the rotation and two-thirds burned. (From Phillips, 1991 after Lovat, 1911).

1. Areas where Bracken is present, unless arrangements are made to control Bracken should it invade the burnt area (see section on Bracken control).
2. Exposed summits and ridges, and areas above 300 m in the north-west or about 600 m in the south and east. Here, heather is kept short by wind clipping and may survive as a prostrate, vegetatively layering mat. Regrowth after burning is very slow, the risk of initiating erosion is high, and the benefits of burning are non-existent.
3. Steep slopes, especially where any scree is present, where the often thin soils may erode if bared by burning and where adequate control of a fire is difficult.
4. Steep, damp, sheltered slopes (often back lying and on north or north east aspects) where heather often maintains itself by vegetative layering (Fig. 11.13) and conditions suitable for burning are infrequent. Many rare mosses and liverworts also grow in such places and are eliminated by burning.
5. Wet blanket bogs on thick peats, especially where pool systems are present (Plate 12). These are of considerable botanical and ornithological value and

Fig. 11.13. Heather can maintain itself by layering (producing stem roots) in damp heaths and bogs where creeping stems become buried by the vigorous growth of mosses or accumulation of grass and sedge litter. This Heather stem was over 2 m long, its point of origin was dead, and it was supported entirely by the roots which had been produced halfway along the stem. Most of the other moorland dwarf-shrubs layer even more prolifically. (A.J. MacDonald.)

are likely to be damaged by burning. It is difficult to achieve successful and controlled burning in these situations. There is always a risk of starting a catastrophic peat fire accompanied by disastrous long-term erosion.

6. Patches of tall heather that provide nesting sites for Merlin *Falco columbarius* and Hen Harrier *Circus cyaneus*. These are often at traditional locations and are usually quite localised in extent.

7. Areas of Juniper *Juniperis communis* or Hawthorn *Crataegus monogyna* scrub which are important for nature conservation and may be severely damaged or destroyed by burning.

If moorland burning is being contemplated it must be properly planned, fires must be properly controlled, and you should first learn something about the effects of fire on the vegetation and the land. Some general rules for muirburn are shown in Box 1. Phillips *et al.* (1993) provide an up-to-date code for muirburn in Scotland and MAFF (1992) have produced a detailed code for heather and grass burning in England and Wales.

Cutting

This is not weather-dependent and there are not the risks associated with fire which may spread out of control. A tractor with a forage harvester or flail mower can cut about 8 hectares per day. The cut material is best removed, however, to facilitate regeneration which is arguably less nutritious (Phillips *et al.*, 1993) (Fig. 11.14). Cutting is about 3 times more expensive than burning and can be practised only where the terrain is suitable for the machinery. There appear to be only small differences between cutting and burning so far as invertebrate diversity is concerned (Usher & Thompson, 1993). Cutting is most useful as an aid to fire management. Firebreaks cut with a tractor mounted swipe, immediately before a fire is set, can make the work of controlling a fire considerably easier, and may double the efficiency of a fire control squad (Fig. 11.15).

Managing grazing sheep

Sheep are crucial in maintaining a balance between dwarf shrubs and grassland and in influencing the cycle of burning and cutting required (Fig. 11.16). Sheep can influence the fertility of the soils, and it is a moot point that they maintain a substantially higher diversity and abundance of invertebrates than would be present in their absence.

The impact of sheep on heather depends on the amount of grassland also present. Grass is much preferred as forage. Consequently, heather is grazed mostly in winter when the availability of live grass leaf is reduced, although some heather will also be eaten in the latter half of the summer before the shoots have lignified and when grass production is beginning to decline. With all-year stocking of hill ground it is essential, if heather loss is to be avoided, that the winter carrying capacity of the ground is not exceeded (excluding any supplementary feeding which may be given). Summer-only stocking rates can be higher, provided that grass production is sufficient to allow the

---

Box 11.1 Some general points to remember about muirburning

Dos   **Don'ts**

- Plan your muirburning
- Make sure you have told your neighbours that you intend to burn
- On Sites of Special Scientific Interest, notify and get permission from the country conservation agencies
- Be extra careful near archaeological features
- Burn across and down the slope
- Know where the fire will be stopped
- Have enough people and equipment to control the fire under all circumstances
- Be very careful in light, variable winds
- Be extra-careful when burning areas containing blow-grass (Flying Bent, Purple Moor-grass)
- Ensure that the fire is completely out when you leave the hill

- Do not burn out of season
- Do not burn within 50 feet (15 m) of a public road or so that smoke causes danger to road users
- Do not attempt to burn wet blanket bogs, where there are normally bog-pools
- Do not burn areas of Bracken or Juniper
- Do not burn traditional nesting sites for Merlin and Hen Harrier
- Do not burn steep slopes
- Do not burn uphill, or into scree, rocks or scrub woodland
- Do not burn when the wind is strong enough to blow bits of heather and dust into your eyes
- Do not burn at night or leave a fire unattended at any time of day
- Do not burn on your own under any conditions

---

stock to obtain most of their forage requirements from grass during the period when they are present. As the intake of grass by sheep increases to around 30%, heather intake increases by 20%, but where the sheep diet consists of more than 30% grass the intake of heather decreases by 1% for every 2% increase in grass. This means that small increases in the quality of a poor diet can result in sheep grazing more, rather than less, heather.

Fig. 11.14. Heather cutting on the East Lomond in Fife, (*a*) immediately after cutting and (*b*) abundant regeneration of Heather, Bilberry and Wavy Hair-grass *Deschampsia flexuosa* after two years. (A.J. MacDonald.)

Fig. 11.15. A firebreak cut with a tractor-mounted swipe immediately before a management fire is set can greatly increase the efficiency of a fire control squad. However, they should not be relied on to definitely stop a fire. (A.J. MacDonald.)

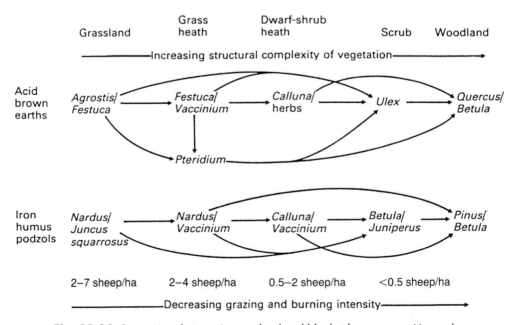

Fig. 11.16. Successional stages in moorland and blanket bog communities under heavy grazing pressure and burning (after Sydes and Miller, 1988).

Fig. 11.17. At this site in the western Cairngorms very heavy grazing by red deer has produced a 'carpet' of suppressed Heather which is at risk of loss. (A.J. MacDonald.)

Where feeding blocks are put out in winter, localised grazing and trampling pressures can be considerable and this has a profoundly suppressive effect on heather. The improved diet quality can lead to increased intakes of heather. Where feeding blocks or hay are provided ('fothering') the actual feeding stations should be rotated. Feeding block sites should be moved at least every three weeks and 20 m from the previous location, and fothering sites should be moved over a similar cycle but at least 250 metres from the previous fothering site (Hudson, 1986; MacDonald, 1993).

The relationship between sheep density and heather cover centres on the proportion of annual shoot production of the heather removed by grazing. As more than 40% of the year's growth is removed there is an increasing probability of long-lasting damage to heather and loss of heather cover. For example, a sustained utilisation rate of 60% over five years would be likely to reduce by half an initial 75% cover of moderately vigorous heather. The precise effect will depend on the age and vigour of the heather, on the vigour of competing plant species, and on other factors such as damaging weather conditions. For old or weak heather the critical offtake may be as low as 10–20% of annual production. Where more than 80% of annual growth is removed, loss of heather is likely even with vigorous heather growing in ideal conditions. Fig. 11.18 shows the effects of heavy grazing on the growth form of heather of various ages.

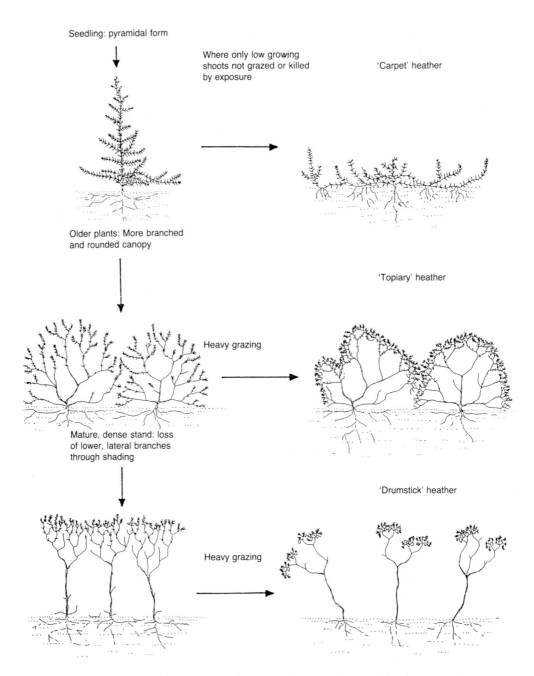

Seedling: pyramidal form

Where only low growing
shoots not grazed or killed
by exposure

'Carpet' heather

Older plants: More branched
and rounded canopy

'Topiary' heather

Heavy grazing

Mature, dense stand: loss
of lower, lateral branches
through shading

'Drumstick' heather

Heavy grazing

Fig. 11.18. The creation of carpet, topiary and drumstick growth forms of Heather in relation to heavy grazing pressure. (After MacDonald, 1993.)

Table 11.2. *Management prescriptions for enhancing moorland grazed by sheep. (1)–(3) gives order of stages for each condition*

Current state		Proposed new prescriptions
Heather condition	Annual average stocking density (ewes per hectare)	Ewes per hectare plus any off-wintering (O–W)[1] in ordered stages
Good	<2.0	(1) 0.75–1.5 and zero O–W (2) 1.5 –2.0 and 50% O–W
Poor	2–3	(1) 0.5–0.75 (for years 1–5) and 100% O–W (2) 1.5–2.0 (for >5 years) and 50% O–W
Suppressed	3–4	(1) Summer graze (for years 1–5) and 100% O–W on grass (2) 0.5–0.75 (for years 6–10) and 50–100% O–W (3) 1.5–2.0 (for years 11–15) and 50% O–W

Note: [1]Off-wintered sheep are put on in-bye land or are housed.
Source: Adapted from Thompson *et al.* 1995a.

Experience and trials indicate that the cover of young, vigorous and dominant heather can be maintained when the sheep stocking rate on the heather is about 2.2 ewes/ha. Where the heather is less vigorous, perhaps because it is old or growing in a bog or competing with other vigorous species such as Purple Moor-grass or Hare's-tail Cottongrass, stocking rates as low as 0.5 ewes/ha are required (Grant, *et al.*, 1978, 1982; Rawes, 1983).

As a simple guide, Table 11.2 provides new sheep stocking prescriptions for good, poor and suppressed heather moorland in the southern/eastern upland areas of Britain (from Thompson *et al.*, 1995a). It is suggested that these are amended in the following ways to adapt them for different situations. On very poor, acid mineral soils and peats the indicated stocking rates should be halved. Where the altitude is greater than 600 m they should be halved, and in going from the southern English uplands to the north west of Scotland the stocking rates should be reduced by two-thirds. If more than one of these factors is present they should be applied multiplicatively.

In many parts of the uplands tall herb species such as Angelica *Angelica sylvestris*, Wood Crane's-bill *Geranium sylvaticum*, Alpine Saw-wort *Saussurea alpina*, Roseroot *Sedum rosea* and Globeflower *Trollius europaeus* and montane willow species are largely restricted to cliff-ledges inaccessible to sheep and deer (Fig. 11.19). Any spread from their rocky refuges would require very substantial reductions in grazing pressures.

Fig. 11.19. Montane willows, often with very restricted distributions, are usually restricted to situations such as this, where there is reduced or no grazing. (A.J. MacDonald.)

Similar problems and opportunities can also exist on lower altitude moorlands (Fig. 11.21). However, other types of herb-rich upland vegetation may require moderate levels of grazing to maintain botanical diversity and their distinctive characteristics of conservation value. This applies particularly to dwarf herb-rich grassland communities on more lime-rich soils where, among a diverse assemblage of species, one or more of the following may be abundant: Alpine Lady's-mantle *Alchemilla alpina*, Moonwort *Botrychium lunaria*, Glaucous Sedge *Carex flacca*, Flea Sedge *Carex pulicaris*, Rockroses *Helianthemum* spp., Gentians *Gentianella* spp., Fairy Flax *Linum catharticum*, Mossy Cyphel *Minuartia sedoides*, Mossy Saxifrage *Saxifraga hypnoides*, Purple Saxifrage *Saxifraga oppositifolia*, Yellow Saxifrage *Saxifraga aizoides*, Sibbaldia *Sibbaldia procumbens*, Moss Campion *Silene acaulis*, Lesser Clubmoss *Selaginella selaginoides*, Wild Thyme *Thymus praecox*, and Mountain Pansy *Viola lutea*. Experience on Ben Lawers and from studies elsewhere suggests that suitable stocking rates for these types of vegetation should be about 0.5 ewes/ha at 900 m altitude increasing to 1.5 ewes/ha at 300 m altitude.

The suggested stocking rates in this section can only be rough guides. It is often best to adopt an 'adaptive' or 'suck-it-and-see' form of management, although this requires conscientious monitoring to be successful.

Other types of damage to heather

A number of other types of heather damage may be confused with damage due to heavy grazing pressure (MacDonald, 1993). Among these are the browning and death of heather shoots due to desiccation. This is most likely to occur in the winter, when the ground is frozen and when there are strong, drying winds (Braid & Tervet, 1937; Watson *et al.*, 1966). Similar effects may be produced by crushing of heather shoots by vehicle passage or trampling by people or livestock, but this is usually quite localised. Heather shoots may also be damaged or killed by fungal attack e.g. by Snow Mould or Heather Rhizomorph Fungus *Marasmius androsaceus*, but again these do not usually affect extensive areas of heather (Braid & Tervet, 1937; MacDonald, 1949; Watson *et al.*, 1966). These types of damage can be distinguished quite easily from the effects of heavy grazing because the heather shoots and leaves remain intact. There is little that management can do to alleviate the effects of weather or fungal attack although burning of affected areas, and reducing grazing pressure, may encourage better heather regeneration.

Heather and Bilberry may also be damaged by insects (Fig. 11.20), most commonly the Heather Beetle *Lochmaea suturalis*, the Winter Moth *Operophtera brumata*, the Magpie Moth *Abraxus grossulariata* and the Vapourer Moth *Orygia antiqua*. In all these cases it is the grubs or caterpillars which do the damage. Populations of these insects tend to show considerable fluctuation in numbers from year to year. Damage tends to be unpredictable in both its location and extent. Large outbreaks causing damage to areas ranging from hectares to square kilometres are possible but rare. Browning of heather as a result of attack by these insects is usually noticeable by late summer, and individual shoots and leaves of the heather have a 'nibbled' or ragged appearance when looked at closely. Management measures to ameliorate this damage are limited. There are no pesticides which can be used legally in this situation. Burning may help to control outbreaks of the moth species since vulnerable eggs, caterpillars or adults may be present in and around affected areas during the legal burning season. However, burning is likely to be ineffective against heather beetle, although it may encourage regeneration of affected heather. The damaging beetle grubs are not present during the legal burning period, and the adults and eggs are well protected in the soil (and their location hard to predict) when burning is permitted.

Bracken control

Bracken is a native species but not of high nature conservation value. It is neither particularly rich nor particularly poor in associated insect species. It can be attractive to some passerine bird species, e.g. Whinchat *Saxicola rubetra*, Tree Pipit *Anthus trivialis*, Yellowhammer *Emberiza citrinella*, Stonechat *Saxicola torquata hibernans*, Meadow Pipit *Anthus pratensis*, and occasionally Nightjar *Caprimulgus europaeus*, but dense stands of

Fig. 11.20. Damage to heather from severe outbreaks of defoliating insects usually affects discrete patches or bands as in this example of Magpie moth (*Abraxus grossulariata*) damage (centre of picture) in western Sutherland. (A.J. MacDonald.)

pure bracken are not very attractive even to these. It is toxic to grazing animals, harbours ticks, and can be a prolific invader, advancing at up to 1.5 m per year. The rate of spread is greater over heavily burned and grazed areas and is often associated with areas where there has been some soil disturbance or improvement in the past, such as on abandoned smallholdings and crofts.

It can be controlled by cutting. Where it is cut once, in late July, there will be a 50% reduction in bracken cover over 3–6 years, with 10–30% left after 10 years; where it is cut twice in June/late July, only 10% will remain after two years (Lowday & Marrs, 1992).

Spraying with Asulam is effective, but needs careful timing. This chemical damages actively growing plant tissues. To be fully effective against bracken it needs to be transported from the fronds down to the rhizomes and the ideal period for this is when the fronds are just fully expanded in late July through to early August. The sprayed fronds will not appear to be affected. If sprayed earlier than this there may be a marked effect on the fronds but the rhizomes will escape unscathed and there will also be an increased risk of damage to other species. If spraying is too late in the season there may be little uptake by the hardened fronds. Care should be taken not to spray other species of ferns which are all sensitive to Asulam. If aerial spraying is undertaken then buffer zones around sensitive areas need to be at least 160 m wide (Marrs *et al.*, 1992).

Whichever control methods are adopted their effectiveness will not last indefinitely. Bracken will quickly reinvade suitable areas if the treatments are stopped.

Drainage and fertilisation

Drainage of moorland areas by open hill drains ('grips') is now actively discouraged because it can cause erosion, flash-flooding and the silting-up of streams. Although grips are usually only cut to a depth of 50 cm they may erode to several metres deep if badly located. The open grips can also act as very effective traps for young birds and lambs. The effects on invertebrate biodiversity, bird density and sheep performance are negligible (Coulson *et al.*, 1990; Stewart & Lance, 1983).

As Hudson (1986) points out where there are boggy flushes these should be regarded as an asset rather than an impediment to management. These can have 10–20 times more arthropods than in surrounding drier moorland and are important in providing a rich source of food and shelter for grouse chicks. Artificial bog flushes are readily constructed by excavating holes to which limestone chippings are added. If the underlying geology is less acidic (e.g. over carboniferous limestone) bog flushes and streams can be impeded to increase the mineral and nutrient richness of the water and immediate vegetation (conservation agencies should be consulted before this is carried out on Sites of Special Scientific Interest).

The juxtaposition of fertilised or reclaimed ground next to moorland has an arguably beneficial impact on the diversity and density of breeding birds (Usher & Thompson,

1993). We caution against the further application of lime and nitrogenous fertiliser to moorland areas but suggest that marginal hill ground dominated by graminoids can be maintained in this way.

Heather restoration

Generally, it is most cost-effective to concentrate effort on improving the condition of heavily grazed Heather rather than trying to restore Heather where it has completely disappeared. When improving the condition of Heather the degree of recovery of Heather cover is not so much related to the cover remaining as to the density of plants remaining. Even Heather plants grazed down to only a few centimetres can recover, and if they occur at more than four plants per square metre then a dramatic increase in Heather cover can occur within five years if grazing pressure is substantially or completely removed. Where Heather has completely disappeared then at least some soil disturbance will be required, and Heather seed collection and sowing may also be necessary.

The soil under heather or vegetation which was recently heathery, contains a large seed bank of a range of heathland species. In acid, peaty soils Heather seed is known to remain viable for decades (Hill & Stevens, 1981). Germination of these seeds requires exposure to light and therefore soil disturbance.

Where no Heather seed bank remains then Heather seed must be collected and introduced. This can be done by collecting the remaining litter and soil surface from recently burned areas of vigorous Heather, either by hand or by using an industrial vacuum cleaner run from a portable generator. The former should produce about 30–40 kg per man-day, the latter about 80–100 kg per machine-day. The collected material should be spread at a rate of about 200 g per square metre. If storage is necessary the material should be dried first. This source of seed has the advantage that a range of heathland species will be introduced and the heat treatment produced by the prior burning may improve germination rates (Hudson 1992).

Heather seed can also be obtained by cutting Heather shoots carrying seed capsules during the period mid October to mid January. Where ground conditions are suitable it may be possible to use a forage harvester or a flail mower and baler, with collection rates of 6000 kg per machine-day possible on moderately productive sites. If storage is necessary the material should first be dried. The rate of application should be about 600 g per square metre.

If restoration applies to areas from which the existing heather vegetation has been removed, e.g. on the sides of a road cutting through heather moorland, restoration can be achieved by using the original topsoil. Preliminary earth-moving should remove the top 5 cm of litter and soil, after burning or flail mowing of the area, which should then be stored before re-spreading to a depth of 2 cm on the ground to be restored. Storage can be for up to a year and should be in shallow heaps, less than 1.5 m high (to avoid compaction and fermentation) on Terram sheet.

Fig. 11.21. At lower altitudes there are frequently areas where woodland and scrub, most often of willows and birch, could be regenerated (*a*) from existing heavily browsed bushes, and (*b*) from suppressed bushes. (A.J. MacDonald.)

Regardless of the type of seed-containing material used it is desirable to prevent loss of the spread material by applying an open covering of forestry brashings or by using an acid soil tolerant, short-lived, nurse grass, e.g. Common Bent *Agrostis capillaris*, Brown Bent *Agrostis vinealis*, Wavy Hair-grass *Deschampsia flexuosa*, or Sheep's Fescue *Festuca ovina*. These should be sown at a rate of about 2 g per square metre and on very infertile mineral or peat substrates an initial application of NPK fertiliser at 50–100 kg per hectare will encourage establishment. There must also be protection from grazing animals for up to five years.

Judging the success of management

The foregoing are just some of the key components to good upland management. Having put these techniques into practice the next stage is to determine your success through monitoring. This can very greatly in time and resources required depending on the complexity of the variables to be monitored and on the degree of precision required. There is always a trade-off between the expense of obtaining precise and comprehensive monitoring data which will enable remedial management to be undertaken at an early stage of change, and the expense of remedial management when triggered at a later stage of grosser change by cheaper and coarser monitoring methods.

Three methods of varying complexity for monitoring Heather cover, and grazing impacts on it, are summarised in Box 11.2.

Two of the methods involve being able to identify Heather stands which are potentially vulnerable to heavier grazing pressures and on being able to identify other signs of heavy grazing pressure. They also depend on being able to assess the intensity of Heather shoot offtake by herbivores. The latter can be done by assessing the percentage of Heather shoot numbers within a stand which have been grazed. If 66% or two-thirds of the shoots show signs of having been grazed then this is equivalent to about 40% of the year's growth of Heather being removed. For vigorous Heather, offtake rates greater than this are likely to lead to long-term loss of the Heather. Grazing pressure on Heather is best assessed in April–May by which time most of the grazing of the previous season's growth will have occurred but there will not have been any significant new growth.

Vulnerable stands of Heather can be identified as follows:

1. young Heather, up to about five years old or about 15 cm tall, regenerating after a recent burn and particularly where it is mixed with a high proportion of grasses
2. older Heather which does not form a continuous canopy but is mixed with a substantial proportion of grasses and herbs

3. Heather within 50 m of areas of more palatable bent/fescue (*Agrostis/Festuca*) grassland
4. old Heather stands in which the branches are beginning to collapse
5. Heather close to feeding stations (within 50 m).

If Heather in vulnerable areas is found to be overgrazed then the possibility of more widespread overgrazing should be checked by returning to the site in July–August and noting if:

1. there is widespread grazing of Heather at this time of year
2. other dwarf-shrubs such as Bilberry are heavily grazed, producing only shoots less than about 8 cm tall (when not suppressed by heather)
3. grassland areas are very closely cropped to a sward height of less than 1.5 cm and there is an abundance of mosses, such as *Rhytidiadelphus squarrosus*, or bare ground within the sward
4. there are noticeable signs of grazing by sheep (or deer) of unpalatable grasses such as Mat-grass *Nardus stricta*, or Purple Moor-grass early in the spring.

If all these indicators are present on a site then it is very likely that substantial areas of Heather, not necessarily identified as vulnerable, will be overgrazed.

Final point

We started by referring to the myths and magic of managing the uplands of Britain. We have provided some simple guidance, but the manager is urged to consult the more comprehensive and detailed accounts listed under References.

Acknowledgements

Many people have contributed to this chapter, notably colleagues in Scottish Natural Heritage, Countryside Council for Wales, English Nature, Joint Nature Conservation Committee, The Game Conservancy, Royal Society for the Protection of Birds and The Heather Trust. It is a pleasure to thank the following for specific comments: Ceri Evans, John Lawton, Alan Stubbs, Colin Tubbs, John Wilson, Michael Usher, Helen Armstrong, Nigel Buxton, Sandy Payne, Ros Smith and Neale Taylor; and to thank Bill Sutherland and David Hill for their encouragement.

Box 11.2 Methods for monitoring grazing impacts on heather and changes in heather cover

Method	Objectives	Data collected	Analysis	Time required per assessment for a site of about 500 ha (man-days)	Repeat period (years)
1. Photographic monitoring of changes in heather cover	(a) Permanent objective visual record of gross changes in cover (b) Record of unexpected changes	Photographs covering whole site from fixed points	Semi-objective assessment of heather cover changes by comparison of photographs	5–8 initially 2–4 subsequently	2–5
2. Rapid assessment of heather utilisation on selected areas with highest risk of overgrazing	(a) Quick, approximate assessment of proportion of most vulnerable stands subject to damaging utilisation (see text)	(a) Map of distribution and extent of most vulnerable stands (b) % shoot numbers grazed in each vulnerable stand	Approximate assessment of extent and degree of damaging utilisation on a site	2–3	1 if overgrazing of vulnerable stands, 5 otherwise

		(c) Observations of extent of different grazing induced heather growth forms (see text) (d) Observations of grazing impacts on other vegetation types (see text)			
	(b) Preliminary assessment of possible overgrazing on non-vulnerable stands				
3. Rapid assessment and mapping of heather utilisation over a whole site	Systematic extension of Method 2 to assess all heather stands on a site	(a) Map of heather stands (b) % shoot numbers grazed in each stand	5	Comprehensive and objective assessment of extent of possible damaging grazing pressures	As Method 2

References

Bardgett, R.D. & Marsden, J.H. (1992). *Heather Condition and Management in England and Wales*. Peterborough: English Nature (Science Directorate).

Bibby, C.J. (1988). Impacts of agriculture on upland birds. In *Ecological change in the Uplands*, ed. M.B. Usher & D.B.A. Thompson, pp. 223–36. Oxford: Blackwell Scientific Publications.

Birks, H.J.B. (1988). Long-term ecological change in the British uplands. In *Ecological Change in the Uplands*, ed. M.B. Usher & D.B.A. Thompson, pp. 37–56. Oxford: Blackwell Scientific Publications.

Braid, K.W. & Tervet, I.W. (1937). Certain botanical aspects of the dying-out of heather. *Scottish Journal of Agriculture* **20**, 365–72.

Clutton-Brock, T.H. & Albon, S.D. (1989). *Red Deer in the Highlands*. Oxford: BSP Professional Books.

Coulson, J.C., Butterfield, J.E.L. & Henderson, E. (1990). The effect of open drainage ditches on the plant and invertebrate communities of moorland and on the decomposition of peat. *Journal of Applied Ecology* **27**, 549–61.

Environmental Advisory Unit, University of Liverpool (1988). *Heathland Restoration: a Handbook of Techniques*. Liverpool: British Gas plc and University of Liverpool Environmental Advisory Unit.

Grant, S.A., Barthram, G.T., Lamb, W.I.C. & Milne, J.A. (1978). Effects of season and level of grazing on the utilisation of heather by sheep. 1. Response of the sward. *Journal of the British Grassland Society* **33**, 289–300.

Grant, S.A., Milne, J.A., Barthram, G.T. & Souter, W.G. (1982). Effects of season and level of grazing on the utilisation of heather by sheep. 3. Longer-term responses and sward recovery. *Grass and Forage Science* **37**, 311–26.

Hill, M.O. & Stevens, P.A. (1981). The density of viable seeds in soils of forest plantations in upland Britain. *Journal of Ecology* **69**, 693–709.

Hudson, P.J. (1986). *Red Grouse: the Biology and Management of a Wild Gamebird*. Fordingbridge: Game Conservancy Trust.

Hudson, P.J. (1992). *Grouse in Space and Time*. Fordingbridge: Game Conservancy Trust.

Lord Lovat (1911). *Grouse in Health and Disease*. Final Report of the Committee of Inquiry on Grouse Disease. London: Smith, Elder & Co.

Lowday, J.E. & Marrs, R.H. (1992). Control of bracken and the restoration of heathland. I. Control of bracken. *Journal of Applied Ecology* **29**, 195–203.

Macaulay Land Use Research Institute. (1988). *Heather Moorland: a Guide to Grazing Management*. Aberdeen: Scottish Agricultural Colleges.

MacDonald, A.J. (1993). *Heather Damage: A guide to Types of Damage and their Causes*. Research & Survey in Nature Conservation No. 28, 2nd edition. Peterborough: Joint Nature Conservation Committee.

MacDonald, J.A. (1949). The heather rhizomorph fungus, *Marasmius androsaceus* Fries. *Proceedings of the Royal Society of Edinburgh* **B63**, 230–41.

Ministry of Agriculture, Fisheries and Food. (1992). *The Heather and Grass Burning Code*. London: HMSO.

Marrs, R.H., Frost, A.J., Plant, R.A. & Lunnis, P. (1992). Aerial applications of asulam: a bioassay technique for assessing buffer zones to protect sensitive sites in upland Britain. *Biological Conservation* **59**, 19–23.

Miles, J. (1988). Vegetation and soil change in the uplands. In *Ecological Change in the Uplands*, ed. M.B. Usher & D.B.A. Thompson, pp. 55–70. Oxford: Blackwell Scientific Publications.

Miles, J., Welch, D. & Chapman, S.B. (1978). Vegetation and management in the uplands. In *Upland Land in England and Wales*, ed. O.W. Heal, pp. 77–95, Cheltenham: Countryside Commission.

Moss, R., Miller, G.R. & Allen, S.E. (1972). Selection of heather by captive Red Grouse (*Lagopus lagopus scoticus*). *Journal of Animal Ecology* **44**, 233–44.

Nature Conservancy Council & Countryside Commission for Scotland (1988). *National Countryside Monitoring Scheme. Scotland. Grampian.* (Available from SNH Publications Section, Battleby, Redgorton, Perth PH1 3EW.)

Orford, N. (1986). Breeding distribution of the Twite in central Britain. *Bird Study* **20**, 51–62.

Pearsall, W.H. (1950). *Mountains and Moorlands.* London: Collins.

Phillips, J. (1991). Heather burning and management: 1911–1991. In Joseph Nickerson Reconciliation Project 7th Annual Report, May 1991.

Phillips, J., Watson, A. & MacDonald, A. (1993). *A Muirburn Code.* Edinburgh: Scottish Natural Heritage.

Picozzi, N. (1986). *Black Grouse Research in North-east Scotland.* Report to the World Pheasant Association, ITE Project 764. Huntingdon: Institute of Terrestrial Ecology.

Ratcliffe, D.A. (1976). Observations on the breeding of the Golden Plover in Great Britain. *Bird Study* **7**, 81–93.

Ratcliffe, D.A. (ed.) (1977). *A Nature Conservation Review.* 2 vols. Cambridge: Cambridge University Press.

Ratcliffe, D.A. (1990). *Bird Life of Mountain and Upland.* Cambridge: Cambridge University Press.

Ratcliffe, D.A. & Thompson, D.B.A. (1988) The British Uplands: their ecological character and international significance. In *Ecological Change in the Uplands*, ed by M.B. Usher & D.B.A. Thompson pp. 9–36. Oxford: Blackwell Scientific Publications.

Rawes, M. (1983). Changes in two high altitude blanket bogs after cessation of sheep grazing. *Journal of Ecology* **71**, 219–35.

Roberts, J.L. & Green, D. (1983). Breeding failure and decline of Merlins on a North Wales moor. *Bird Study* **30**, 193–200.

Rodwell, J.S. (1991). *British Plant Communities.* Vol. 1, *Woodland and Scrub.* Vol. 2, *Mires and Heaths.* Cambridge: Cambridge University Press.

Rodwell, J.S. (1992). *British Plant Communities.* Vol. 3, *Grasslands and Montane Communities.* Cambridge: Cambridge University Press.

Rowan, W. (1921). Observations on the breeding habits of the Merlin. *British Birds* **15**, 122–9.

Savory, C.J. (1986). Utilisation of different ages of heather on three Scottish moors by red grouse, mountain hares, sheep and red deer. *Holarctic Ecology* **9**, 65–71.

Scottish Natural Heritage (1992). *National Countryside Monitoring Scheme. Scotland. Dumfries and Galloway.* (Available from SNH Publications Section, Battleby, Redgorton, Perth PH1 3EW.)

Simmons, I.G. (1990). The mid-Holocene ecological history of the moorlands on England and Wales and its relevance for conservation. *Environmental Conservation* **17**, 61–9.

Stewart, A.J.A. & Lance, A.N. (1983). Moor draining: a review of impacts on land use. *Journal of Environmental Management* **17**, 81–99.

Sydes, C. & Miller, G.R. (1988). Range management and nature conservation in the British uplands. In *Ecological Change in the Uplands*, ed. M.B. Usher & D.B.A. Thompson, pp. 323–38. Oxford: Blackwell Scientific Publications.

Thirgood, S.J. & Hewson, R. (1987). Shelter characteristics of mountain hare resting sites. *Holarctic Ecology* **10**, 294–8.

Thompson, D.B.A., Hester, A.J. & Usher, M.B. (eds) (1995a). Heaths and Moorland: cultural landscapes. Edinburgh and London: HMSO.

Thompson, D.B.A., MacDonald, A.J., Marsden, J.H. & Galbraith, C.A. (1995b). Upland heather moorland in Great Britain: a review of international importance, vegetation change and some objectives for nature conservation. *Biological Conservation* **71**, 163–78.

Thompson, D.B.A. & Sydes, C. (in press). The British Uplands. In *Our Natural Heritage*, ed. I.F.G. McLean. Peterborough: Joint Nature Conservation Committee.

Thompson, D.B.A., Stroud, D.A. & Pienkowski, M.W. (1988). Afforestation and upland birds: consequences for population ecology. In *Ecological Change in the Uplands*, ed. M.B. Usher & D.B.A. Thompson, pp. 237–60. Oxford: Blackwell Scientific Publications.

Usher, M.B. & Thompson, D.B.A. (eds) (1988). *Ecological Change in the Uplands*. Special Publication of the British Ecological Society No. 7. Oxford: Blackwell Scientific Publications.

Usher, M.B. & Thompson, D.B.A. (1993). Variation in the upland heathlands of Great Britain: conservation importance. *Biological Conservation* **66**, 69–81.

Watson, A., Miller, G.R. & Green, F.W.H. (1966). Winter browning of heather *Calluna vulgaris* and other moorland plants. *Transactions of the Botanical Society of Edinburgh* **10**, 195–203.

Watson, D. (1977). *The Hen Harrier*. Berkhamsted: T.&A.D. Poyser.

Welch, D. (1984). Studies in the grazing of heather moorland in North-east Scotland. II. Response of heather. *Journal of Applied Ecology* **21**, 197–201.

Williams, G.A. (1981). The Merlin in Wales: breeding numbers, habitat and success. *British Birds* **74**, 205–14.

Woodland and scrub   12

ROBERT J. FULLER AND GEORGE F. PETERKEN

Introduction

The composition and structure of the woodlands we have today have been determined as much by their history of management as by climate and soils. These factors in turn influence how we value each wood for nature conservation and the choice of management strategy and practice. Although the historical perspective has been strongly emphasised in the last 25 years, it is far from new; perhaps the first modern measure in woodland conservation was the protection of the 'Ancient and Ornamental' woodlands under the 1877 New Forest Act.

Ancient woods (those which have existed since before 1600) are now recognised as the most important for nature conservation (Spencer & Kirby, 1992). Many comprise modified remnants of Britain's original natural vegetation and include communities of plants and animals which we suppose they inherited directly from the Atlantic forests and have sustained for millennia on or about their present sites. However, even secondary woodlands have a structural complexity which sets them apart from other vegetation formations and may develop into rich habitats. Whilst ancient woodlands are the priority, woodland managers should also seize every opportunity to diversify secondary woods. In a country which still has only 11% of its land area under trees, most landscapes would be enriched by establishing more woodland.

Whilst there is a good case for setting aside some woods as strictly unmanaged reserves, most woods need to be managed, both to maintain their wildlife and to yield timber and other material products. Our woodland wildlife has been heavily influenced by past management systems – notably coppice and wood-pasture – and we can most surely maintain it by continuing, or by gently adapting, the traditional systems. The high forest systems which prevail in modern forestry offer different opportunities, particularly for wildlife associated with forest interior conditions. Above all, management enables the wildlife-rich open spaces to be maintained and renewed, and a full range of age-classes to be perpetuated within the stands.

This chapter therefore embraces ancient and more recent woods, though conservation objectives differ markedly between them. We first make some general points about objectives and priorities, then summarise management in each of the main woodland types as defined by management system and structure. We also consider the management of open spaces, which often makes the difference between a rich and an average woodland. Finally, we consider how the various management systems might be combined within a single wood or group of woods. Throughout we have three questions in mind; (i) is management always needed?, (ii) what general prescription is appropriate for the site?, and (iii) how should general prescriptions be modified in recognition of the special features of a wood and its surroundings?

Woodlands and their management are so complex that we cannot present management prescriptions for individual species, nor go into detail about silvicultural systems. Readers are referred to Kirby (1984), Marren (1990), Mitchell and Kirby (1989), Peterken (1993), Rackham (1980, 1990) and Watkins (1990) for more information on the woodland ecosystem. Matthews (1989) describes the various silvicultural systems and Evans (1984) describes the technicalities of managing broadleaved woodlands in Britain.

Setting the objectives of management

Successful conservation management in woodland demands thoughtful planning based on an understanding of the habitat needs of species, coupled with commitment and vision. The potential variety of management manipulations is endless, so long-term objectives are essential. Seldom will the management of a site be driven by the needs of just one species, or even of one taxonomic group. More typically, management will seek to cater for a range of species, each of which will respond uniquely to any treatment.

Change within woodlands is relatively slow, so a decade of neglect or uncertainty about the objectives and strategy of management is rarely disastrous. The main exception is active coppice where cessation of management can lead to the rapid disappearance of early successional species. Management strategies should be planned to last for centuries, even though the detailed plans may be modified occasionally: short-term tactics should be based on long-term vision. Woodlands can even be managed at the scale of individual trees, which affords a degree of management selectivity which is denied to other habitats.

The perspective of conservation in ancient and recent woodlands is very different. In ancient woods, especially on nature reserves, commercial considerations should be subordinated to those of conservation, whereas timber production is a major objective in most recent woodland. In ancient woodlands, the accepted priorities are generally to maintain existing habitats, especially where there has been continuity of coppicing or wood-pasture, or to re-establish habitats previously present by resuming traditional treatments. The objective is to maintain the existing community by forestalling further extinctions of species with demanding habitat requirements. Within conifer forests and recently established woods, the objective of conservation management is more often to enhance species diversity. One exception to this general rule is that some British conifer plantations established in nutrient-poor areas carry important populations of nationally scarce bird species (e.g. Capercaillie *Tetrao urogallus*, Nightjar *Caprimulgus europaeus*, Woodlark *Lullula arborea*). In some cases management aimed at the habitat needs of these species may be a higher priority than management aiming to maximise the variety of species. Large-scale introduction of broadleaves into such sites may well enrich the plant and animal communities present but it would probably be to the detriment of the scarce species associated with the conifer forest.

Management objectives may differ strikingly from one ancient wood to another. The two main traditions of woodland management in Britain are coppicing and wood-pasture which have created complementary suites of habitats for woodland plants and animals. Coppicing entails periodic cutting of trees with regeneration by natural growth from the cut stumps (stools). The rotations typically vary from 5 to 30 years according to the tree species, growth rates and markets. In wood-pasture, animals are grazed within the woodland and trees are managed for timber or for large poles by pollarding. Parkland is an extreme form of wood-pasture. Structurally, the two systems are very different. Wood-pasture has allowed the survival of more mature trees and dead wood than are found in most coppice. As a result of heavy shade and grazing, the wood-pasture field layer is relatively sparse, whereas managed coppice is typified by a continuity of open space and vigorous ground vegetation. Consequently, the conservation interest of ancient woods tends to diverge between species associated with open, early successional habitats (coppice) and species requiring mature, late successional habitats (wood-pasture). Maintaining this fundamental difference of ecological character is a vital conservation objective in ancient woodlands.

Five general considerations should be borne in mind when planning the management of ancient woods:

1. Woods often act as reserves for the whole landscape, especially in intensively arable regions. Many, for example, should be regarded as grassland reserves, as well as woodland reserves, because ancient rides sometimes carry relic semi-natural grassland. Management of open spaces within the woods may, therefore, be at least as important as management of the tree stands themselves. Many woods are still linked to a network or mosaic of habitats in their surroundings, effectively a 'landscape-scale reserve'.

2. Management should aim to sustain all species now present within the wood. Most woods, especially small woods in intensively arable districts, are now so isolated that species with limited mobility cannot readily recolonise after local extinction (see Chapter 1).

3. Managers should base silvicultural treatments on the native tree and shrub

species already in the wood. Maintenance of each species at its present level of abundance is not necessary, provided all species are retained. Maintenance of their present distributions within a wood is desirable, but minor adjustments are usually acceptable, especially if these come about by natural regeneration.

4. The long-term aim should be a balanced age-structure within each wood or cluster of small woods. Ideal treatments should aim to (i) maintain a continuous supply of young growth through regular felling and (ii) protect and enhance mature features, such as large trees and dead wood.

5. In the great majority of ancient woods, management should either continue or revive the management which was traditional in the wood, or incorporate the main habitat features of traditional management within high forest systems. The creation of near-natural woodland is, however, an entirely valid objective for some ancient woods (see below).

In existing secondary woods and new woods principles (1) and (4) are important. Whilst it is always preferable to create more native woodland (3), more flexibility is acceptable in management systems and choice of tree species within secondary woodlands. In commercial forests, the success of nature conservation measures is determined largely by the amount and treatment of open spaces, and the age-class distribution of the whole forest.

When combining these considerations into a coherent plan, it is essential to consider the constraints. Idealised plans which fail are inferior to realistic plans which are carried out. Whilst coppicing is often regarded as the ideal form of conservation management in ancient woods with a history of coppicing, financial and practical considerations often rule it out (see Box 12.1). It must also be recognised that woodland management often demands skilled professional input, though volunteers can tackle many small-scale tasks especially in cutting small-scale coppice, scrub, glades and rides.

Woodland communities

Woodland is so rich in species that management cannot hope to cater individually for each one. So certain groups of plants and animals have come to be focal points for woodland conservation management. Some of these groups are vulnerable because they have specialised requirements and may be confined to certain types of woodland which have become scarce. Others are convenient management indicators which utilise a range of habitats: targeting the management on them would simplify management planning and should benefit many other species. Key conservation issues and management objectives for these groups are outlined below. For examples of species and for more information on their ecology and habitat requirements, refer to the publications cited after the headings.

Tree communities (Rackham, 1980 & 1990; Rodwell, 1991; Peterken, 1993)

Amongst ancient woods there are fascinating patterns in tree communities, often related to soil type, which possibly reflect patterns in the composition of the primeval forest as revealed by pollen analysis. Genetically, many trees within ancient woods may be direct descendants of those in the original forest. Rackham (1990) has listed 54 types of tree community in ancient woodlands. The Highland Scots Pine *Pinus sylvestris* and Birch *Betula* spp. woods are especially significant as representatives of the boreal forests. In the English lowlands, woods containing Small-leaved Lime *Tilia cordata* are interesting because they may be relics of the time when lime was the commonest tree in the primeval forest. An aim of management should be to maintain semi-natural communities of trees, wherever possible by using natural regeneration, but if this is not practicable then by planting using stock of local provenance.

Ground flora (Rackham, 1980; Rodwell, 1991; Peterken, 1993)

The flora of ancient woods, particularly coppices, tends to be richer than that of recent woodland and certain plants are concentrated in such woods. The list of plants strongly associated with ancient woodland is both long and variable from district to district. Where there has been unbroken continuity of woodland cover, the flora may have descended from that of the primeval forest. Since 1945 the fabric of the British countryside has profoundly changed. Ancient woods (and other semi-natural habitats too) have become increasingly isolated with the consequence that species with poor mobility are less able to colonise them, or to re-colonise, should they become extinct. This applies particularly to many plants and invertebrates. It is all the more important, therefore, that management should ensure that species do not become extinct as a result of habitat deterioration. The continuity of young growth and open spaces is particularly critical for light-demanding and grassland species.

Epiphytic flora and fungi (Harding & Rose, 1986)

Mature trees in wood-pasture in those parts of Britain with a reasonably unpolluted atmosphere often carry interesting communities of epiphytic lichens, bryophytes and fungi. The richness and luxuriance of bryophytes and lichens increases towards the west of Britain. Amongst the richest assemblages of epiphytes are those in the New Forest and along the western seaboard. At sites where mature trees have been continuously available since the original establishment of tree cover these communities may resemble those that lived in the primeval forest (Fig. 12.1). In the lowlands alone there are more than 70 species of lichens that are seldom, if ever, found outside old wood-pasture. The main requirement for epiphyte conservation is that there should be continuity of mature trees, of the appropriate species, at the key sites. The trees with the richest lichen floras are oaks *Quercus* spp., Ash *Fraxinus excelsior*, Beech *Fagus sylvatica* and elms *Ulmus* spp.

Saproxylic invertebrates (Harding & Rose, 1986; Warren & Key, 1991; Kirby, 1992)

These are animals that depend on dead, dying or living wood, or associated fungi, at some stage in their life cycle. Perhaps more than a fifth of the fauna of woodland depends on dead and decaying trees, the main groups being beetles and flies (Fig. 12.2). Wood-pasture, even individual old trees, can be important for populations of such species. Continuity of the specialised habitat of mature trees, has allowed a relic fauna to persist at certain sites. The main management needs are to protect these sites and their old trees, to prevent removal of dead wood and to maintain or introduce management that will ensure the persistence of old trees and their associated micro-habitats. The continuity of prime condition dead wood is important because many saproxylic invertebrates need old wood in a specific condition, for example moisture content can be critical. At these sites it is vital that consideration is given to the long-term recruitment of pollards and large trees.

Fig. 12.1. *Lobaria virens* growing on a large oak in Great Wood, Borrowdale, Lake District. The *Lobaria* assemblage is presumed to have been abundant in primitive, old-growth forests, but is now reduced to relic locations throughout most of Europe. (P. Wakely, English Nature.)

Fig. 12.2. (*a*) *Armillaria mellea* (Honey Fungus) growing on the remains of a fallen beech pollard in Vinney Ridge, New Forest, Hampshire. Large fallen logs and dead wood on and within large trees form an important habitat for fungi, and beetles such as this rare click beetle *Lacon quercus* (*b*) which lives in red heart rot only in the very oldest of oaks. This type of habitat is much reduced in managed forests. (*a*, P. Wakely, English Nature; *b*, R. Key.)

Butterflies (Thomas, 1991; Fuller & Warren, 1991; Warren & Key, 1991;
Kirby, 1992; Greatorex-Davies *et al.*, 1993)

Most species of woodland butterflies live in open areas such as rides, glades and young growth where the food plants for their larvae grow and nectar sources are available. Massive declines have occurred in the majority of these species over the present century, even within many nature reserves. Comparable declines have occurred in other groups of early successional invertebrates. Two factors have been particularly important. First, the great reduction of coppicing as a commercially viable system has deprived these insects of large areas of habitat. Second, the increasing isolation of woods (see above), and of suitable patches within woods, coupled with weak ability to colonise, has rendered populations increasingly vulnerable. Continuity of suitable conditions has been essential to the survival of populations at particular sites. In coppiced woods the habitat may be suitable for some species for merely the first 2 or 3 years after cutting so that a very short break in a coppice rotation could place a population in jeopardy. See below for further information on effects of coppicing and ride management (see Boxes 12.1, 12.2).

Small mammals (Gurnell & Pepper, 1988; Bright & Morris, 1989;
Gurnell *et al.*, 1992)

Two mammal species are of particular concern in woods: the Dormouse *Muscardinus avellanarius* and the Red Squirrel *Sciurus vulgaris*. Both have declined this century and are locally distributed. Their habitats and conservation needs are very different. Dormice live in broadleaved woods with vigorous field and shrub layers. Actively coppiced woods with a variety of tree and shrub species, and particularly Hazel *Corylus avellana*, offer a good habitat, especially the middle stages of growth where the canopy offers continuous arboreal pathways. Red Squirrels in Britain are mainly confined to large conifer plantations, especially Scots Pine, though there are populations still surviving in mixed and broadleaved woods where Grey Squirrels *Sciurus carolinensis* are absent, on islands and in the north. Where these species are known to be present, management should aim to maintain suitable structures of habitat on appropriate scales (see below).

Birds (Petty & Avery 1990; Fuller, 1990; Fuller 1992; Fuller 1995)

Many woodland managers wish to manage sites to benefit bird populations. A high diversity of bird and mammal species can often be encouraged by maintaining a mixture of age classes. Birds are different to most other groups in that they are extremely mobile and are generally good colonists. The need to conserve local populations is less urgent than for many other groups but there are exceptions. Firstly, those rare species associated closely with the Scottish native pinewoods, notably Scottish Crossbill *Loxia scotica*, Crested Tit *Parus cristatus* and Capercaillie. Secondly, large segments of the

British populations of Nightjar and Woodlark are associated with recently replanted clear-felled areas in lowland conifer plantations. Continued provision of suitable habitat in their main breeding strongholds is important. Thirdly, some scarce species can potentially be helped to spread by ensuring that their current sites are protected and managed sensitively, and that opportunities are taken to create new suitable habitat. Golden Oriole *Oriolus oriolus* is a good example, appearing to prefer in Britain fairly open poplar *Populus* spp. plantations. Fourthly, the disappearance of the Nightingale *Luscinia megarhynchos* from many sites in southern Britain, probably has a basis in habitat deterioration. The species depends on dense closed-canopy scrub or thickets. Nightingales have not colonised many coppices where suitable habitat has been re-created, suggesting that it may not be a strong colonist and that appropriate management is desirable to help populations survive in their current locations.

Conservation management in woodland stands

Prescriptions for conservation management are of two broad types: those that are concerned with the tree stand itself and those that focus on other parts of woods such as rides and glades. Here we consider management of stands in relation to various silvicultural systems and stand types, both traditional and contemporary. The types of woodland and circumstances where each is most appropriate are considered, the major variants of management are discussed, and guidelines are given for their implementation. The term 'coupe' is used to mean a felled patch in which the trees subsequently grow as an even-aged stand.

Coppice

Until the turn of this century coppicing was the usual treatment for the majority of lowland ancient woods and for much upland Sessile Oak *Quercus petraea*. It had been intensively practised over such a long period that it must have been a profound influence on the British woodland fauna and flora. Coppicing is widely regarded as a priority management for ancient woodland, for many of our scarcer plants and animals are adapted to the conditions it generates. Not only can coppicing serve to maintain populations of rare species, but it creates a variety of habitat structure supporting a high diversity of species. Many species are adapted for living at one particular stage of coppice growth so there is usually a rapid turnover in species at any one place as the coppice grows. Coppicing also offers an appealing cultural contact with the past and an open woodland structure which is attractive to many people, but not to all. For more information on coppice management read Buckley (1992) and Fuller & Warren (1993).

The highest priority should be given to coppicing in ancient woods where it survives as a living practice. Coppicing is also a priority where it has recently ceased, though it is likely that species dependent on the earliest stages of growth will have been lost unless open rides or glades have provided refuges. Careful thought should be given to sites where coppicing has long been abandoned. Any species which require young coppice will already have vanished. There may be a low density of vigorous stools which will lead to poor coppice regrowth. Furthermore, coppicing is labour intensive so it must be asked whether a long-term programme can be sustained. The markets for coppice underwood are few and specialised (e.g. thatching spars, hurdles, charcoal) and much woodland may not generate sufficient marketable wood to interest professional cutters. On top of all this, deer are causing increasingly severe problems for woodland regeneration.

Though the basic principle of coppicing is straightforward, there is a spectrum of variants from scrubby woods, managed on short rotations without standard trees, to ones with many standards, resembling high forest. It is easy to manipulate several components of coppice systems: the rotation length, the size and location of coupes, and the density of standard trees. It is also possible to alter the density of coppice stools and the tree species composition by tree removal, planting or layering (see below). Hence, a coppiced wood is a highly controlled environment.

The conservation interest of coppice is concentrated into the early years of growth – the period before and just after the canopy closes. Once the vegetation has thinned out in the lower 2 m or so, which typically has occurred by 10 years after cutting, the habitat is often of lower conservation interest, though Dormice may use it and a range of scarcer moths and molluscs are to be found in older coppice. Most of the spectacular flowers and rare invertebrates are associated with the very youngest growth, well before the canopy closes. The more interesting birds, notably warblers and Nightingales, depend on middle-aged growth when the shrub layer is at its thickest. Therefore, for conservation purposes it is important to *maintain large areas of coppice of all ages up to 10 years*. In much mixed coppice a reasonable rotation would be 15 years, where the aim was to cut a fifteenth of the wood per annum or perhaps a fifth every three years. Alternatively, a more complex 'split' rotation could be adopted whereby part of the wood was cut on a short rotation and part on a long rotation. The ideal is to *cut at least one coupe each year*. If it is not feasible or cost-effective to cut every year then a large coupe (at least 0.5 ha) cut every two or three years will suffice.

Small coupes, less than one third of a hectare (approximately 60 m × 60 m), are best avoided because they can be too heavily shaded and are more prone to heavy deer browsing. *Aim to cut coupes of at least half a hectare in size* (70 m × 70 m). An exception may be where Dormice are present because large coupes of young coppice may form significant barriers to these animals. This may not be too much of a problem, however, provided substantial blocks of middle-aged and older coppice are always available. *Cut successive coupes adjacent to one another*. The reasons for doing this are two-fold. Firstly, it will be easier for weakly mobile species to colonise new habitat. Secondly, this will

Fig. 12.3. Mixed coppice-with-standards at Swanton Novers Great Wood, Norfolk. The underwood of Hazel, Small-leaved Lime, Ash, Maple *Acer campestre*, Alder *Alnus glutinosa*, Bird Cherry *Prunus padus* and other species has continued to be cut regularly on a short rotation, supplying bundles of brushwood for river bed stabilisation. The Oak and Ash standards are all old; the succession should have been renewed by reserving saplings at recent coppicings. (G.F. Peterken.)

create larger blocks of suitable habitat for species with relatively large space requirements, particularly birds. When starting a new coppice regime in a wood, concentrate the coupes in one area rather than scattering them about the wood. Present knowledge is inadequate to make detailed recommendations about the ideal shape and juxtaposition of coupes of different ages. We suggest varying the shapes of coupes but avoiding very narrow ones which could be vulnerable to shading.

A common cause of shading and poor coppice regrowth is too many large standard trees. Whilst these trees may provide habitats not available elsewhere in coppice, they should not be allowed to overwhelm the main conservation interest which is the species of young growth. The conservation of species associated with mature trees is best achieved through wood-pasture, high forest and natural woodland (see below). As a general rule, leave *no more than 15 large standards (crown diameter exceeding 10 m) per hectare*, though more small trees should be present to replace these when felled (Figs. 12.3, 12.4). When thinning or felling standards, try to leave a variety of tree species. Any old trees at the edge of the wood should be left and perhaps managed as pollards (see wood-pasture). Consider creating new pollards at the wood edge which, in time, will become a source of dead wood.

Fig. 12.4. Coppice-with-standards in Bradfield Woods, Suffolk. Here too, coppicing has continued without a break and ancient coppice stools retain full vigour. Furthermore, the succession of standards has been maintained by cutting larger trees and reserving a few 'wavers' to replace them. (P. Wakely, English Nature.)

Fig. 12.5. A 'dead hedge' protecting recently cut coppice from deer browsing. Whilst this may not be 100% effective in excluding deer, it may deter them enough during the first two years of tree growth to allow the coppice sprouts to grow out of reach. (P. Wakely, English Nature.)

Even with a low density of standards, poor regeneration is common for two reasons. Firstly, there may be a low density of viable stools leading to a patchy canopy and dense bramble. This can be avoided either by not cutting poorly stocked areas in the first place, or by planting out new stools, or by layering. The latter involves partially cutting through a coppice stem, bending it over, and pegging it along the ground until it sends out new shoots (BTCV, 1980). The second problem is deer. Numbers of deer are now so high that they are jeopardising the future of coppicing in many woods (Fig. 12.5). Several actions can be taken to reduce damage from deer but none is entirely satisfactory. The first is to cut the coupes as large as possible. Deer tend to stay close to cover so it is often the edges of open coupes within the wood that receive the heaviest browsing. Unfortunately, even coupes of 1 ha will not have a great effect in reducing overall damage where deer populations are large. Individual stools can be protected to some degree by heaps of brushwood or, more effectively, by fences of brushwood or wire (Fig. 12.5). It is more efficient, however, to fence the entire coupe but this is not straightforward. The most secure fencing would be 2 m deer fence, but this is expensive, unsightly and creates a collision hazard for large birds. Electric fences can be used to deter the smallest species of deer but should not be used where Roe *Capreolus*

Box 12.1 Bradfield Woods: an example of successful coppice management

Bradfield Woods, Suffolk, the celebrated example of more or less continuous coppicing since the Middle Ages, exhibits many features of good coppice practice (see Fig. 12.4):

- large coppice coupes (average size 1.3 ha in 1987)
- large areas of young, open growth (enhanced by use of a 'split' rotation with 70% of the wood cut on a 20–25 year rotation and 30% on a 10–12 year rotation)
- proximity of young coupes (the young growth forms fairly discrete blocks)
- low density of large standard trees
- high density of stools (maintained by layering where necessary)
- protection of young growth from deer by carefully constructed brushwood fences, coupled with reduction of deer numbers
- maintenance of an intricate ride system with linear coppice along some

As a result of the low density of standards, high density of stools and their protection from deer, the coppice regeneration is vigorous. Typically, the regrowth becomes impenetrable some five years after cutting by which time it is an excellent habitat for breeding warblers and Nightingales.

---

capreolus, Fallow *Dama dama* or Red Deer *Cervus elaphus* are present. Flexi-netting should not be used at all. Dense, tall brushwood fences around freshly cut coupes have proved effective in some woods (Fig. 12.5, Box 12.1). The only long-term solution is to conduct sustained campaigns of deer culling which demand professional skills and the full co-operation of landowners over substantial areas. To reduce the population to a level compatible with tree growth requires a sustained effort over a large area including land adjacent to the wood.

Derelict coppice

Large areas of coppice now lie neglected. Some of these woods are assuming the appearance of high forest but many retain the uniform multi-stemmed structure characteristic of old coppice and there is little point in seeking to retain this structure. The options are either to restore coppicing, to convert them to managed high forest or to treat them as natural forest (Fuller, 1990). Old coppice lacks species associated with the young, open growth. It is usually poor in species that live in old growth too, though this is less true of derelict oak coppice. Even when neglected for decades, much old coppice rarely acquires a rich saproxylic fauna. This is partly a problem of colonisation but also important is the fact that old coppice is often dominated by Hazel and can take a very long time to change its structure towards mature woodland.

As discussed above, coppicing is not always the most sensible approach. Often there will be a stronger case for converting coppice into broadleaved high forest. This can be achieved either by felling and replanting, preferably in small groups over a period of time to create a varied age structure, or by thinning. The exact approach chosen will depend on the type of coppice involved. In the case of pure Hazel, for example, there may be little alternative to replanting. The usual form of thinning in old coppice is singling, in which all the stems but one on each stool are removed to promote a single-stemmed structure. The thinning may need to be repeated after a few years to further open up the canopy and promote growth of individual trees. This approach does, however, need to be adopted with care. In areas where the windthrow hazard is high (and this includes much derelict Oak coppice on steep valley sides in western Britain) it is not sensible to single extensive blocks of old coppice. A better solution here may be to undertake patchy recoppicing with a view to subsequent singling to create a habitat of less uniform structure.

Wood-pasture and grazing in woodland

In the lowlands, examples of extant wood-pasture systems are rarer than those of coppice. The utmost priority must be given to protecting surviving sites from changes in land use, including conversion to other forms of woodland management (Harding & Rose, 1986). The aim should be to ensure the survival of ancient trees with their associated microhabitats (hollow trunks, dying and dead wood in various stages of decay). At many sites there is little need for much active management provided that mature and overmature trees are maintained.

The most extensive wood-pastures in Britain are now those upland Birch and Sessile Oak woods which are open to deer and sheep (Mitchell & Kirby, 1990) (Fig. 12.6). There has long been concern that intense grazing threatens the very survival of many of these woods, but in fact most pasture woods have regenerated intermittently when grazing pressure has been temporarily reduced (e.g. in the New Forest). Even if grazing is not yet threatening the wood's survival, many groups of plants and animals would benefit from some relaxation of grazing, which would allow an underwood to develop and produce varied structure. Unfortunately, it is by no means straightforward to realise this laudable aim.

The special conservation interest of some wood-pasture could actually be damaged by relaxation of grazing. For example, some epiphytic lichens would be excessively shaded if the undergrowth became dense. Another problem in upland oakwoods is that their unique bird communities, characterised by high densities of Pied Flycatchers *Ficedula hypoleuca*, Wood Warblers *Phylloscopus sibilatrix* and Redstarts *Phoenicurus*

Fig. 12.6. Johnny's Wood, Borrowdale, Cumbria. Like most western oakwoods, this is open to grazing by sheep or deer. The undershrubs have vanished, except from outcrops; regeneration is limited; woodland herbs are partially replaced by grasses; but bryophyte swards have developed vigorously. (P. Wakely, English Nature.)

phoenicurus, would be unlikely to persist in woods with dense undergrowth.

Complete exclusion of animals is, therefore, unlikely to be satisfactory because this is likely to damage existing conservation interest. A better approach would be to manipulate grazing in such a way that heavily grazed patches remain within a mosaic of grazed and ungrazed woodland. It may be possible to achieve this through low, intermittent stocking. Rotational fencing of patches within the wood is another possibility but this would be costly and labour demanding. Unfortunately, it is impossible to give exact advice because the effects of stocking rates on the ecology of woodland are so poorly understood. Far more research is needed on regeneration and the responses of different groups of plants and animals across the continuum from complete absence of grazing to severe overgrazing. The effects will probably vary according to the types of animals involved, the density of stocking, the seasonal pattern of grazing, the time elapsed since relaxation of grazing and the type of woodland. Furthermore, reducing grazing may not be the best solution at all. In western oakwoods, adequate regeneration and habitat patchiness may be produced by thinning or felling small blocks of trees, the extra light being sufficient stimulus to trigger local regeneration. This solution would only work where the density of herbivores was not excessively high and in any case would be inappropriate in lowland wood-pasture.

Pollarding is now rarely carried out within woodland (Fig. 12.7). From a conservation viewpoint this practice is important because it tends to prolong the life of a tree and it has been responsible for producing many of the magnificent ancient Beeches and Oaks in lowland wood-pasture which are rich in epiphytes, fungi and saproxylic invertebrates. Continuation of the pollarding tradition in lowland wood-pasture is, therefore, highly desirable. Initiating a programme of pollarding on young or middle-aged trees is easy enough if the trees are not heavily shaded. Re-pollarding an ancient tree which has not been cut for many decades is a different matter, though it can be successful if carefully planned. In view of the effort and risk involved to the tree one must ask whether it really is worth pollarding such old trees. Those interested in taking this further are strongly urged to consult Mitchell (1989) and Read (1992) who give much detailed practical advice on this subject.

Fig. 12.7. Beech pollards in Mark Ash Wood, New Forest, Hampshire. These trees have not been lopped for two centuries and now many are falling apart. Whenever grazing pressure has been reduced, gaps have been filled by groups of saplings. (P. Wakely, English Nature.)

High forest

Modern forestry relies on high forest systems. The trees are grown from seedlings, which can be naturally regenerated, but are usually planted. The main product is large timber so the rotations are much longer than those used in coppice. High forest offers habitats complementary to those of traditional management, and some of the desirable features of traditional systems can be retained within high forest. High forest is so extensive in Britain that incorporating sound conservation practices within productive forestry would greatly enhance the regional populations of many species. Here we discuss broad principles of conservation management; aspects of high forest management that are exclusive to conifer forests are considered in the following section. For more detail see Evans (1984), Kirby (1984), Mitchell & Kirby (1989), Forestry Commission (1990), Matthews (1989), Peterken (1993).

Four broad silvicultural systems can be recognised:

1. *Clear-fell systems*. Trees are cut in large coupes, typically at least 1 ha, often much larger. Restocking is usually by planting. This creates a patchwork of even-aged stands, though it may take many decades to establish a varied age structure where large areas have been afforested in a short period. Nearly all British conifer stands are treated as clear-fell systems.

2. *Shelterwood systems*. Involves cutting a stand in at least two stages with restocking by natural regeneration. The first cut heavily thins the canopy, leaving sufficient trees to provide seed and shelter for seedlings. The remaining canopy trees are finally cut when the young growth is adequately established. This system creates even-aged stands, mixed with 2-storey stands undergoing regeneration. The size of coupes is variable but often several hectares in France, where the system is far more common than in Britain.

3. *Group-fell systems*. A smaller scale version of clear-felling, sometimes referred to as 'group selection'. Groups of trees, typically no larger than 0.3 ha, are felled. Restocking is either by natural seeding or by planting. Group-felling creates a finer-grain mosaic of even-aged stands than clear-felling.

4. *Selection systems*. Management is at the level of the individual tree rather than in coupes. The overall appearance of the forest remains constant because felling and regeneration are widely spread and frequent. Regeneration is usually natural. Practicable only for shade-bearing species, such as Beech and Small-leaved Lime. Selection systems are rare in Britain.

It would be interesting to see more examples of selection forestry in British woodlands. It creates extremely complex structures which may support rich communities of birds and other animals, though species needing open spaces or a vigorous ground vegetation would probably be scarce. Selection systems are, therefore, less appropriate than group-felling for ancient sites with a recent history of coppice management.

The other systems, like coppice, create open space within the forest, so potentially they also benefit species dependent on temporary gaps and young-growth. However, rotation lengths are longer than in coppice, though they may be as short as 45 years in conifers, so the proportion of land under young growth is smaller. Cutting of coupes is often widely spread, both in space and time, so for some species, suitable habitat may not be constantly available within a wood, or within the same part of a wood. Species with limited mobility may not survive such discontinuities in the availability of open spaces. Particular attention needs to be given to this problem where high forest has been established on former ancient woodland sites (Fuller & Warren, 1991). Here relic populations of, for example, Pearl-bordered Fritillary *Boloria euphrosyne*, may be associated with a rich ground flora in the young plantations. In such situations it is unlikely that it will be possible to cut adjacent new coupes to provide alternative habitat, but it may be possible to manage rides as refuges for such species (see below). Group-felling results in more contiguity of open habitats than clear-cutting but the smaller coupes associated with group-felling may be more shaded and hence less suitable for some butterflies. The young stages of shelterwood are more shaded than those of either group- or clear-felling, thus butterflies and other species demanding open warm conditions will be at a disadvantage unless open rides or glades are maintained.

Looking at the other end of the woodland cycle, any relic ancient trees, such as pollards that often survive along boundaries, should be protected because these may harbour rare species that could colonise future old growth if this becomes established nearby. Indeed, it is highly desirable to leave all old or dead wood to decay naturally within the wood and to develop a policy for ensuring continuity of a certain amount of old and rotting trees. Dead wood should not be imported from other woods.

Rather little is known about the effects of coupe size and shape on biological communities. We can be sure, however, that the relatively coarse-grained patchwork created by clear-fell systems will favour different species to the finer-grained mosaic found within group-fell systems. Species that avoid compartment edges, or which need open unshaded habitats, will generally benefit from large coupes, whereas ones associated with edges are more likely to benefit from small coupes. Where conservation is an important objective, decisions about the size of coupes will rest upon the type of community that is desired. As a working rule, larger coupes are more appropriate in extensive conifer forests, especially in the uplands, and group-felling is more appropriate in broadleaved woods. One possible undesirable effect of large coupes in conifer forests is that the vigorous growth of grass within them usually leads to a huge increase in voles which attract predators, such as Fox *Vulpes vulpes*, which in turn may impact on birds, such as Capercaillie, nesting in the surrounding forest.

Finally, some thoughts are specific to broadleaved woods. In general, natural

Fig. 12.8. Mixed broadleaved woodland, Trench Woods, Worcestershire. This stand may have been thinned several years ago allowing the development of a vigorous understorey which is beneficial to a wide range of bird species. The stand would now benefit from further thinning. Group-felling systems are appropriate for such lowland broadleaved woods. (D.M. Green.)

regeneration is preferable to planting. Trees are quite capable of establishing themselves provided browsing pressure and shading are not too great. Where planting is necessary, use species native to the site, preferably trees of local provenance. Conifer plantations on former ancient broadleaved sites should generally be gradually restored to broadleaved woodland but other changes in tree species composition should only be undertaken after very careful thought. Sycamore *Acer pseudoplatanus*, for example, is widely perceived as a major conservation problem because it is a highly invasive, non-native tree which casts a heavy shade. In fact, it is by no means certain that it will form persistent large monocultures unless planted, though its spread within ancient woods should be resisted where possible on the grounds that it may disrupt semi-natural communities of trees. More serious are dense thickets of *Rhododendron* which suppress ground flora and create woods of extremely unnatural structure. Two effective methods of controlling *Rhododendron* are the winching of cut stumps and roots from the ground, and the drilling and treatment of stumps with ammonium sulphamate (Evans and Becker, 1988).

Conifer forests

Extensive conifer forests managed by clear-felling can provide habitats for species associated with different stages of woodland succession because they are harvested rotationally in coupes. The general aim, therefore, should be to create a full range of age classes, ensuring that at any time there are substantial areas of young and mature growth. Many large first-generation conifer plantations contain extensive tracts of even-aged trees but at the end of the first rotation there comes an opportunity to break up this uniformity. To achieve this restructuring requires that some areas will be felled either before or beyond their normal economic felling age in shapes designed to fit in with the landscape.

Economic forestry involves harvesting trees as soon as growth rates decline. Within managed conifer forests trees seldom develop the natural features of old growth, even less so than in broadleaved woods. Retaining patches of woodland to grow well beyond their economic felling age is, therefore, highly desirable for conservation. Such areas, termed retentions, need not remain unmanaged, but rather can be treated as long-rotation stands. It may be desirable to retain small, permanently uncut core areas (Peterken *et al.*, 1992) although windblow is a problem in some areas. Retentions may develop habitats, particularly dead wood, which are unavailable elsewhere in the forest and which will benefit invertebrates and birds such as Capercaillie, which need old stands, and hole-nesters (Currie & Bamford, 1982). Dead wood in large conifer forests could be provided both through a network of retentions and by leaving scattered isolated dead trees within the shorter rotation stands, though retentions are likely to have the greatest conservation advantages.

Many commercial forests are essentially monocultures of spruce or pine. These forests could certainly be improved for birds, and perhaps other wildlife, by diversifying the tree species. Small adjustments may have large benefits. Creation of small broadleaved patches – even the planting of scattered oaks – may benefit a variety of insectivorous songbirds (Petty & Avery, 1990). Pure stands of Sitka Spruce *Picea sitchensis* can potentially be enriched as bird habitats by the addition of small amounts of Scots Pine, Norway Spruce *Picea abies* and Larch *Larix* spp. to contribute 5–10% of the total planting. The main benefit is that annual cone production will be less variable for seed-eating species such as Crossbill and Siskin *Carduelis spinus*. Young conifer plantations with much natural regeneration of Birch are likely to be richer in birds than ones that are largely devoid of non-crop trees.

Some species of conifer high forest merit particular discussion. In Scottish pinewoods, Capercaillies may benefit from thinning to ensure that a field layer of *Calluna* and *Vaccinium* develops beneath the canopy to provide a rich food source and nest cover for the birds. This may also help to generate a denser shrub layer of Juniper *Juniperus communis* which will probably increase the density of songbirds. Avoiding too high a density of pine trees may also benefit Crossbills which seem to prefer feeding on

Fig. 12.9. Thinned Oak stand in Wyre Forest, Shropshire. Thinning has let in more light, but has temporarily generated a uniform stand with no underwood. However, the remaining trees will grow faster and larger, and an underwood is likely to develop as shrubs and saplings regenerate providing that grazing pressure is not too severe. (D.M. Green.)

trees which are well spaced out. It would be undesirable to manage all native pinewoods by thinning because non-intervention is appropriate for many areas (see below) and Red Squirrels prefer areas with a close tree spacing, presumably because this aids their mobility through the canopy. The best conifer habitats for Red Squirrels are thought to be those offering extensive mosaics of thicket stages, for cover, and pole stage growth, for food. Mixtures of pine and spruce also help to provide a continuous supply of seed.

Manipulating tree density by thinning may be a valuable management tool in other types of high forest (Figs. 12.8, 12.9). Provided the risk of wind damage is not too great, thinning could be used to create a balance between mature trees and an open canopy permitting the development of a dense understorey or ground flora. These stands would potentially carry very diverse bird communities including species requiring a dense shrub layer, such as several summer visitors, as well as birds of mature forest, notably hole-nesters. There is likely to be an optimum density of mature trees, though research is needed before more guidance can be given on ideal tree spacing in different woodlands.

Scrub succession

Scrub is a transient formation which is commonly ignored by conservationists, and often seen as a threat to grassland, heathland and fen. Yet some scrub is interesting in its own right and can be rich in invertebrates and provide important breeding, feeding and roosting sites for birds. Most forms of scrub need management if they are to persist. The rate of habitat change varies with the shrub species, soil type and altitude.

Examples of scrub types of particular conservation interest are:

1. *Mixed calcareous scrub.* Typically composed of Hawthorn *Crataegus monogyna*, Buckthorn *Rhamnus catharticus*, Privet *Ligustrum vulgare*, Dogwood *Cornus sanguinea*, Roses *Rosa* spp, Whitebeam *Sorbus aria*, Wayfaring Tree *Viburnum lantana*. This type of scrub supports many species of insects and the berries are eaten by a wide range of birds.

2. *Juniper scrub.* Increasingly scarce in southern England. Ways are currently being sought of regenerating Juniper.

3. *Blackthorn* Prunus spinosa *scrub in ancient woodland.* In the Midland clay belt, Blackthorn is important as the food plant of the Black Hairstreak *Strymonidia pruni*. These insects are extremely localised, even within apparently suitable habitat. Bushes at least 20 years old must be present and an appropriate management would be cutting very small patches (10 m × 10 m) on a rotation of perhaps 30 years. The Brown Hairstreak *Thecla betulae*, on the other hand, needs growths of young, dynamic Blackthorn.

4. *Submontane scrub.* At the upper limit of tree growth on British mountains, scrub consisting of Sallow *Salix* spp, Junipers and Birches would once have been the characteristic vegetation, though now it is extremely rare. Birch scrub is a natural vegetation at lower altitudes in many upland areas though it is often suppressed by burning or browsing. Expansion of Birch scrub, especially on Scotland's hills, is desirable.

Hawthorn is the commonest pioneer scrub species where agricultural land is abandoned, or grazing pressure relaxed. Pure Hawthorn scrub can be rich in insects and birds in its early and middle stages of growth. Extensive areas of old 'leggy' scrub should not be allowed to develop because many of the more interesting insects and birds, such as warblers and Nightingales, live in the earlier stages of growth. Scrub that has recently closed canopy, but which still has fairly dense foliage extending down to near ground level, can carry extremely high densities of birds as well as affording good cover for roosting birds, and berries for thrushes. Blackthorn scrub spreads rapidly by suckering and also creates a dense structure that is favourable to many birds.

Maintaining large areas of scrub is an entirely appropriate aim of conservation management on those chalk downs and commons (often former wood-pasture) where it does not conflict with other aims, such as conserving a rich grassland flora. The most obvious approach is to use rotational cutting to maintain patches of scrub at different stages of growth from freshly cut to closed canopy. This will provide conditions for a spectrum of successional communities. Rotation lengths and patch sizes will depend on site conditions such as rates of regrowth and pressure from deer (see coppice). It may be appropriate to cut much smaller patches than in coppice, creating an intimate mixture of scrub and grassland, with structural gradation between. Such habitat can suit both grassland and scrub species, as well as those using edges. It can also provide warm, sheltered conditions for insects.

It may prove easier to maintain habitat mosaics by grazing rather than cutting. Existing rabbit and deer populations may do the job, but low intensity grazing by sheep, cattle or ponies should be considered (Fig. 12.10). The choice of grazing stock will depend on the type of vegetation that is being sought, for all animals graze in different ways inevitably leading to changes in the tree and scrub composition. Large herbivores have been used more widely in The Netherlands (van Wieren, 1991) than in Britain as a management tool for creating habitat mosaics. This approach may, however, have considerable applications in those British heaths, grasslands and sand dunes where the aim is to allow *some* scrub to grow up. We cannot give guidance on stocking rates. Much depends on the particular characteristics of each site which the habitat managers

Fig. 12.10. Cattle grazing willow scrub at Cabin Hill, Merseyside. Scrub will either become impenetrably thick, or pass into woodland if it is not periodically grazed, cut, or burned. (P. Wakely, English Nature.)

themselves must assess. Success in managing patchy vegetation requires a skill which has to be honed through trial and error. One advantage that grazing holds over cutting is that it creates habitats of more natural appearance, with subtle gradations and patchiness. Furthermore, scrub maintained by cutting will turn into coppice, whereas intermittent or patchy pasturage may perpetuate scrub as a stage in a woody succession from unwooded vegetation.

Natural woodland

There is a strong case for allowing natural woodlands containing old growth stands to develop within some British nature reserves. Such sites (i) develop valuable habitats for species dependent on mature trees and dead wood, (ii) permit scientific research and educational opportunities for understanding the natural dynamics of woodland, (iii) can form a baseline for monitoring widespread environmental change, and (iv) generate a wilderness atmosphere missing from British woodlands.

The main attributes of natural temperate woodlands in Britain would be (Peterken, 1991, 1992):

1. *Patchiness*. Spatially, the forest would consist of patches at different stages of growth. Small gaps would be created periodically through the death of single trees, larger gaps less frequently through storm damage. In Britain, wind rather than fire, would be the main agent of catastrophic disturbance, though fire was probably important in the native pinewoods and in districts with thin, nutrient-poor soils.
2. *Massiveness*. Some individual trees would grow extremely large.
3. *Dead wood*. An abundance of dead wood as standing snags, dead limbs, fallen trunks and branches, and decaying cores of large, living trees. Dead wood will be present in all stages of decay. Its volume will be greatest in the oldest stands and young stands developing after catastrophic disturbances.

Britain has no virgin forest, so the issue is: which sites would be most appropriate for natural woodland to be re-created? Fairly mature stands offer the best starting point, including long-neglected coppice and semi-natural high forest. Ideally, the proposed sites should be large (at least 10, preferably 100 ha) to minimise edge effects and enable a full range of growth phases to be permanently present. Some species of birds, notably Capercaillie, would benefit from areas of old growth, in this case pine, extending over several square kilometres.

Adopting a natural woodland option does not necessarily imply total non-intervention (Peterken, 1991). Because the characteristics of old-growth forest will develop very slowly, the process might be accelerated by initial intervention, perhaps thinning or felling small patches to break up the uniformity of the stand. Generally, however, we do not advocate this kind of 'priming', feeling that the natural processes are more interesting. The focus of interest is the dynamic relationship between native species, so the usual policy is to accept any changes which take place in the amount and distribution of native trees and shrubs, whilst removing the exotics. On a few sites, however, it would be interesting to adopt complete non-intervention and permit whatever changes occurred, including colonisation by exotics. At all natural woodland sites, any urge to tidy up, say, the 'debris' arising from storm damage must be totally resisted.

It is not possible to recreate truly natural forest in Britain. Even large tracts of woodland will inevitably be influenced by their surroundings (invasions of alien and edge species, pesticide drift, air pollution) and we have lost most of our large mammals.

Fig. 12.11. Windthrow in Norsey Wood, Essex, after the storm of October 1987. Gaps so produced naturally diversify the woodland structure and composition, and generate pit-and-mound microtopography. If the site manager prefers tidiness, then the limbs can be cut, the rootstocks can be rocked into place and the result will be a small patch of coppice. Most of the toppled trees will re-sprout rapidly even if they are not so treated. (P. Wakely, English Nature.)

Fig. 12.12. Ride in Ham Street woods, Kent. Open rides with a varied marginal structure of tall grasses, herbs and shrubs are richest for wildlife. (P. Wakely, English Nature.)

We do not know what levels of grazing would have been sustained in our natural woodland (see wood-pasture). None of these problems should deter us from seeking to re-establish near-natural woodlands on suitable sites.

Management of open spaces, edges and watercourses

Rides, glades, external edges and watercourses all make a contribution to the diversity of the woodland flora and fauna that is disproportionate to their area. This is particularly true in large conifer plantations, though the rides and glades in ancient woodland are also foci of diversity. Petty and Avery (1990) recommended that some 10% of the total area of conifer forests should be given over to a network of habitats that do not form part of the productive forest area.

Conservation management in woods often focuses on rides and glades (Figs. 12.12, 12.13). These can support large numbers of light-demanding species which may be absent elsewhere in the wood. Well-managed rides and glades can provide vital habitats for scarce grassland plants and large numbers of grassland and scrub invertebrates. These open habitats are important in providing a large range of flower resources (nectar and pollen) for insects, and can serve as population sources from which newly created

Fig. 12.13. Narrow ride in Great West Wood, Lincolnshire. Sandwiched between tall, semi-natural Birch-Ash-Lime woodland on the left, and rapidly growing plantations on the right, this ride is already too narrow and will soon be overshaded and impoverished. (P. Wakely, English Nature.)

young coppice or plantation can be colonised. A good network of rides acts as a link between different patches of young growth within the stands. Particularly in high forest, rides and glades can be essential refuges for many species. Prescriptions are outlined in Box 12.2 but readers are urged to consult Warren & Fuller (1993) and Greatorex-Davies (1991) for more detail.

Box 12.2 Ride and glade management

It is essential to keep shading to a minimum because the majority of species dependent on rides and glades demand warm, sunny conditions. Three factors influence shading of rides: ride width, height of adjacent trees and orientation of the ride. The taller the trees, the wider must be the ride. As a general rule, *rides should be 1.5 times as wide as the height of the bordering crop trees* so that where trees reach 20 m, rides should be at least 30 m wide (width is the distance between the crop trees).

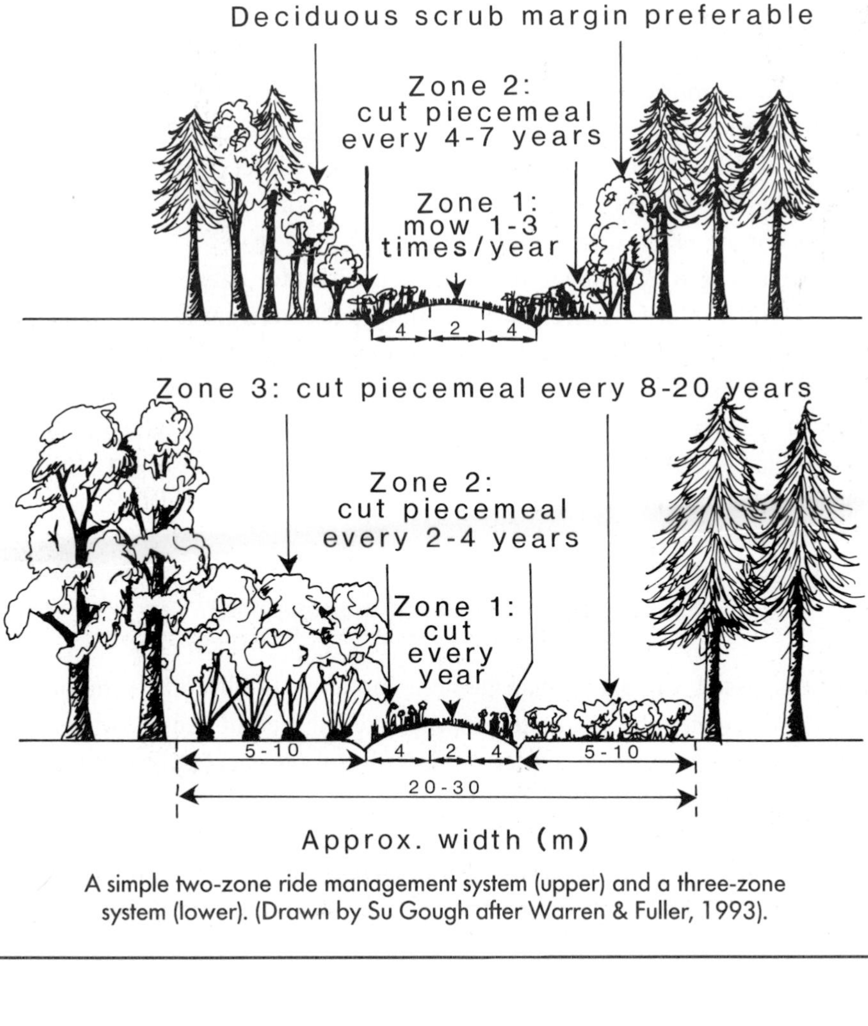

Deciduous scrub margin preferable

Zone 2:
cut piecemeal
every 4-7 years

Zone 1:
mow 1-3
times/year

4 | 2 | 4

Zone 3: cut piecemeal every 8-20 years

Zone 2:
cut piecemeal
every 2-4 years

Zone 1:
cut
every
year

5 - 10 | 4 | 2 | 4 | 5 - 10

20 - 30

Approx. width (m)

A simple two-zone ride management system (upper) and a three-zone system (lower). (Drawn by Su Gough after Warren & Fuller, 1993).

Rides aligned east–west receive more sunlight than do ones running north–south, so the latter ideally need to be wider than the former. Whilst many rides benefit from widening, this should be undertaken cautiously. For example, in some conifer plantations the only broadleaved shrubs and trees may exist along ride sides and their simultaneous removal would be disastrous. Ride-widening is best carried out gradually over several years. In some woods, old ride systems are best left unmanaged, especially in semi-natural high forest and wood-pasture where the conservation interest lies mainly with the old growth species. Attention should also be given to any sunny banks that exist within the wood for these sites are important for solitary bees and wasps, and other invertebrates. Such banks need to be kept free of vegetation.

For rides, the ideal management is to maintain linear parallel belts of herbaceous and shrubby vegetation (see Warren & Fuller, 1993). The most complex prescriptions, involving three separate belts each cut on different rotation lengths, ensure a continuity of habitats ranging from open grassland to closed-canopy scrub or coppice. Glades can be created in conjunction with ride systems. One method is to create 'box-junctions' at ride intersections by removing the corners between each ride (see Warren & Fuller, 1993). Each of the four segments can then be managed on a rotation which may or may not differ between segments. Glades can also be created in the form of scallops at intervals along rides. Where ancient glades, perhaps old deer launds, survive within a wood, every effort should be made to maintain their open character through an appropriate cutting or grazing regime.

The external edges of woods often provide habitats for a large number of invertebrates and birds (Fuller & Warren, 1991). Such edges often are rich in food resources, especially flowers and fruits, and the vegetation structure can be more complex than within the wood itself. Management should seek to perpetuate any scrub and old trees that exist at woodland edges. In upland forests, watercourses form linear features containing relatively natural, wildlife rich habitats for which the Forestry Commission has developed management guidelines (Forestry Commission, 1991).

Integrated woodland management

We have discussed woodland management in terms of discrete components and systems, but these must be assembled appropriately if the full benefits of conservation management are to be realised. Different silvicultural systems can be combined within single woods. The combinations chosen for ecologically similar woods need not be identical. In fact, we must consider management at all scales from parts of woods to the management pattern for the woods of a whole region (Fuller, 1990; Fuller and Warren, 1991; Peterken, 1993). Long-term strategies are desirable which seek to create a complementary range of habitats and associated communities. Box 12.3 exemplifies this approach.

The particular combinations which are optimal depend on the individual features of each wood. We have already stressed the importance of maintaining wood-pasture and active coppice where it has a long history. Where rare species are present (of any wildlife group), their needs will take priority, but such species are often very localised within individual woods and some woods have no rare species. However, where there are no such overriding special needs and where woods are sufficiently large, it is desirable to include both compartments of coppice management and compartments of high forest or unmanaged woodland, for this combines the important young and old stages of growth (see Kirby, 1992). The actual distribution of different systems within a wood will be partly determined by the inherited stand conditions, but one should avoid an intimate mixture of two systems to ensure that species with requirements for young or old growth are not presented with excessively fragmented habitats. The contribution of open spaces must be related to the management of stands: for example, in the long rotations of high forest systems it is desirable to have wider rides and to ensure that felling is out-of-step on the two sides of any ride.

Integration is also desirable in another sense. Woodland nature reserves should be complementary to ordinary forestry: they provide the conditions which are not otherwise ensured within management for timber and landscape (Peterken, 1991). Furthermore, they are no longer functionally separate from other woodlands, for all forests are subject to Nature Conservation Guidelines (Forestry Commission, 1990). In purely quantitative terms, reserves protect only a minority of woodland wildlife. Accordingly, although conservation management of woodlands is a valid concept, it is only part of a wider spectrum of environmentally sustainable management which should be applied to all woodlands and forests.

Acknowledgements

Rob Fuller's work on woodlands was funded by the Joint Nature Conservation Committee on behalf of English Nature, Scottish Natural Heritage and the Countryside Council for Wales. We thank the following for their comments on the manuscript: David Bullock, Robin Buxton, Fred Currie, Pete Fordham, Robin Gill, Ted Green, Keith Kirby, Matthew Oates, Ron Summers, Martin Warren, and the editors and consultants. Sophie Foulger did a huge amount of work on various drafts of this chapter.

Box 12.3 Strategic management of ancient woodlands: an example

The Berkshire, Buckinghamshire and Oxfordshire Naturalists' Trust (BBONT) owns 10 ancient woods of total area 358 ha. Planning the management of these woods has been carried out in two stages:

(1) *A strategic review* of the occurrence of different stand types across the 10 woods and the identification of suitable long-term stand treatments for all compartments so as to (a) identify treatments in keeping with the existing character and conservation interest of each wood and, (b) to create a series of woodland structures supporting a variety of plant and animal communities. This review also states the policy objectives for BBONT's woods, including many of the management recommendations contained in this chapter.

(2) *Formulation of conventional management plans* for each wood, detailing precise treatments for compartments (e.g. rotation length, thinning regimes) and for glades, rides and other edges. These management prescriptions build on the framework established in the strategic review. Hence, the strategic review sets a coherent policy for the management of BBONT's woods and ensures that the management plans for individual woods are not entirely independent.

The aim was to ensure that each wood complements the rest by formulating plans which were (i) appropriate to the interest and current condition of each compartment of each wood, but which also (ii) ensure that no two woods are managed identically, and (iii) include in the woods as a whole a reasonable amount of each of the main management options.

A summary of the long-term stand types proposed for the woods is given in the table below.

	Short rotation coppice	Long rotation coppice	Broadleaved high forest	Mixed high forest (broadleaves and conifers)	Natural woodland (i.e. indefinitely unmanaged)	Scrub
Wood A (12.9 ha)		2.1	10.8			
Wood B (20.2 ha)	1.2	3.7	12.6		2.7	
Wood C (85.0 ha)	10.0	11.0	38.0	11.0	15.0	
Wood D (43.0 ha)	4.0		28.0		11.0	
Wood E (22.4 ha)			22.4			
Wood F (42.5 ha)			42.5			
Wood G (9.5 ha)			9.5			
Wood H (23.1 ha)	3.5	8.9	9.6		1.1	
Wood I (41.9 ha)	10.6		15.7		13.0	2.6
Wood J (57.3 ha)	3.7	0.8	39.2		10.4	3.2
Total	33.0	26.5	228.3	11.0	53.2	5.8
	(9.2%)	(7.4%)	(63.8%)	(3.1%)	(14.9%)	(1.6%)

Significant proportions of the broadleaved high forest will be derived from derelict coppice (38 ha) and from conifer plantations (53 ha). The overall balance of habitats will be: coppice 60 ha (17% of total area), high forest 239 ha (67%), natural woodland 53 ha (15%), scrub 6 ha (2%). BBONT manages substantial areas of chalk scrub which were not included within this review though it would be possible to undertake a similar exercise for grassland and scrub habitats.

References

Bright, P. & Morris, P.A. (1989). *A Practical Guide to Dormouse Conservation*. Occasional publication of the Mammal Society II. London: The Mammal Society.

British Trust for Conservation Volunteers. (1980). *Woodlands: a Practical Conservation Handbook*. Wallingford: BTCV.

Buckley, G.P. (ed.) (1992). *Ecology and Management of Coppice Woodlands*. London: Chapman & Hall.

Currie, F.A. & Bamford, R. (1982). The value to birdlife of retaining small conifer stands beyond normal felling age within forests. *Quarterly Journal of Forestry* **76**, 153–60.

Evans, C.E. & Becker, D. (1988). Rhododendron control on RSPB reserves. *RSPB Conservation Review* **2**, 54–6.

Evans, J. (1984). *Silviculture of Broadleaved Woodland*. Forestry Commission Bulletin 62. London: HMSO.

Forestry Commission. (1990). *Forest Nature Conservation: Guidelines*. London: HMSO.

Forestry Commission. (1991). *Forests & Water: Guidelines*, revised edition. Edinburgh: Forestry Commission.

Fuller, R.J. (1990). Responses of birds to lowland woodland management in Britain: opportunities for integrating conservation with forestry. *Sitta* **4**, 39–50.

Fuller, R.J. (1992). Effects of coppice management on woodland breeding birds. In *Ecology and Management of Coppice Woodlands*, ed. G.P. Buckley, pp. 169–92. London: Chapman & Hall.

Fuller, R.J. (1995). *Bird Life of Woodland and Forest*. Cambridge: Cambridge University Press.

Fuller, R.J. & Warren, M.S. (1993). *Coppiced Woodlands: Their Management for Wildlife*, second edition. Peterborough: Joint Nature Conservation Committee.

Fuller, R.J. & Warren, M.S. (1991). Conservation management in ancient and modern woodlands: responses of fauna to edges and rotations. In *The Scientific Management of Temperate Communities for Conservation*, ed. I.F. Spellerberg, F.B. Goldsmith & M.G. Morris, pp. 445–71. Oxford: Blackwell Scientific Publications.

Greatorex-Davies, J.N. (1991). Woodland edge management for invertebrates p. 25–30. In *Edge Management in Woodlands*, ed. R. Ferris-Kaan. Forestry Commission Occasional Paper 28. Edinburgh: Forestry Commission.

Greatorex-Davies, J.N., Sparks, T.H., Hall, M.L. & Marrs, R.H. (1993). The influence of shade on butterflies in rides of coniferised lowland woods in southern England and implications for conservation management. *Biological Conservation* **63**, 31–41.

Gurnell, J., Hicks, M. & Whitbread, S. (1992). The effects of coppice management on small mammal populations. In *Ecology and Management of Coppice Woodlands*, ed. G.P. Buckley, pp. 213–32. London: Chapman & Hall.

Gurnell, J. & Pepper, H. (1988). Perspectives on the management of red and grey squirrels. In *Wildlife management in forests*, ed. D.C. Jardine, pp. 92–109. Edinburgh: Institute of Chartered Foresters.

Harding, P.T. & Rose, F. (1986). *Pasture-Woodlands in Lowland Britain*. Huntingdon: Institute of Terrestrial Ecology.

Kirby, K.J. (1984). *Forestry Operations and Broadleaved Woodland Conservation*. Focus on Nature Conservation No. 6. Peterborough: Nature Conservancy Council.

Kirby, K.J. (1992) Accumulation of dead wood: a missing ingredient in coppicing? In *Ecology and Management of Coppice Woodlands*, ed. G.P. Buckley, pp. 99–112. London: Chapman & Hall.

Kirby, P. (1992). *Habitat Management for Invertebrates: a Practical Handbook.* Sandy: Royal Society for the Protection of Birds.

Marren, P. (1990). *Woodland Heritage.* Newton Abbott: David & Charles.

Matthews, J.D. (1989). *Silvicultural Systems.* Oxford: Clarendon Press.

Mitchell, F.J.G. & Kirby, K.J. (1990). The impact of large herbivores on the conservation of semi-natural woods in the British uplands. *Forestry* **63**, 334–53.

Mitchell, P.L. (1989). Repollarding large neglected pollards: a review of current practice and results. *Arboricultural Journal* **13**, 125–42.

Mitchell, P.L. & Kirby, K.J. (1989). *Ecological Effects of Forestry Practices in Long-established Woodland and their Implications for Nature Conservation.* Oxford Forestry Institute Occasional Paper No. 39. Oxford: Oxford Forestry Institute.

Peterken, G.F. (1991). Ecological issues in the management of woodland nature reserves. In *The Scientific Management of Temperate Communities for Conservation*, ed. I.F. Spellerberg, F.B. Goldsmith & M.G. Morris, pp. 245–72. Oxford: Blackwell Scientific Publications.

Peterken, G.F. (1992). Conservation of old growth: a European perspective. *Natural Areas Journal* **12**, 10–19.

Peterken, G.F. (1993). *Woodland Conservation and Management*, second edition. London: Chapman & Hall.

Peterken, G.F., Ausherman, D., Buchenau, M. & Forman, R.T.T. (1992). Old-growth conservation within British upland conifer plantations. *Forestry* **65**, 127–44.

Petty, S.J. & Avery, M.I. (1990). *Forest Bird Communities: a Review of the Ecology and Management of Forest Bird Communities in Relation to Silvicultural Practices in the British Uplands.* Forestry Commission Occasional Paper 26. Edinburgh: Forestry Commission.

Rackham, O. (1980). *Ancient Woodland: its History, Vegetation and Uses in England.* London: Edward Arnold.

Rackham, O. (1990). *Trees and Woodland in the British Landscape*, third edition. London: Dent.

Read, H.J. (ed.). (1992). *Pollard and Veteran Tree Management.* Corporation of London.

Rodwell, J.S. (ed.) (1991). *British Plant Communities.* Volume 1: *Woodlands and Scrub.* Cambridge: Cambridge University Press.

Spencer, J.W. & Kirby, K.J. (1992). An inventory of ancient woodland for England and Wales. *Biological Conservation* **62**, 77–93.

Thomas, J.A. (1991). Rare species conservation: case studies of European butterflies. In *The Scientific Management of Temperate Communities for Conservation*, ed. I.F. Spellerberg, F.B. Goldsmith & M.G. Morris. pp. 149–97. Oxford: Blackwell Scientific Publications.

van Wieren, S.E. (1991). The management of populations of large mammals. In *The Scientific Management of Temperate Communities for Conservation*, ed. I.F. Spellerberg, F.B. Goldsmith & M.G. Morris. pp. 103–27. Oxford: Blackwell Scientific Publications.

Warren, M.S. & Fuller, R.J. (1993). *Woodland Rides and Glades: their Management for Wildlife*, second edition. Peterborough: Joint Nature Conservation Committee.

Warren, M.S. & Key, R.S. (1991). Woodlands: past, present and potential for insects. In *The Conservation of Insects and their Habitats*, ed. N.M. Collins & J. Thomas. London: Academic Press.

Watkins, C. (1990). *Woodland Management and Conservation.* Newton Abbott: David & Charles.

13 Urban areas

CHRIS BAINES

Introduction

The world is becoming increasingly urban. Already nine out of ten people in Britain live in towns. By the year 2020, it is expected that nine out of ten people in the rest of the world will do the same. The humdrum pressure of urban living makes human contact with the natural world more important than ever, so our towns and cities need to provide very easy access to relatively wild, green landscapes. For the old, and the very young in particular, wildlife on the doorstep is almost the only wildlife that counts. Ironically, almost all the commitments to habitat management, from central government funding and statutory protection, to practical action by enthusiasts on the ground, is directed towards remote rural landscapes, keeping rare species in protective custody, for the pleasure of the privileged few. If these exclusive habitats are to survive, then they need championing by the urban majority, who in turn must be inspired through familiarity with the wildlife they can see every day of the week.

The land resource

There is a myth that urban landscapes are concrete jungles, paved wall to wall with tarmac – a hostile environment where nature struggles to survive. In fact even the most densely built-up places have abundant open space, and in a typical western town or city, greenspace is in the majority. In Leicester, for example, 25% of the city's land area is occupied by private gardens, and the 'official' green open space of sports fields, public parks, school playing fields, hospital and college grounds, road verges and golf courses occupies almost as much again. There is a third category of green open space in most urban areas too, and certainly in older industrial areas it can be very extensive indeed. This is the 'unofficial' wildspace, often labelled derelict, where nature has re-colonised the ruins of past industrial activity. The combined resource of surplus railway land, worked out mineral quarries, demolished factories and tipped land has created a complex landscape of disturbed ground, mixed mineral substrates, and varied vegetation which is often rich in the wildlife that thrives best in pioneering communities. The extent of this unofficial wildspace is immense, and is continually expanding and maturing too. In the West Midlands borough of Sandwell, the Black Country Urban Forestry Unit compared aerial photographs from 1977 and 1989, and found that woodland cover had almost doubled in that 12 year period, and that almost half the new woodland had emerged unaided, simply through natural colonisation of neglected 'derelict' land.

Existing management

The management of almost all the green open space in urban areas falls into one of two categories. 'Official' landscapes are clinically managed under intensive regimes where machines and chemicals sterilise the environment. Typically, grasslands are close-mown at frequent intervals through the growing season, fallen leaves are swept away, and often burned or buried for the sake of tidiness, and native weeds are ousted in favour of exotic horticultural alternatives. This neat and tidy style of grounds maintenance is applied to most private gardens too. The result is a massive commitment to expensive management cycles, and a relatively lifeless range of habitats.

The alternative to horticultural sterility is a style of management based on neglect. Here the quality of habitat is determined by a combination of ground condition and availability of colonisers, combined with such biotic factors as nutrient boosting through fly tipping, casual burning, firewood-cutting and erosion through simple human pressure. For the most part these landscapes are the wildspaces of dereliction, and almost the only active habitat management is likely to be scrub clearance in response to community pressure for tidiness, increased access and improved security.

The resources available for managing urban habitat are enormous – far greater than those available in the rural countryside. The difficulty is in establishing the need to be creative with those resources. For example, each year a city the size of Birmingham or Manchester will spend in excess of £10 million (1990 prices) exclusively on mowing the municipal grass. Nationally the urban grass-cutting bill exceeds £1000 million. As well as the vast fund committed annually to urban landscape management, there are other huge resources available as a result of the constant change which urban spaces undergo. Wherever new roads are built, or other site development is undertaken it is possible to channel money and effort into positive landscape change. The one aspect of management which is difficult to finance, is that of people management. The social pressure on urban habitats is relentless. Such things as children's play, fouling by dogs, motor-cycling, fly-tipping, malicious vandalism, toxic waste dumping and airgun shooting combine to pressurise wildlife communities. Their impact can be greatly reduced through skilled management: adopting a countryside ranger style of supervision which combines policing with play leadership and education. Financing this kind of labour force generally is only possible if some other aspect of revenue expenditure is eroded, and that often proves difficult.

One further management resource which is especially significant in the urban context is the community, local people. Here the critical concern is that this volunteer resource is used appropriately. Whilst there may be technical skills available, the real strength of a community input to management lies in such areas as site surveillance, site wildlife recording, and such labour intensive 'fun' events as haymaking, seed gathering and pond planting.

Urban grasslands

Occasionally there are relic grasslands surviving in urban areas, containing fragments of diverse fauna and flora. For the most part, though, the grasslands of towns and cities are relatively modern, established with a sowing of just two or three species of grass, and colonised by a very limited range of other species. They are mostly managed under an intensive regime of summer mowing, with the finer playing pitches – golf greens,

Fig. 13.1. Formal parks and public open spaces are full of opportunities (and financial resources) which could create new wildlife habitat. (C. Baines.)

bowling greens and cricket squares – cropped as much as three times each week, cleared of the clippings, and then heavily fertilised to maintain the nutrient levels. The majority of other grasslands are cut every 10 to 14 days, using either a flail, a rotating blade, or cylinder mower, leaving the clippings to return to the soil. This regime builds up fertility and creates a mat of dead grass 'thatch' and tends to favour a small range of well adapted species. Rosette plants such as Dandelions *Taraxacum officinale*, Daisies *Bellis perennis* and Catsears *Hypochaeris radicata* thrive, along with creeping species such as Selfheal *Prunella vulgaris*, and Yarrow *Achillea millefolium*. Craneflies and earth-worms are the principal macro-invertebrates, and Starlings *Sturnus vulgaris* are the most common vertebrates.

To increase species diversity, there is a need to address three key aspects of the habitat. Firstly, there is a need to extend the length of time between cuts, to allow for plants to flower and seed, and the timing of these gaps will be critical to certain species. Secondly, there is a need to reduce fertility, if species diversity is to be increased. In rich soils a few vigorous plants tend to thrive, and squeeze out most of their less aggressive competitors. The simplest way of achieving this is to remove the clippings, though that has very significant ramifications so far as machinery requirements are concerned. The organic waste that results can, of course, be used an an ingredient in compost-making, or applied directly to the surface of the soil as a mulch in newly planted tree and shrub areas. Thirdly, there will generally be a need to introduce additional species of plants, since urban sites are relatively isolated from sources of natural re-colonisation.

Hay meadow management is the most popular regime adopted to enhance species diversity in urban grasslands. The cycle generally involves a mid to late summer mowing, ideally using a blade cutting machine rather than a flail so that the vegetation is cut cleanly, rather than chewed and shredded. This makes removal of the hay far easier. On large areas an agricultural hay cutter bar is ideal, though it is critically important to ensure that this is properly guarded at all times as this machine is very dangerous. Urban meadows are a favourite place to dump obstacles such as bicycle wheels, fence posts, lengths of wire and shopping trolleys. Just one such obstacle can seriously damage the cutter bar, and so it is important to walk the meadow immediately before mowing begins. On smaller areas, the ideal machine for mowing meadow-length grass is an auto-scythe, a machine with a pair of reciprocating cutter blades similar to a mechanical hedge trimmer. (A traditional hand scythe does the same job but is much harder work and a good deal slower.) If this kind of machine is used, then the hay can be left to dry in the sun, and tossed to keep it aerated, before being raked up and removed (see Fig. 13.3). This allows the seed to fall. If a flail or covered rotary mower is used, then the grass will be chopped and shredded in the process, and it is then very difficult to dry and harvest later. In this case it is far better removed as an integral part of the mowing process, rather than being left to become soggy. On a large scale, agricultural forage harvesters can be used, though there may need to be two consecutive cuts to cope

Fig. 13.2. These native sallow leaves offer essential food to dependent invertebrates. (C. Baines.)

Fig. 13.3. Haymaking can be a fun way of involving local people in habitat management. Here the hay is raked and dried on temporary timber frames before being harvested for local guinea-pigs. (N. Jowes.)

with long grass. For smaller areas, or in grassland made inaccessible by obstacles such as trees, there are smaller more manoeuvrable harvesters now available, but the capacity of their grass containers is very limited, and frequent emptying is necessary.

There is also great merit in leaving some areas of grassland completely uncut. Mowing every three or four years will help control scrub invasion, but the rank, tussocky grass which develops is very important overwintering habitat for a good many invertebrates, for newts, and for small mammals.

Traditional rural meadows are generally grazed for part of the autumn and winter season. The re-growth which emerges after the summer mowing (known traditionally as the aftermath) provides grazing for cattle and/or sheep, and the livestock serve the double purpose of cropping the vegetation close, so allowing the spring meadow flowers improved light for the following growing season, and also poaching or cutting up the soil surface, pressing in the fallen seed, and providing a diverse range of niches both for plant establishment and for specialist birds. Snipe *Gallinago gallinago*, for instance, feed through the winter in the soft mud of poached meadows, and ground-nesting birds may lay their eggs in hoof-depressions. In the urban context, grazing is rarely practical and so there is a need to substitute the livestock with a second mechanical cut, either in late autumn, or early spring. Some trampling may be possible too, if people can be persuaded to play the part of cattle.

Boosting the floral diversity of urban grasslands is only worth attempting once a sympathetic management regime has been successfully established, and conditions are suitable for a wider range of species. New species are much easier to introduce successfully as established plants than as seed. The more aggressive meadow wild-flowers can be introduced by direct seeding, but only if some means can be found of penetrating the sward and the thatch, and enabling the seed to germinate in the soil below. Mechanical scarifying with a spiked harrow is one possibility, and there has been great success with the use of agricultural slot-seeders which plant the individual seed, and suppress the immediately adjacent grasses with a squirt of contact herbicide (see Fig. 13.4). This latter technique has been used with spectacular success for Oxeye Daisy *Leucanthemum vulgare* and Cowslip *Primula veris* by Miriam Rothschild, on her farm in Northamptonshire, but is not successful as a means of establishing plants with finer seeds.

Adding new species

Species introduction using established seedlings is far more expensive and time consuming, but is more likely to lead to success. If soil conditions are poor, the sward is thin, and competition is weak, then small plugs of soil, each containing a single rooted seedling, can be successfully pushed down into the surface of the soil. However, if the existing grassland is at all vigorously competitive, then larger plants should be planted, to make sure that they can survive the early establishment stage.

Since the prime conservation value of urban grassland habitat is in providing the 'wildflower meadow experience' rather than re-creating complex botanical communities, it is advisable to adopt a simplistic approach, and to begin diversification by introducing relatively tough meadow species such as Oxeye Daisy, Black Knapweed *Centaurea nigra* and Yellow Rattle *Rhinanthus minor*. Since the Yellow Rattle is parasitic on grasses, it is an important ingredient as it helps suppress the grasses, and so reduces the competition for other desirable species. Knapweed is an important nectar plant for butterflies, and as the grasses become less competitive, caterpillar food plants such as Birdsfoot Trefoil *Lotus corniculatus* and Lady's Smock *Cardamine pratensis* can be encouraged or introduced.

Urban grazing

Whilst grazing is difficult to manage well in urban areas – both sheep and cattle are vulnerable to human abuse, and to dog worrying – there is grazing livestock available. Geese and ponies are both frequent features of town grasslands, but both produce grassland communities which are of poor species diversity, and relatively unattractive.

Populations of Canada Geese *Branta canadensis* are reaching plague proportions. Geese graze extremely closely, turning their heads on one side to scissor away the entire

Fig. 13.4. Mechanical slot-seeding has been used successfully to introduce Oxeye Daisy to this Rye Grass meadow. (C. Baines.)

grass sward. This, combined with the fertility effect of their droppings and the trampling by wet webbed feet produces a smeared surface where grasses eventually disappear, and Daisies are almost the only plants which survive. Attractive lake-side grasslands are very difficult to maintain unless geese are very vigorously suppressed, an unpopular management policy amongst tax paying bird feeders. Pricking the eggs or coating with liquid paraffin are the preferred techniques, but most city geese move out of town in pairs to breed, and once dispersed this option is impractical.

Ponies and horses are equally unsatisfactory as grazers. They graze coarsely and selectively, and unless stocking levels are extremely low, they will leave a rough pasture where unpalatable broadleaved plants such as Meadow Buttercup *Ranunculus acris* and Marsh Marigold *Caltha palustris* gradually take over. In urban paddocks the stocking level is almost invariably far too high, and this leads to heavy soil erosion, and mud baths, particularly around gates, sheltering hedges and points where the public can feed and stroke the animals. Overgrazing, and the lack of trace elements in impoverished pasture, often results in the horses and ponies browsing on hedgerows and trees, and frequently killing trees in paddocks by stripping the bark and eating it.

Finally, horses and ponies adopt specific latrine areas where manure accumulates. This in time builds up soil fertility, boosts nitrogen levels, and since these fouled areas are not grazed by the animals, the vegetation becomes rank, dominated by Stinging Nettles *Urtica dioica*, and quickly gives the grassland a very scruffy appearance. The nettles are valuable as larval plants for several species of butterflies, and the dung supports a rich fauna of beetles and other insects, but otherwise the horse pasture generally provides poor habitat for invertebrates.

The floral diversity of urban grassland can be improved by restricting horse grazing intensity, by supplementing with other kinds of grazing livestock, by feeding mineral supplements and by resting the pasture and managing it for hay every three to four years. Tethered grazing is a common means of managing urban livestock, and if carefully controlled and rotated, it can help produce a particularly rich mosaic of varied grassland habitat.

Urban woodlands

There are very few treeless towns. The species list of typical town birds bears witness to that: Blackbirds *Turdus merula*, Wrens *Troglodytes troglodytes*, Robins *Erithacus rubecula*, Song Thrushes *Turdus philomelos* and the numerous tits are all essentially woodland species. However, it is true to say that very few urban trees grow in habitats which would be recognised as woodland. Ironically, the most successful woody communities from the point of view of wildlife diversity, are those naturally regenerated scrub woodlands which have sprung up in the disturbed land and mineral deposits of railway sidings, abandoned gravel pits and demolished buildings. For the most part, though, these latter-day pioneer woodlands are dominated by wind-borne colonisers such as Silver Birch *Betula pendula*, Sycamore *Acer pseudoplatanus* and various willows *Salix* spp. and by definition they lack any of the old trees, deep leaf litter-rich soils or spring wildflowers which are a feature of traditional woodlands. By contrast, the more formal green spaces in towns are often graced with large, mature trees, but here there is rarely any shrubby understorey or herb layer, and the urban preoccupation with sterile tidiness ensures that dead branches, hollow trunks and autumn leaves are all removed before they can make a positive contribution to the habitat.

Success in managing urban woodland must begin with a massive public rediscovery of the true nature of woodland. Whilst there is an impressive commitment to trees and tree planting, with residents chaining themselves to threatened giants, and community tree planting a mainstay of practical conservation, there is precious little evidence of a popular enthusiasm for the less glamorous qualities of woodland, such as decay.

In a project such as Operation Greenheart, undertaken in the parks of Walsall Metropolitan Borough Council by the local authority and the Black Country Urban Forestry Unit, the emphasis is very much on raising public awareness of the special qualities woodland has to offer, highlighted through the work of sculptors, dawn chorus concerts, nest-box schemes and seed gathering.

Fig. 13.5. Urban wasteland like this is ecologically diverse, species-rich and very accessible – all it lacks is skilled and sensitive management. (D. Woodfall.)

Death and decay

The greatest challenge for the would-be urban woodlander, is to re-establish death and decay as a foundation for life. Grass cutting is such an all-pervasive feature of urban open space management, that to drive it out from beneath the canopy of town trees demands a major rethink in terms of budgets, bonus schemes and training. The broader concerns for the environment are beginning to help the argument. 'Green' Councils are increasingly sensitive to the argument that excessive machine-use burns unnecessary fossil fuels, adds to global warming and acid rain and increases noise and air pollution. The environmental and financial cost of landfill or incineration waste disposal is also of serious concern. Mountains of leaves are disposed of this way each autumn.

As a first move towards rebuilding woodland habitat beneath the canopy of urban trees, grass cutting can realistically be replaced by the spreading of leaf-sweepings. As playing fields, and other formal grasslands are cleared in autumn, the fallen leaves can be transferred to the nearest woodland floor. Road sweepings are slightly suspect, because these leaves contain an accumulation of chemical pollutants, but woodland is probably one of the safest places to absorb them.

Ecologically, the value of dead standing timber, diseased branches and hollow trunks is immense – but in towns there is an understandable concern for public safety, and the common practice is to remove any diseased limb, or fell any unstable tree at the earliest opportunity. There is considerable evidence that hollow trees are inherently *more* stable. However, whilst this change in perspective is slowly absorbed by tree surgeons, arboricultural officers and others, there are a number of ways of achieving a compromise which will benefit conservation.

Firstly, where potential woodland cover is extensive, the management plan should encourage principal public access routes where paths are clearly marked, always passable and free from hazards. Along these routes, and around woodland car parks, outdoor classrooms and other heavily used areas, a high level of arboricultural rigour should be maintained.

By contrast, other areas of the woodland should be actively promoted as undisturbed sanctuary areas, where public access is made difficult, and where hollow trees and dead branches can be conserved. Sanctuary areas in an otherwise accessible woodland will greatly increase the public's enjoyment, by strengthening a foodchain with increases in fungi and invertebrates, and so increasing songbird and small mammal populations. Short side paths which lead to comfortable seats will allow visitors good views into the richer parts of the wood without risk either to themselves or to the wildlife they are hoping to enjoy.

As a further compensation for the over-zealous tidying, artificial nest sites and bat roosts can be provided, and their manufacture and monitoring offers an excellent opportunity for community involvement in woodland management. All the hole-nesting bird species can be persuaded to nest in boxes, and wildlife conservation bodies such as the Royal Society for the Protection of Birds and The Wildlife Trusts produce practical guides to house type and fixing requirements.

Finally, where branches and trees do have to be lopped or felled, every effort should be made to keep the timber in the woodland. Whilst dead branches in the canopy of trees play a very particular role in woodland, they still have enormous value if they are allowed to decay amongst the leaf litter and wildflowers of the woodland floor. The larger baulks of timber can be placed beside paths. In the sun they will serve as casual seating, and in the shade they will support a wide range of plants and animals.

Carpets of wildflowers

Woodland wildflowers are not good colonisers. In the wild they tend to spread mainly by vegetative means – Bluebell *Endymion non-scriptus* bulbs and Yellow Archangel *Lamiastrum galeobdolan* runners are typical – or by seeds which need a good built-in reserve of energy to grow up into the light, and therefore are too heavy to be blown or carried far. Developing a new woodland, even on the boundary of an existing one, will fail at floor level if simply left to nature. However, woodland wildflowers are relatively easy to propagate under nursery conditions. Best results are achieved in the protected half-shade of a netted bed, and if possible, the seed compost should be made of woodland leafmould, and the seed should be collected locally. The lighter dappled shade of woodland path sides is the best habitat for most of the more colourful wildflowers, and again concentration of such species as Primroses *Primula vulgaris*, violets *Viola* spp., Red Campion *Silene dioica*, Foxglove *Digitalis purpurea* and Wood Anemone *Anemone nemorosa* here will best serve the dual purpose of ecology and recreation. To maximise the insect populations, the aim should be to establish a continuous succession of pollen and nectar flowers, with plants such as thistles and Cow Parsley *Anthriscus sylvestris* being particularly important. Sallow *Salix caprea* and Sloe *Prunus spinosa* provide early spring flowers and Ivy *Hedera helix* provides late autumn nectar. Some of the best examples of artificial woodland wildflower colonies are to be found in the parks of Amstelveen, close to Schiphol airport, on the edge of Amsterdam. Here, the local authority parks department has created a range of beautiful wildflower parks from scratch in less than 40 years.

Exotics

One further feature of many urban woodlands which distinguishes them from most of the conservation woodland of the countryside, is the frequent presence of exotic species. This is particularly true in the shrub layer, where species and cultivars with their origins in the temperate forests of China, Japan, North and South America have been planted as 'ornamental' alternatives to our native Holly *Ilex aquifolium*, Hazel *Corylus avellana*, Box *Buxus sempervirens* and Rose. Many of the trees are exotic too:

acacias, Tulip Trees and chestnuts are frequently more common than the Oak *Quercus robur*, Lime *Tilia cordata*, and Ash *Fraxinus excelsior* of native woodlands. Even at ground level, it is often exotics that carpet the woodland floor – spring bulbs from the eastern Mediterranean, and evergreen ground cover plants. These plants are an important part of our culture. The plant collectors of the nineteenth century were heavily subsidised by British imperialism. The rhododendron walks, azalea days and daffodil festivals need to be celebrated as an integral part of urban woodland management, but there is a need to alert the captive rhododendron audience to the significance of native species too. The food chains which originate with food-plant-specific caterpillars lead eventually to the more glamorous Sparrowhawks *Accipiter nisus*, Tawny Owls *Strix aluco*, Badgers *Meles meles* and bats which make up the acceptable, marketable face of woodland wildlife. Native plants are the fundamental platform of the diverse life-support system. Generally it is possible to ease in the natives gradually, beginning with the popular, colourful spring flowers, and involving the community wherever possible, in the planting of the Hazels, Hollies and Honeysuckles that will enrich the rest.

Naturally regenerating scrub woodland

Giant 'lollipop' trees and close-mown grass are one kind of urban habitat that can be converted to more complete woodland by encouraging the techniques outlined above. The other key resource is the natural scrubland, which has colonised so much of post-industrial urban Europe. Here, the challenge is to underpin the benefits of neglect with sensitive, creative management. In many cases, as for instance with the 'Beeching forest' of 25 year old rail closures, the first wave of Silver Birch and willow pioneers is already making way for long-term woodland trees (Fig. 13.6). As the leaf litter builds a rudimentary soil, the more mobile insects quickly colonise, birds move in to roost in the twiggy canopy, and seed is introduced to the sheltered habitat of the woodland floor. Jays *Garrulus glandarius* bring in and bury acorns, Blackbirds deposit Hawthorn *Crataegus monogyna*, Bramble *Rubus fruticosus* and Elder *Sambucus nigra* seed. Ash and Sycamore seeds blow in and begin to take over too. There is rarely any need to plant the trees, and many of the shrubs will colonise new woods eventually. These pioneer communities are often extremely rich in colonising fungi too – but once again the woodland wildflowers are absent. There is nowhere for them to colonise from, their seed is not dispersed by wind, and they must be introduced.

The manager of this kind of woodland faces a dilemma. It is probably at its more attractive and 'user friendly' at the stage where the first generation is just being taken over by the second. As the wind-borne pioneers are, by definition, light demanding, they are not sustainable as a community without very dramatic interference. If a light, open canopy of Silver Birch is to be maintained, then there is a need to adopt a rotational programme of thinning and/or coppicing and site disturbance, to allow new Birch to germinate in open clearings. In busy urban landscapes where the established woodland is popular, great care must be taken to explain such seemingly destructive management, and every effort should be made to re-direct busy footpaths so that visitors always enjoy the woodland in its most popular 'light romantic' phase.

The woodland network

There are rarely very extensive areas of woodland in towns, although big blocks do occur in some European cities. However, the combination of the two kinds of woodland – unofficial pioneer scrub and official mature parkland – often combine to form a very extensive network, or web of interconnecting woodland corridors. Where new planting is contemplated, or natural regeneration is to be actively encouraged, a strategic urban forestry policy is invaluable, with the clear aim of strengthening the integrity of this woodland network. So often, new planting takes place either as completely isolated islands, or as an addition to existing trees. Far better to use the extra resources to link existing pockets of woodland together and increase the opportunity for woodland species to move freely around town.

Functional woodland

The prime function of trees and woodland in towns has always been considered to be 'amenity', trees to give pleasure to people. In the last decade or so, the attendant

Fig. 13.6. The 'Beeching' forest. After 25 years of abandonment, the natural pioneers of birch and willow woodland have successfully colonised the nation's redundant shunting yards. (C. Baines.)

pleasure of woodland wildlife has been recognised, and so habitat management has become recognised as a function of urban woodland management. Increasingly, though, trees and woodland in towns are being valued for more mercenary functional values. They undoubtedly clean the air by filtering out dust and absorbing noxious chemicals. The shelter they provide reduces wind speeds, and this in turn helps reduce the heating costs of buildings in their lee. A wooded setting for a building is acknowledged to be more attractive than an open windswept one, and since this is reflected in the higher price of houses in leafy neighbourhoods, new planting is slowly being promoted as a cost-effective way of boosting development land values and attracting inward investment. This last consideration is leading to the establishment of a quite new kind of urban woodland: one which is installed partly as a short-term stop-gap which can green the vacant land but be sacrificed, at least in part, when built development does eventually take place. Here pioneer ecology yields more appropriate techniques than horticulture. There is no need for expensive top soil, since the hardcore of the development site is an ideal medium for establishing trees. In the short-term, ripping the ground with heavy machinery to improve aeration and some supplementary irrigation offer the best support for establishing the small seedling trees, and in the longer term, if the woodland is retained, then it will follow the development pattern of the 'Beeching forests' that already cover so much of our abandoned railway land.

There is one further refinement of the short-life woodland, and that is the production of biomass fuel crops from coppiced plantations. Most of the research into fast growing woody species has been aimed at agricultural soils, and certainly if energy yields are to be significant, then the willows, poplars and other fast-growing trees involved will demand a rich fertile growing medium. There are, however, urban circumstances where soil fertility is rich enough to yield an energy crop. Here it is possible to provide the environmentally attractive benefits of both shelter and renew-able fuel, by growing biomass coppice as a green land use between the buildings of industrial estates, or around heat-hungry schools and hospitals. Ecologically, the biomass coppice prototypes are very disappointing. A preoccupation with energy yield has led to monoculture plantations of clonally selected willows or poplars which are at their most efficient when they soak up all the sunlight. These are extremely sterile landscapes. However, if coppice cycles are extended from three to five or six year intervals, and the species mix is varied to include such alternatives as Wild Cherry *Prunus avium*, Ash and Alder *Alnus glutinosa*, and the native willows and poplars, then these new kinds of woodland could be productive in both renewable energy and wildlife.

Urban wetlands

Much of the surface water has been squeezed out of urban areas. Rivers have been confined between engineered floodbanks, streams and ditches have been piped underground, and most of the marshy riverside land was drained and built over long ago. However, there are still some wet areas, and there is certainly great scope for increasing the sensitivity of their management.

Urban wetlands tend to be seen as repositories for casual waste disposal. Everything from supermarket trolleys to waste sump oil is dumped into canals, ponds, lakes and rivers, and a good deal of the essential habitat management of urban wetlands revolves around dealing with this problem. Oil is particularly damaging, and so inflows should be fitted with oil traps as a matter of course. Where spillages do occur, floating booms and suction pumps may help contain the problem. Prevention of dumping is best achieved through a combination of things: restricting vehicular access to the water's edge, clearly signing to dissuade, making potential commercial polluters such as garages more aware, keeping on top of rubbish removal and involving local people from the surrounding neighbourhood in the clean-up.

Chemical pollution is more difficult to prevent as so often it is invisible. The establishment of a network of voluntary wetland watchdogs from within the local

Fig. 13.7. Urban areas are criss-crossed by significant ecological corridors such as this canal in the Black Country. (C. Baines.)

Fig. 13.8. An artificial wetland in its third year after construction and planting.
Camley Street, London. (C. Baines.)

community may help alert officials to problems. Pesticides are a common cause of chemical pollution too, and it is vital that a rigorous code of practice is adopted within the vicinity of wetlands. This applies to recreation managers, but should also be effectively applied to allotment holders, and to riverside industries such as timber yards which may well be using pesticides.

The best protection for urban wetlands lies in fostering public access and popularity. However, public access brings with it the risk of accident, and there is often political pressure to fill in wetlands, or to force people out. Easy access must be backed by on-site interpretation, attention to detail where there are hard waterside surfaces and furnishings, and maximum visibility to ensure that children in particular can be seen clearly, long before they get into difficulties.

Wetlands are extremely dynamic, rapidly changing ecosystems, and skilful habitat management needs to accommodate this. Shallow ponds and canals in particular rapidly colonise with the more invasive emergent aquatic plants such as Unbranched Burr-Reed *Sparganium simplex*, Reed Mace *Typha latifolia* and Flag Iris *Iris pseudacorus*. Regular removal is important, but it should be possible to phase this very disruptive operation, and in any event it is best carried out between November and the end of February.

There is also a management problem which is the reverse of this. Wetland vegetation in urban areas is frequently destroyed by two factors, both of which are difficult to control. Angling is the first. Bank erosion, cutting of waterside trees and 'weed' clearance frequently destroy the wildlife potential and the beauty of heavily fished waters. One remedy is to make very positive provision for anglers, installing convenient fishing platforms, whilst at the same time prohibiting fishing from the open bank. In this way sanctuary zones can be maintained, where the full complement of aquatic vegetation can be encouraged, and of course the quality of fishing will improve, since these conservation areas will provide secure breeding sites for the fish themselves.

Wildfowl are the other major damagers of wetland habitat. Hand-feeding boosts the duck and goose population, pollutes the water with stale bread, and extra droppings, and the increased bird activity stirs up the mud. In addition, when the artificial food runs out, the birds inevitably feed on the aquatic plants. This problem has recently been aggravated in some areas by the explosive population growth of Canada Geese. These large birds were originally introduced as ornamental additions, have adapted well, and now thousands of them spend the winter on urban park lakes, and move out of town each spring to breed. Their numbers need to be vigorously reduced and controlled, but this is difficult since few of them nest in urban sites, where egg pricking would be convenient, and they are long-lived. Direct culling is technically relatively easy, since they moult each year and can be rounded up whilst flightless, but they are popular with townspeople, and consequently culling is an extremely sensitive issue.

Short-life disturbance sites

Disturbance is a key characteristic of urbanisation. Sometimes land is disturbed on a dramatic scale, as with war damage, major mineral extraction or industrial tipping. There is a more modest scale of disturbance too, which takes place constantly as soil is cultivated, buildings are demolished or roads are widened.

The wildlife communities which occupy these 'new' landscapes are the most dynamic of all. A combination of up-turned dormant seed from within the site, and incoming wind-dispersed ruderal plants very quickly re-vegetates any disturbed site which is free from all but the most toxic of pollution. These plant communities are often extremely colourful, fast-growing and very productive; characteristically they produce colourful flowers for insect pollination and clouds of wind-borne seed within a few weeks of their own germination. Many of them have the additional ability to spread vegetatively, usually by shallow rhizomes, and so stabilise the surface of the site. The attendant animal life is extremely colourful and varied too. Flocks of seed-eating birds, scavenging mammals and brightly coloured pollinating insects make these fascinating landscapes for bringing nature and people closer together.

Fig. 13.9. A wet wildflower meadow, created from scratch in a suburban housing area by Amstelveen parks department, The Netherlands. (C. Baines.)

The most essential aspect of their management is establishment of a respectable image, to deter abuse, and encourage respect. Neatly maintained margins, well-signed footpaths, seats and welcoming site labels all help to protect these habitats. Giving a temporary site a name, and explaining its value for nature can dramatically improve the public's attitude towards it.

The most colourful, diverse early stage of revegetation passes very quickly. Many of the pioneering annuals flower and seed for just a single generation, and then blow away to new sites, or lie as dormant seed beneath the suppressing canopy of more permanent plants. If this very special kind of habitat is to be maintained, then frequent re-disturbance is essential. By stirring up the surface, new dormant seeds are exposed, a niche is revealed for wind borne incomers, and the colourful colonisation can begin again. In a vigorous urban economy the natural process of redundancy and redevelopment generates a constant stream of new sites for disturbance habitat; but if the economy stagnates and activity grinds to a halt, it may be necessary to initiate new sites just to maintain the mobile population of thistles, willowherbs and ragworts that depend upon disturbance for survival.

Here again, as with the urban woodlands, exotic species play a prominent role in these communities. Many sites are colonised quite naturally by garden plants with windborne seeds. Michaelmas Daisies *Aster novi-belgii* and Golden Rod *Solidago virgaurea* are classic examples. Human dispersal is another key factor. The dumping of excess garden plants is a major source of plant invasion, and urban 'wastelands' are frequently carpeted with colonies of such exotic plants as Lupins *Lupinus* spp., Montbretia *Crocosmia* x *crocosmiiflora* and Gardener's Garters *Phalaris arundinacea*, plants which typically outgrow their gardens and are dumped around the edge of vacant land. Whilst these exotics may not always offer food for leaf-eating larvae they can boost the nectar supply, for example, Buddleia *Buddleja davidii* is a river-gravel coloniser from China which attracts clouds of butterflies to rubble sites in August. The seeds and fruits of these exotics help feed the flocks of birds in winter too.

In conclusion, there is much about the management of urban habitats which is a simple repeat of the rural experience. However, in almost every case, there is an extra dominant factor. People are what make the urban situations so different. They provide the extra pressure, but they also provide the critical support, and through their appreciation of the nature on their doorstep, fostered through the skilful management of local habitats, the popular, political backing is created for the positive conservation of wildlife habitats everywhere else.

References

Baines, C. (1986). *The Wild Side of Town*. London: BBC Publications.

Baines, C. & Smart, J. (1991). *A Guide to Habitat Creation*. Chichester: Packard Publishing.

Black Country Urban Forestry Unit (1991). *Woodland Planting Guidelines for the Urban Forest*. Sandwell: Black Country Urban Forestry Unit.

Buckley, G.P. (1989) *Biological Habitat Reconstruction*. Lymington: Belhaven Press.

Emery, M. (1986). *Promoting Nature in Cities and Towns*. London: Croom Helm.

English Nature (1989). *The Establishment and Management of Wildflower Meadows*. Focus on Nature Conservation No. 21. Peterborough: English Nature.

English Nature (South-West Region) (1991). *The Greater Bristol Nature Conservation Strategy*. Taunton: English Nature.

Gilbert, O.L. (1989). *The Ecology of Urban Habitats*. London: Chapman & Hall.

Johnson, J. (1990). *Nature Areas for City People*. London: London Ecology Unit.

Leicester City Council (1989). *Leicester Ecology Strategy*. Leicester: Leicester City Council.

Leicester City Council (1990) *Open Space Management for Nature Conservation*. Leicester: Leicester City Council.

Street, M. (1989). *Ponds and Lakes for Wildfowl*. Fordingbridge: The Game Conservancy.

Tregay, R. & Gustavson, K. (1983). *Oakwood's New Landscape*. Warrington: Warrington and Runcorn Development Corporation.

Wells, T., Bell, S. & Frost, A. (1981). *Creating Attractive Grasslands Using Native Plant Species*. Peterborough: Nature Conservancy Council (English Nature).

Some useful addresses

Statutory bodies

English Nature Northminster House, Peterborough PE1 1UA (the statutory body for conservation in England).

Countryside Council for Wales Plas Penrhos, Bangor, Gwynedd LL57 2LQ (the statutory body for conservation in Wales)

Scottish Natural Heritage 12 Hope Terrace, Edinburgh EH9 2AS (the statutory body for conservation in Scotland)

Department of the Environment, Northern Ireland Calvert House, 23 Castle Place, Belfast BT1 1FY (the statutory body for conservation in Northern Ireland)

Joint Nature Conservation Committee Monkstone House, City Road, Peterborough PE1 1JY (carries out research and gives advice to the four bodies above)

Countryside Commission John Dower House, Crescent Place, Cheltenham, Gloucestershire GL50 3RA (statutory body, giving advice and providing funding on conservation and access)

Forestry Authority (England) Great Eastern House, Tennison Road, Cambridge CB1 2DU

Forestry Authority (Scotland) Portcullis House, 21 India Street, Glasgow G2 4PL

Forestry Authority (Wales) North Road, Aberystwyth, Dyfed SY23 2EF

Forestry Service (Northern Ireland) Deparment of Agriculture, Dundonald House, Upper Newtownards Road, Belfast BT4 3SB

Environment Agency 30–34 Albert Embankment, London SE1 7TL (statutory body giving advice on waterbodies and watercourses)

Other organisations

Agricultural Development and Advisory Service (Headquarters) Oxford Spires Business Park, The Boulevard, Kidlington, Oxfordshire OX5 1NZ

Biological Records Centre c/o Institute of Terrestrial Ecology, Monk's Wood, Abbot's Ripton, Cambridgeshire PE17 2LS

Botanical Society of the British Isles c/o Department of Botany, Natural History Museum, Cromwell Road, London SW7 5BD

British Dragonfly Society 1 Haydn Avenue, Purley, Surrey CR8 4AG

British Ecological Society 26 Blades Court, Deodar Road, Putney, London SW15 2NU

British Herpetological Society c/o London Zoo, Regents Park, London NW1 4RY

British Organic Farmers and Growers Association 86 Colston Street, Bristol BS1 5BB

British Trust for Conservation Volunteers 36 St Mary's Street, Wallingford, Oxfordshire OX10 0EU

British Trust for Ornithology The Nunnery, Nunnery Place, Thetford, Norfolk IP24 2PU

British Wildlife Magazine Lower Barn, Rook's Farm, Rotherwick, Basingstoke, Hampshire RG27 9BG

Butterfly Conservation PO Box 222, Dedham, Colchester, Essex CO7 6EY

Council for the Protection of Rural England Warwick House, 25 Buckingham Palace Road, London SW1W 0PP

Farming and Wildlife Advisory Group National Agricultural Centre, Stoneleigh, Kenilworth, Warwickshire CV8 2RX

The Game Conservancy Trust, Fordingbridge, Hampshire ST1 1EF

Groundwork Trust Bank House, 8 Chapel Street, Shaw, Oldham OL2 8AJ

The Heather Trust (formerly **The Joseph Nickerson Reconciliation Project**) The Secretary, Arngibbon, Arnprior, Kippen, Stirlingshire FK8 3ES

Institute of Terrestrial Ecology Monks Wood, Abbot's Ripton, Cambridgeshire PE17 2LS

Landscape Institute 12 Carlton House Terrace, London SW1Y 5AH

Macaulay Land Use Research Institute Craigiebuckler, Aberdeen, AB9 2QJ

The Mammal Society 15 Cloisters Business Centre, 8 Battersea Park Road, London SW8 4BG

Plantlife c/o The Natural History Museum, Cromwell Road, London SW7 5BD

Royal Society for the Protection of Birds The Lodge, Sandy, Bedfordshire SG19 2DL

The Tree Council 35 Belgrave Square, London SW1X 8QN

Urban Wildlife Group 11 Albert Street, Birmingham, B4 7UA
Wildfowl and Wetlands Trust Slimbridge, Gloucestershire GL2 7BT
The Wildlife Trusts The Green, Witham Park, Waterside South, Lincoln LN5 7JR
The Woodland Trust Autumn Park, Dysart Road, Grantham, Lincolnshire NG31 6LL
WorldWide Fund for Nature Panda House, Weyside Park, Catteshall Lane, Godalming, Surrey GU7 1XR

Index of species by common name

Note: page numbers in italic refer to tables and figures

Plants

Subject index

Note: Page numbers in italic refer to tables and figures